The Search for New Elementary Particles: Status and Prospects

Proceedings Supplements
International Journal of Modern Physics A (1993)

Published volumes:

Physics and Experiments with Linear Colliders
Saariselkä, Finland, September 9–14, 1991 (ed. R Orava, P Eerola & M Nordberg)

XVth International Conference on
High Energy Accelerators
Hamburg, Germany, July 20–24, 1992 (ed. J Rossbach)

Forthcoming volumes:

International Conference on
Differential Geometric Methods in Theoretical Physics
Nankai, China, June 5–9, 1992 (ed. C N Yang & M L Ge)

International Workshop on Theoretical Physics
String Quantum Gravity and Physics at the Planck Energy Scale
Erice, Sicily, June 21–28, 1992 (ed. N Sanchez)

Batavia Meeting
1992 Meeting of the Division of Particles and Fields
of the American Physical Society
Batavia, Illinois, USA, October 13–17, 1992 (ed. R Raja & J Yoh)

Subscription Information

For 1993 there will be five volumes (ten issues). The total subscription price is US$580. The special price of US$470 is available to subscribers of the International Journal of Modern Physics A, D and E. For postage and handling, add US$80. Subscription orders may be sent to any one of the following World Scientific offices:

USA: World Scientific Publishing Co. Inc., Suite 1B,
1060 Main Street, River Edge, NJ 07661, USA

UK: World Scientific Publishing Co., 73 Lynton Mead,
Totteridge, London, N20 8DH, England

Singapore: World Scientific Publishing Co. Pte. Ltd.
Farrer Road, P O Box 128, Singapore 9128

Int. J. Mod. Phys. A (Proc. Suppl.) 3B (1993)

INTERNATIONAL CENTRE FOR THEORETICAL PHYSICS

INTERNATIONAL ATOMIC ENERGY AGENCY

UNITED NATIONS EDUCATIONAL, SCIENTIFIC AND CULTURAL ORGANIZATION

Proceedings of the Trieste Workshop on

THE SEARCH FOR NEW ELEMENTARY PARTICLES: STATUS AND PROSPECTS

Trieste, Italy 20 – 22 May 1992

Edited by

Gregor Herten
Universität Freiburg
Freiburg, Germany

Lydia Beers
SLAC, Stanford, CA, USA

Martin L. Perl
SLAC, Stanford, CA, USA

World Scientific
Singapore • New Jersey • London • Hong Kong

Published by

World Scientific Publishing Co. Pte. Ltd.
5 Toh Tuck Link, Singapore 596224
USA office: 27 Warren Street, Suite 401-402, Hackensack, NJ 07601
UK office: 57 Shelton Street, Covent Garden, London WC2H 9HE

British Library Cataloguing-in-Publication Data
A catalogue record for this book is available from the British Library.

Proceedings of the Trieste Workshop on
THE SEARCH FOR NEW ELEMENTARY PARTICLES: STATUS AND PROSPECT

ISBN-13 978-981-02-1236-0
ISBN-10 981-02-1236-4
ISBN-13 978-981-02-1271-1 (pbk)
ISBN-10 981-02-1271-2 (pbk)

PREFACE

In the last decade we have seen many important advancements in particle physics. They have been achieved in a close interplay between experimental data, theoretical modelling and predictions, and further specific experiments which reduce the wide range of models to a few selected successful theories. Generally one can say that the distinction between various theoretical models was either obtained by precise measurements or by the discovery of new particles and new interactions. Examples for the first method are the precise confirmation of QED by g-2 measurements, the observation of P and CP violation and more recently the precise tests of the Standard Model from the study of Z^0 decays at LEP. There are many examples that the observation of new particles had an important impact on the development of physics, e.g. the discovery of the Ω^-, J/ψ, τ, Υ, W, Z. Presently both roads are pursued to test the Standard Model and/or to find evidence for new physics.

The aim of this workshop about new particle searches is to review the status of present particle searches and the prospects for future searches at new accelerators and laboratories. The goal is to bring physicists from various fields together, who study similar effects and particles with different approaches and techniques. Examples are high energy physicists working with e^+e^-, pp, $p\bar{p}$, ep or neutrino beams and underground experimenters searching for proton decay, dark matter, monopoles and solar neutrinos. Dark matter candidates could be observed through their interaction in highly precise underground detectors or through their production in Z^0 decays at LEP.

Many present and future experiments focus on the Higgs boson. Its discovery would yield an experimental confirmation of the symmetry breaking mechanism and thus an understanding of the generation of particle masses. The signature of the Higgs boson depends strongly on its mass and also on details of the model, e.g. the minimal Standard Model or a two doublet Higgs models as required in SUSY theories. Are all Higgs masses accessible with present or planned experiments? What are the prospects to detect supersymmetric Higgses and other SUSY particles? What new detectors and accelerators are needed to observe these new particles and to determine their properties. These are a few of the questions which will be addressed at this workshop.

In preparing this meeting we have profited greatly from the support and suggestions of the staff of the International Centre for Theoretical Physics. We are especially grateful to the director of the centre A. Salam, and to H. R. Dalafi, L. Bertocchi and M. Bowman.

Editors:
G. Herten
L. Beers
M. L. Perl

CONTENTS

The Search for New Elementary Particles: Status and Prospects

SEARCH FOR THE TOP QUARK AT FERMILAB: STATUS AND PROSPECTS

F.D. SNIDER
High Energy Physics Department, The Johns Hopkins University
Baltimore, MD 21218

ABSTRACT

We review the current status of efforts to find the top-quark, starting with mass limits set by experiments at e^+e^- and hadron colliders. Focussing on the hadron colliders, we then discuss recently published results, including searches for standard model charged current decays, decays to charged Higgs bosons, and decay-mode independent searches. After reviewing the planned experimental program at the Fermilab Tevatron Collider, we estimate the sensitivity of the CDF and D0 experiments. Based upon the expected luminosity and background rates, we conclude these experiments have a high probability of finding a standard model top with a mass less than 150 GeV/c^2 during the next running period. Much higher mass regions will become accessible over the next five to seven years provided background rates are controlled.

1 Introduction

The existence of the top-quark (t), the weak-isospin partner of the bottom-quark (b), is a central prediction of the standard model [1,2] (SM). Although the top-quark has not yet been observed,[3–12] numerous indirect observations point to its existence within the framework specified by the SM. Primary among these is the absence of flavor-changing neutral currents in b decays [13] and the forward-backward asymmetry in $e^+e^- \to b\bar{b}$. [14]

While the existence of the top-quark is mandated by the SM, the top-quark mass (m_t) is not and at present remains one of the few unmeasured free parameters of the SM. Fixing m_t in principle fixes the Higgs mass and allows better predictions of such phenomena as rare K-decay branching ratios, the CP violating parameter ϵ'/ϵ and $B^0 - \overline{B^0}$ mixing, all of which may be sensitive to physics beyond the SM. [15] Other properties of the top-quark such as the decay rates and branching ratios are well specified by the SM and represent further probes into possible new physics. It is within this context that the Collider Detector at Fermilab [16] (CDF) and D0 [17] collaborations and the Fermilab Tevatron collider program have attached primary importance to the search for the top-quark.

The goal of this talk is to evaluate the prospects for the top search efforts of the CDF and D0 collaborations. We begin by reviewing the current status of experimental and theoretical mass constraints, then exploring the methods and results of the most recent searches relevant to the upcoming search efforts by the Fermilab experiments. Next, we will discuss the recently completed detector and accelerator upgrades at Fermilab, and focus on the impact these changes are expected to have on the ultimate sensitivity of these experiments.

2 Current m_t limits

Numerous experiments have placed lower bounds on m_t through direct searches for top decay products, or indirect searches that exploit quantities sensitive to top production. The most general limits come from the e^+e^- experiments at LEP and SLAC. Searching for top production via the reaction $e^+e^- \rightarrow Z^0 \rightarrow t\bar{t}$, these experiments arrive at the lower bound $m_t > 46$ GeV/c^{2}at 95% C.L. [3] This limit, derived from a fit to the Z^0 line shape, is independent of the top-quark decay mode.

Experiments at the CERN and Fermilab hadron colliders have mounted both direct and indirect searches. In a direct search for SM decay signatures with muons, UA1 has excluded $m_t < 60$ GeV/c^2 at 95% C.L. [4] The UA2 collaboration searching for electron and missing transverse energy ($\not{E}_t$) signatures has disallowed $m_t < 69$ GeV/c^2 at 95% C.L. [5] More recently, CDF has published a result excluding $m_t < 91$ GeV/c^2 at 95% C.L. [6] based upon searches in several channels involving electrons and muons. All of these results assume that SM charged currents dominate the top decay rate.

The same three experiments have reported top-decay-independent independent limits based upon a measurement of the decay width of the W boson.[7–9] The most restrictive of these is $m_t > 44$ GeV/c^2 at 95% C.L. set by CDF [9] (discussed in Section 6).

In addition to searches for SM decays, both UA1 and UA2 have conducted direct searches for the decay of top into the charged Higgs boson. These searches, the topic of Section 5, assume the existence of two Higgs doublets and exclude regions in the combined parameter space of m_t, Higgs boson mass and ratio of Higgs vacuum expectation values. [10, 11]

The top mass also enters radiative corrections to various electro-weak parameters such as the boson masses.[18–20] This dependence allows theoretical constraints on the top and Higgs boson masses by requiring consistency between weak-neutral-current and boson-mass data. [20] These studies typically favor top masses between 100 and 180 GeV/c^2, depending upon the Higgs boson mass, with central values between 130 and 160 GeV/c^2 for Higgs masses between 10 and 1000 GeV/c^2.

Table 1: Decay modes and branching fractions (BR) for $t\bar{t}$ events.

$t\bar{t}$ decay mode				BR	$t\bar{t}$ decay mode				BR
$q\bar{q}$	b	$q\bar{q}$	$\bar{b}$	36/81	$e\nu$	b	$\tau\nu$	$\bar{b}$	2/81
$q\bar{q}$	b	$l\nu$	$\bar{b}$	12/81	$\mu\nu$	b	$\tau\nu$	$\bar{b}$	2/81
$q\bar{q}$	b	$\mu\nu$	$\bar{b}$	12/81	$e\nu$	b	$e\nu$	$\bar{b}$	1/81
$q\bar{q}$	b	$\tau\nu$	$\bar{b}$	12/81	$\mu\nu$	b	$\mu\nu$	$\bar{b}$	1/81
$e\nu$	b	$\mu\nu$	$\bar{b}$	2/81	$\tau\nu$	b	$\tau\nu$	$\bar{b}$	1/81

3 Standard model top decay signatures

The overall topology of top events depends upon both the production and decay mechanisms. At the Tevatron, top production [21,22] proceeds primarily through $\bar{p}p \to t\bar{t}$ as shown in Figure 1. The direct production of $t\bar{b}$ via W-gluon fusion, [23] although always dominated by $t\bar{t}$ production, contributes more than 20% of the total production cross section for $m_t > 150$ GeV/c^2(Figure 2). In the following, we will consider only $t\bar{t}$ production.

The SM predicts that the top-quark will decay through the process $t \to W^{+*}b$, where W^{+*} is a real or virtual W^+ depending upon the top mass. (Charge conjugate reactions are implied throughout the text.) The decay mode of the W boson pair in a given event therefore defines the top decay signature. Table 1 shows the various possible decay modes and branching fractions.

Although fully hadronic decays dominate the branching fraction, the six jet event signature is difficult to distinguish from the higher rate QCD background. Events in which one of the top-quarks decays semi-leptonically represent the next largest branching fraction. A high transverse momentum (p_t) electron or muon provides an easily identified and reliable trigger signal. The dominant background to these events is W plus jets, which, although tractable, can greatly reduce the sensitivity of this channel. Finally, events in which both top-quarks decay semi-leptonically provide the cleanest signature (excluding τ's) at the price of the smallest branching fraction.

Lepton channels involving a $\tau^{\pm}$ contribute significantly to the total decay rate. Efforts are underway at both CDF and D0 to include analyses into certain $\tau^{\pm}$ channels for the upcoming collider runs.

4 Search for standard model decays at CDF

The CDF collaboration has conducted extensive searches for top decays. [6,24,25] These studies have culminated in the highest published mass limits subject to the assumption that top decays via the SM charged current (published in Reference [6] since the

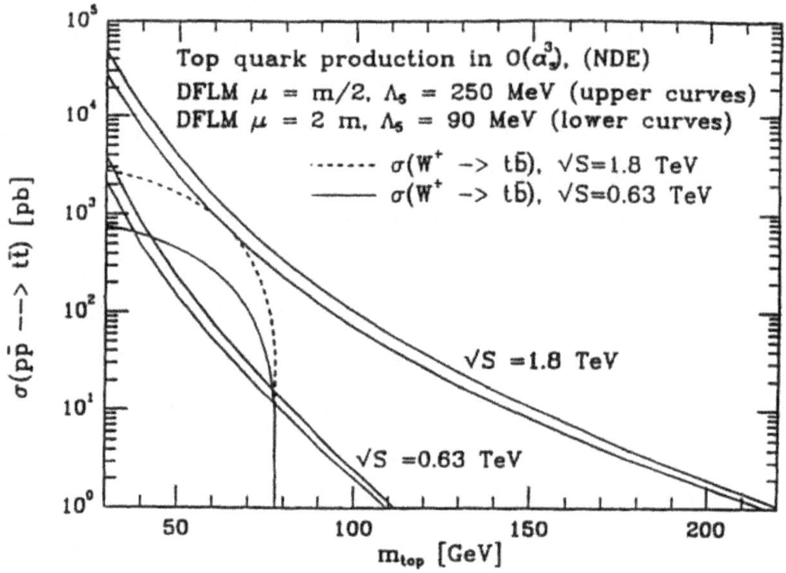

Figure 1: Top-quark production cross section [22] in $\bar{p}p$ collisions at $\sqrt{s}$ = 630 GeV/c^2 and 1800 GeV/c^2.

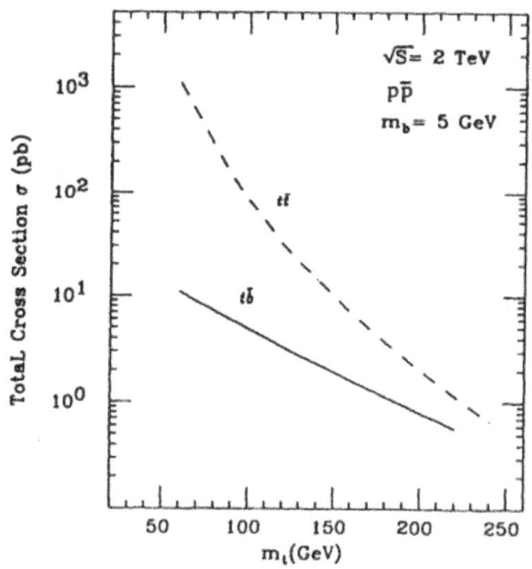

Figure 2: Direct $t\bar{b}$ production cross section through W-gluon fusion [23] in $\bar{p}p$ collisions at $\sqrt{s}$ = 2000 GeV/c^2.

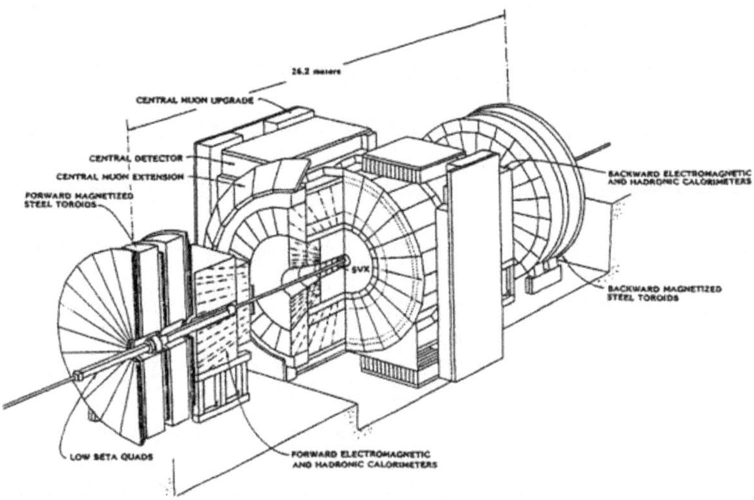

Figure 3: The CDF detector in the 1992 run configuration.

original presentation of this talk). The principal features of future searches at the Tevatron should closely resemble those of the latest CDF analysis.

Before discussing the analyses, we present a brief overview of the CDF detector,[16] shown in Figure 3. The CDF detector consists of tracking chambers within a 1.4 T magnetic field, surrounded by electromagnetic and hadronic calorimeters. For tracks with $|\eta| < 1.2$, ($\eta = -\ln(\tan(\theta/2))$, where θ is the polar angle with respect to the beam axis) the Central Tracking Chamber (CTC) allows measurement of charged particle momentum. The calorimeters provide electron triggering for $|\eta| < 2.3$. Muon chambers, located outside the calorimeters in the central region and forward regions, provide muon identification within $|\eta| < 0.63$ and $2 < |\eta| < 3.6$, and triggering within $|\eta| < 0.63$.

4.1 Di-lepton search

Although the ee, $e\mu$ and $\mu\mu$ channels comprise only 4/81 of the total decay rate, highly efficient triggers and lepton identification algorithms produce a larger total acceptance than in the more copious $e/\mu+$ jets channel.

The data for this analysis were collected using single electron and muon triggers. The electron trigger identified high transverse energy (E_t) showers in the electromagnetic calorimeter associated with a high p_t track within $|\eta| < 1.0$, or simply a high E_t electromagnetic energy deposition for $1.1 < |\eta| < 2.3$. A total integrated luminosity of 4.1 pb^{-1} was collected using this trigger.

The muon trigger required a track in the muon chambers associated with a high

p_t track. This trigger was used to collect an integrated luminosity of 3.5 pb^{-1}.

Event samples were defined based upon offline lepton identification, lepton isolation and event topology requirements. For details of this procedure, the reader should consult Reference [6].

4.1.1 Event selection

Selected events were required to have two leptons, each with $p_t > 15$ GeV/c. Since the decay leptons tend to be well isolated from the hadronic fragmentation products of the top and decay bottom-quarks, the authors required $E_t^{cone} < 4$ GeV, where E_t^{cone} is the total E_t (excluding the lepton) within a cone of radius 0.4 in $\eta - \phi$ space around the lepton, and ϕ is the azimuthal angle.

Two sets of topological requirements were imposed. The first, applied to ee, $\mu\mu$ and $e\mu$ events, demanded that the azimuthal separation between leptons be less than 160°. This requirement reduces a small background in the $e\mu$ channel from $Z \to \tau^+\tau^-$, $b\bar{b}$ and particles mis-identified as muons, and a large background in the ee and $\mu\mu$ channels from Drell-Yan processes. The second requirement, applied to ee and $\mu\mu$ events, demanded that $\not{E}_t > 20$ GeV and that the di-lepton mass (M_{ll}) lie outside the range $75 < M_{ll} < 105$ GeV/c^2. The former criterion reduces background from Drell-Yan, while the latter eliminates $Z \to l^+l^-$ decays.

A single $e\mu$ event survived the selection procedure with an expected background of 2.7±0.9 events. This result is combined with the following lepton plus jets search to arrive at the final limit on the top production cross section.

4.2 Lepton plus jets search

A large background from the process $\bar{p}p \to W$+jets presents a significant impediment to extracting a clean signature in the lepton plus jets channel. CDF has published an analysis sensitive to $m_t < M_W + M_b$, in which the transverse mass distribution of the lepton and $\not{E}_t$ was used to distinguish between the W plus jets background and the virtual W in top decays. [24]

For $m_t > M_W + M_b$, the W boson is real, so the transverse mass distribution is no longer a good discriminant. A second CDF analysis [6] used, instead, a soft muon to tag events with a b-quark in addition to the W plus jets signature. The semi-leptonic branching fraction for $b \to \mu + X$ is approximately 10%, and about 7.5% for the decay chain $b \to c \to \mu + X$. These muons were identified using tracks in the muon chambers with $2 < p_t < 15$ GeV/c and well separated from the two highest E_t jets. For $m_t < 120$ GeV/c^2, these two jets are usually associated with the decay of the $W^\pm$ boson.

Starting from the same set of single lepton triggers as used for the di-lepton analysis, the authors selected events with an electron or muon with $p_t^{e/\mu} > 20$ GeV/c, $\not{E}_t > 20$ GeV, at least two jets with $E_t > 10$ GeV, and a soft muon candidate. For

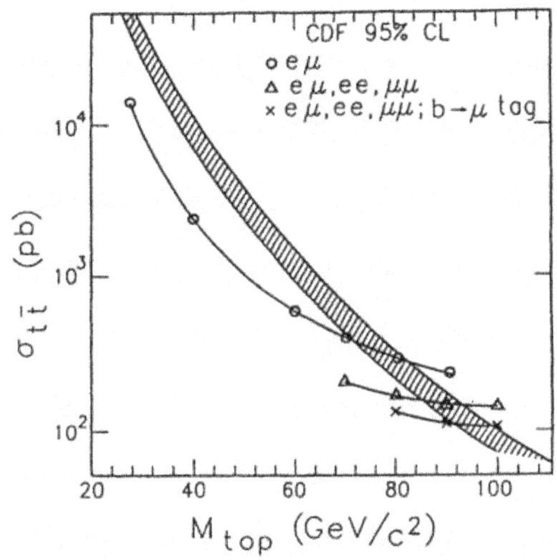

Figure 4: Experimental top cross section limit and predicted $t\bar{t}$ production cross section as a function of top mass. The mass limit corresponds to the point of intersection between the two curves.

the high p_t lepton, the authors required $E_t^{iso} < 1.5$ GeV, where E_t^{iso} is the total E_t in the calorimeter towers adjacent to the tower the lepton entered. (The calorimeter towers in the central region subtend the angles $\Delta\phi = 15\,\text{deg}$ and $\Delta\eta = 0.1$.)

No candidate events were observed where a background of 0.9 was expected.

4.3 Combined top mass limit

The results of the analysis outlined above were used to set a 95% C.L. limit on the top-quark production cross section as a function of m_t. The intersection of the experimental limit curve with the predicted production cross section yields the 95% C.L. bound on the top mass (Figure 4). For the combined CDF analyses, the limit is $m_t > 91$ GeV/c^2 at 95% C.L.

5 Direct search for non-standard model decays

Perhaps the most intriguing aspect of the top search is the possibility of discovering a decay mode that is outside the SM. Such a finding would likely place immediate constraints on the possible form of physics beyond the SM.[15] Moreover, should one of these decay modes dominate the branching fraction, then the mass limits from *all* the direct searches may well be invalid.

Numerous possible decay modes have been explored in the literature. A partial list include decays into supersymmetric partners,[26] charged Higgs bosons,[26,27] and flavor-changing neutral currents[28] arising from the mixing of the t-quark with a new heavy particle. Although many of these processes may have observable signatures, the only results reported recently pertain to searches for the decay of top into the charged Higgs boson, conducted by UA1[10] and UA2.[11] Such searches will take on added significance in the event that no SM top signal is established during the next collider runs.

A charged Higgs boson exists in models with two Higgs doublets,[27] an interesting possibility in part because it is the most simple extension to the minimal Higgs sector of the SM, and also because it arises naturally in minimal supersymmetry (SUSY) models. As discussed by other speakers at this conference,[29] fits to SM parameters favor SUSY extensions over certain others. Both the UA1 and UA2 searches have assumed the minimal SUSY prescriptions. Under these assumptions, one can show that for $m_t < m_W + m_b$, the SM decays are *always* suppressed relative to $t \to H^+b$. For $m_t > m_W + m_b$, the SM decays may or may not be suppressed.

The event signature for $t \to H^+b$ will depend upon the decay mode of the H^+, which, for $M_H < m_t$, will be dominated by $H^+ \to \tau^+\nu_\tau$ or $H^+ \to c\bar{s}$ depending upon the value of $\tan\beta$, the ratio of vacuum expectations, and various masses. Given the difficulty of isolating a $H^+ \to c\bar{s}$ signal from the large QCD background, most direct searches are likely to exploit the $H^+ \to \tau^+\nu_\tau$ channel. The UA1 and UA2 searches employed two different methods to identify τ's and have excluded slightly different regions of m_t-$\tan\beta$ space.

5.1 UA1 search for $t \to H^+b$

The technique used by UA1 involved searching for muons from the decay of the τ or the b-quark in the processes:

$$\begin{aligned}
t &\to H^+b, \\
& \quad H^+ \to \tau^+\nu,\ \tau^+ \to \mu^+\nu\nu,\ b \to X; \text{ and} \\
t &\to H^+b, \\
& \quad H^+ \to \tau^+\nu,\ \tau^+ \to \mu^+\nu\nu,\ b \to \mu X.
\end{aligned}$$

The first process produces the signature μ plus two jets plus $\not{E}_t$ where the muon is isolated, while the second appears as $\mu\mu$ plus jet where one of the muons is not isolated. The authors then define a likelihood function based upon Monte Carlo studies of $t \to H^+b$ and exclude regions of $\tan\beta$-M_H-m_t space. The results are shown in the next section along with the UA2 results.

5.2 Ua2 search for $t \to H^+b$

UA2 exploits the large hadronic branching fraction of the τ by using the jet shape to discriminate between hadronic τ decays and QCD jets. The decay chain

$$\begin{aligned} t &\to H^+b, \\ & H^+ \to \tau^+\nu, \tau^+ \to \text{hadrons} + \nu \end{aligned}$$

produces an event signature with a high E_t jet from the τ decay and $\not{E}_t$ from the neutrinos. Events were selected requiring a missing transverse momentum greater than 20 GeV/c, a leading jet with $E_t > 17$ GeV and no jet with $E_t > 10$ GeV opposite the leading jet. The lateral shape of the leading jet was parameterized by:

$$\rho = (E_m + E'_m)/E_{tot}, \quad (1)$$

where E_{tot} is the total jet energy, and $E_m + E'_m$ is the sum of energies in the highest energy calorimeter cell and its highest energy neighbor. A good statistical separation of τ from hadronic background follows from the requirement $0.75 < \rho < 0.90$, as shown in Figure 5.

The number of τ events is extracted by fitting the ρ distribution to the expected shape from τ plus QCD background, then compared to the rate predicted from W production. The limits on the $t \to H^+b$ rate in $\tan\beta$-m_H-m_t space are shown in Figure 6.

6 Indirect search for top production

Search techniques that depend only on top production play an extremely important role in exploring for non-SM decays of the top-quark. A positive signal obtained using an indirect technique where direct searches have failed would represent strong evidence for a dominant non-SM decay mode. The measurement of the full decay width of the W boson, Γ_W, provides such a solution at hadron colliders for the case $m_t < M_W - M_b$. Until the start of LEP-200, this measurement will provide the only decay-mode independent search technique into the region $m_t \lesssim m_W$.

The actual measurement indirectly determines the W width using the relation:

$$\begin{aligned} R &= \frac{\sigma(p\bar{p} \to W \to l\nu)}{\sigma(p\bar{p} \to Z^0 \to l^+l^-)} \\ &= \frac{\sigma(p\bar{p} \to W)}{\sigma(p\bar{p} \to Z^0)} \frac{\Gamma(Z^0)}{\Gamma(Z^0 \to l^+l^-)} \frac{\Gamma(W \to l\nu)}{\Gamma(W)} \\ &= \frac{N_W}{N_Z} \frac{A_Z}{A_W} \frac{\epsilon_Z}{\epsilon_W}, \end{aligned} \quad (2)$$

where N_i is the observed number of events of type i, A_i is the geometrical acceptance and ϵ_i is the detection and reconstruction efficiency. The value of R is measured

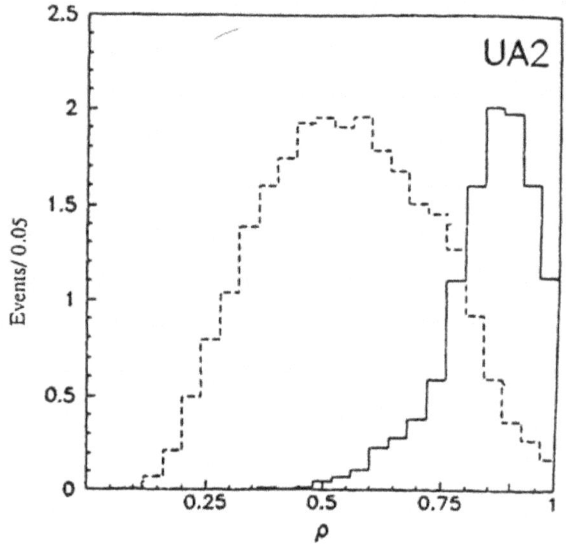

Figure 5: Jet profile parameter, ρ, for di-jet events (dashed) and Monte Carlo τ's (solid) from UA2.

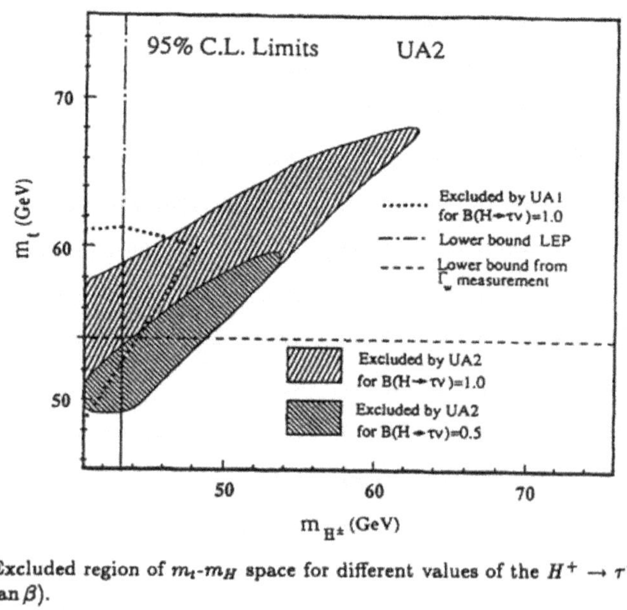

Figure 6: Excluded region of m_t-m_H space for different values of the $H^+ \to \tau^+\nu_\tau$ branching ratio (effectively $\tan\beta$).

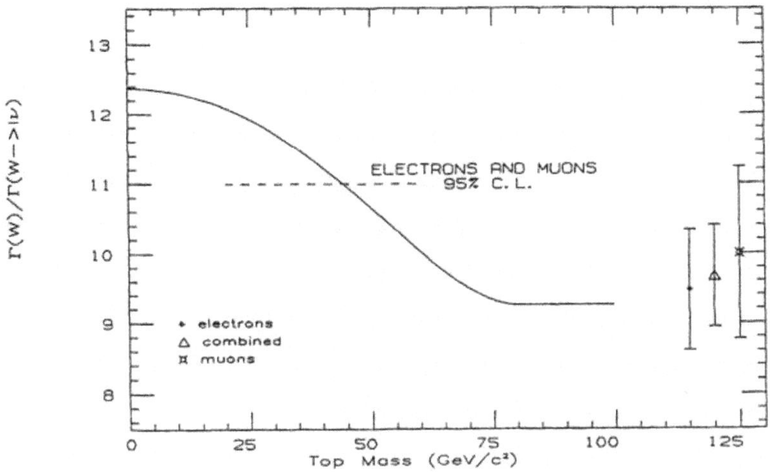

Figure 7: Predicted inverse branching fraction vs. m_t. The experimental result shown corresponds to the limit $m_t > 44$ GeV/c^2 at 95% C.L.

directly, while $\sigma(W)/\sigma(Z^0)$ is predicted, $\Gamma(Z^0)/\Gamma(Z^0 \to l^+l^-)$ is measured precisely at LEP, and $\Gamma(W \to l\nu)$ is predicted by the SM. Solving for Γ_W yields a result more precise than would be obtained from a direct measurement of the width at the hadron colliders.

The value of $R = 10.0 \pm 0.7$ measured by CDF translates into the inverse branching fraction $\Gamma_W/\Gamma(W \to l\nu) = 9.6 \pm 0.74$, where both electron and muon results are combined. The 95% C.L. limit on the top mass, $m_t > 44$ GeV/c^2, can then be obtained using the theoretical inverse branching fraction show in Figure 7.

7 Top search prospects at the Tevatron

The physics program at the Fermilab Tevatron collider is now completing the first phase of a multi-step upgrade program for both the accelerator and the detectors. Perhaps foremost among the changes for the run now in progress at Fermilab is the successful commissioning of the D0 detector. This new detector (see Figure 8) features a hermetic, highly segmented uranium-liquid argon calorimeter fully surrounded by a muon toroid system. The detector should provide excellent electron and muon identification, and electron energy resolution.

The accelerator has planned a series of luminosity upgrades over the next four to six years (see Table 2). The first two stages anticipate a factor of five increase in the instantaneous and integrated luminosities for the 1992 run compared to that of the 1988-1989 run, followed by another factor of two in 1993. Increasing the number of proton and anti-proton bunches in the machine from 6 to 36 will allow substantial

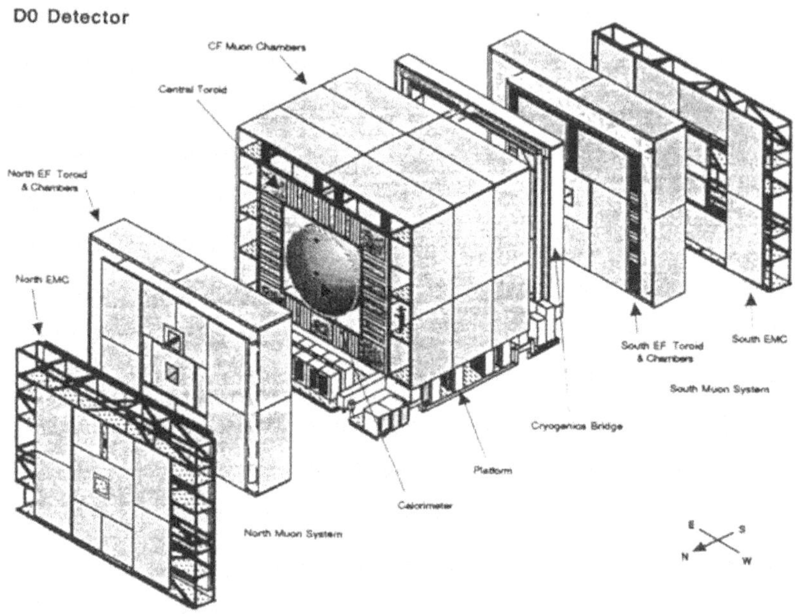

Figure 8: The D0 detector.

increases in instantaneous luminosity while maintaining the number of collisions per crossing at a manageable level. The program culminates in the proposed Main Injector that promises integrated luminosities in the range of 500 pb^{-1} per year.

7.1 Detector upgrades

In order to keep pace with the increasing luminosity of the machine, both the CDF and D0 detectors will undergo a series of upgrades designed to allow operation with significantly increased collision rates, as well as to improve the physics capabilities of the detectors. Below we outline the major elements of these upgrades that may have an impact on the top-quark search.

7.1.1 CDF upgrade schedule

The primary upgrades to the CDF detector will occur in two stages, the first of which is already completed and in operation for the current run (IA). A few of the most significant improvements include a silicon micro-strip vertex detector (SVX);[30]

Table 2: Tevatron upgrade and luminosity goals through the 1990's.

Date	Run	$\mathcal{L}$ $(\times 10^{30})\text{cm}^{-2}\text{s}^{-1}$	$\int \mathcal{L}dt$ (pb^{-1})	Changes to accelerator
1992	Run IA	5.7	25	Electrostatic separators, $\bar{p}$ source improvements
1993	Run IB	10	75	LINAC upgrades
1995	Run II	11	50	36 bunch operation, Tevatron cooling, major CDF/D0 upgrades
1996	Run III	570	500/yr	Main Injector

1.8 times more muon coverage in the trigger; and more muon absorber steel in the central region to reduce hadronic punch-through to the muon chambers. The addition of the SVX should greatly improve the b-quark tagging capabilities of the experiment and offer enhanced background rejection in the lepton plus jets channel. The SVX features an intrinsic resolution of about 10 μm in four layers of single-sided silicon detectors. A number of b tagging studies have been conducted. Most yield tagging efficiencies between 15–30% per event for top decays with $m_t > 110$ GeV/c^2. The tagging rates for background are typically $\lesssim 1\%$.

The second round of upgrades, scheduled to be ready for Run II in 1995, include an improved silicon micro-strip detector (SVXII), an upgraded endcap calorimeter, and filling of the remaining gaps in the muon coverage. These changes will offer significant improvements in the overall top acceptance.

7.1.2 D0 upgrade schedule

The main features of the present D0 detector [17] include non-magnetic tracking in the central and forward regions; transition radiation detectors (TRD) for electron identification in the central region; uranium-liquid argon calorimeters with projective $\Delta\eta = \Delta\phi = 0.1$ towers (four depth segments in the electromagnetic compartment, four in the hadronic); and 13-20 interaction lengths of muon absorber material in front of full-coverage muon detectors and toroids. The muon system offers momentum measurements with $\Delta p_t/p_t \approx 20\%$, and triggering down to $|\eta| = 2.4$. In general, the overall lepton coverage is superior to that of CDF, as is the rejection of muon background, although the muon momentum measurement is worse up to about 100 GeV/c. The combination of TRD's and dE/dx in the tracking chambers should allow an electron identification comparable to that of CDF obtained using an E/p selection criterion.

For the start of Run II, the D0 collaboration intends to extend the muon trigger below $|\eta| = 2.4$. By the start of Run III, D0 has proposed installing a 1.5 T solenoid

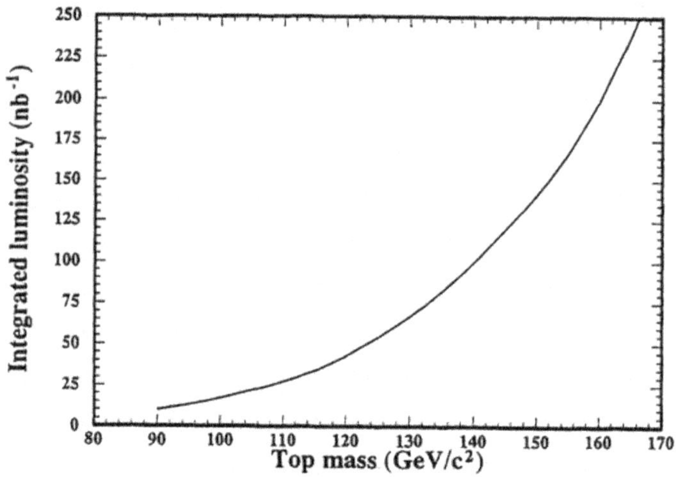

Figure 9: The integrated luminosity required to observe 10 di-lepton events as a function of m_t. We use the lower limit on the predicted value of $\sigma_{t\bar{t}}$ from Ref. [21], and the detection efficiency from CDF for $m_t = 90$ GeV/c^2.

and a silicon tracker in the central detector.

7.2 Projected top sensitivity

A conservative approach toward estimating the future performance of the Fermilab detectors is to assume the same top acceptance as found for $m_t \approx 90$ GeV/c^2 in the CDF searches. Using the lower bound on the theoretical production rate, [21] we should arrive at a somewhat pessimistic prediction. A more detailed analysis should await real data.

Figure 9 shows the integrated luminosity required to observe 10 di-lepton events, $10/(\sigma_{t\bar{t}} BR(t\bar{t} \to ll)\epsilon_{CDF}$, as a function of m_t where ϵ_{CDF} is the efficiency for observing top with a mass $m_t = 90$ GeV/c^2 in the CDF detector from the most recent analysis. The efficiency combines both electron and muon channels.

From Figure 9, we observe that to reach sensitivities at the $m_t = 150$ GeV/c^2 level requires reducing background rates by a factor of about 10. Experience obtained during the CDF di-lepton analysis suggests several approaches:

1. Require $\not{E}_t > 20$ GeV in the $e\mu$ analysis. This will reduce $Z \to \tau^+\tau^-$ background by a factor of 4, $b\bar{b}$ by more than an order of magnitude, and the fake rate by a factor of 3.

2. Increase lepton E_t requirements to 20 or 25 GeV. A 20 GeV requirement reduces $Z \to \tau^+\tau^-$ by a factor of 3 and sharply reduces the rate of fakes.

3. Tighten lepton isolation requirements. Background from $b\bar{b}$ and fakes should improve with tighter isolation cuts.

4. Demand one or more jets with $E_t > 10$ GeV. The ee and $\mu\mu$ channels are especially sensitive to Drell-Yan, while all channels will have problems with diboson production for $m_t \gtrsim 150$ GeV/c^2. A jet requirement is probably the best way to reduce these sources with minimum loss of acceptance.

Based upon the last run, we should see on the order of 2000 W plus jet such events during the next run. This yield requires that a b-quark tag have a background contamination at the 0.1% level. The problem of reaching this level of background suppression is still under investigation, but results from Monte Carlo studies and early data from run IA appear promising. The high tagging efficiency using the SVX allows some room for further improvements. For the "worst" case, CDF expects the acceptance to be comparable to that of the previous run, but with a much higher level of background suppression. D0 does not have a vertex detector, and has pursued other techniques such as likelihood methods to discriminate background from a top signal in this channel. [31]

Finally, the decay independent search will also gain sensitivity with increased luminosity. The statistical uncertainty in the present R measurement is dominated by the number of $Z \to l^+l^-$ events, while the systematic uncertainties are dominated by the uncertainty in lepton finding efficiencies. Since these efficiencies were extracted directly from the data, the uncertainties should improve with additional data. By the end of Run IA, the overall uncertainty in R should be at the ± 0.4 level, which corresponds to a limit in the mid to upper 50 GeV/c^2 range provided the central value does not change. Eventually, the systematic uncertainty becomes dominated at the 2-3% level by uncertainties in the parton distribution functions that enter the acceptance calculation. In all probability, by the time the Fermilab experiments have adequate statistics to push the R measurement to its ultimate precision, the LEP-200 experiments will have completed decay independent searches up to $m_t = 90$ GeV/c^2.

8 Conclusions

We have reviewed the current state of experimental limits on the top mass. The various techniques explored yield limits that depend upon different assumptions about the dominant top decay modes. Mass limits set by experiments at LEP and SLAC have ruled out top-quarks with $m_t < 44.8$ GeV/c^2, while an indirect search at CDF has excluded $m_t < 44$ GeV/c^2 at 95% C.L. independent of decay mode. Direct searches at CDF have excluded top for $m_t < 91$ GeV/c^2 assuming SM decay rates. Additional direct searches for the decay $t \to H^+b$ followed by the decay $H^+ \to \tau^+\nu_\tau$ performed by UA1 and UA2 have excluded regions of parameter space available in two Higgs doublet models.

We have also reviewed the most recent searches at CDF and used the experiences obtained there to estimate the probable top mass sensitivity of CDF and D0 in the upcoming runs. Over the next two runs, these experiments should be able to discover or exclude top masses up to $m_t = 150$ GeV/c^2. Indeed, given the current state of the detectors and their upgrade programs, it seems probable that a SM top anywhere within the allowed range of masses will be found at Fermilab. It is with great eagerness and anticipation that we await just this event.

Acknowledgements

The author wishes to thank M. Contreras, D. Hedin, R. Raja and J.E. Skarha for helpful discussions in preparing this talk.

References

[1] S.L. Glashow, Nucl. Phys. **22** (1961), 579; S. Weinberg, Phys. Rev. Lett **19** (1967), 1264; A.Salam, in *Elementary Particle Theory*, ed. by N. Svartholm (Almqvist and Wiksell, Sweden, 1968), p. 367.

[2] S.L. Glashow, J. Illiopoulos and L. Maiani, Phys. Rev. D. **2** (1970), 1285; M. Kobayashi and M. Maskawa, Prog. Theor. Phys. **49** (1973), 652.

[3] P. Abreu *et al.*, Phys. Lett. **B242** (1990), 536; D. Decamp *et al.*, Phys. Lett. **B236** (1990), 551; M.Z. Akrawy *et al.*, Phys. Lett. **B236** (1990), 364; G.S. Abrams *et al.*, Phys. Rev. Lett **63** (1990), 2447.

[4] C. Albajar *et al.*, Z. Phys. **C48** (1990), 1.

[5] T. Åkesson *et al.*, Z. Phys. **C46** (1990), 179.

[6] R. Abe *et al.*, Phys. Rev. D. **45** (1992), 3921.

[7] C. Albajar *et al.*, Phys. Lett. **B253** (1991), 503.

[8] J. Alitti *et al.*, Phys. Lett. **B276** (1992), 365.

[9] F. Abe *et al.*, Phys. Rev. Lett **69** (1992), 28.

[10] C. Albajar *et al.*, Phys. Lett. **B257** (1991), 459.

[11] J. Alitti *et al.*, Phys. Lett. **B280** (1992), 137.

[12] I. Adachi *et al.*, Phys. Rev. Lett **60** (1988), 97; M. Althoff *et al.*, Phys. Lett. **138B** (1984), 441; C. Berger *et al.*, Phys. Lett. **86B** (1979), 413.

[13] A. Bean *et al.*, Phys. Rev. D. **35** (1987), 3533.

[14] W. Bartel *et al.*, Phys. Lett. **146B** (1984), 437.

[15] For a review, see G.L. Kane, invited lecture at the Workshop on High Energy Phenomenlogy, Mexico City, Mexico, July 1-10, 1991, UM-TH-91-32, Dec. 1991.

[16] F. Abe *et al.*, Nucl. Inst. and Meth. **A271** (1988), 387.

[17] S. Aronson *et al.*, *Design Report: an Experiment at D0 to Study* $\bar{p}p$ *Collisions at 2 TeV* (Fermi National Accelerator Laboratory, Batavia, IL, 1984).

[18] LEP Collaborations, Phys. Lett. **B276** (1992), 247;

[19] D. Bardin *et al.*, Z. Phys. **C44** (1989), 493; Nucl. Phys. **B351** (1991), 1; Phys. Lett. **B255** (1991), 290.

[20] P. Langacker and M. Luo, Phys. Rev. D. **44** (1991), 817; U. Umaldi *et al.*, Phys. Rev. D. **36** (1987), 1385.

[21] R.K. Ellis, Phys. Lett. **B259** (1991), 492.

[22] G. Altarelli *et al.*, Nucl. Phys. **B308** (1988), 724; P. Nason, S. Dawson and R.K. Ellis, Nucl. Phys. **B303** (1988), 607.

[23] C.P. Yuan, Phys. Rev. D. **41** (1990), 42.

[24] F. Abe *et al.*, Phys. Rev. Lett **64** (1990), 142.

[25] F. Abe *et al.*, Phys. Rev. Lett **64** (1990), 147; F. Abe *et al.*, Phys. Rev. D. **43** (1991), 664.

[26] K. Hidaka, Y. Kizukuri, T. Kon, Phys. Lett. **B278** (1992), 155.

[27] J. Reid *et al.*, Z. Phys. **C51** (1991), 395; R.M. Godbole, D.P. Roy, Phys. Rev. D. **43** (1991), 3640; V. Barger, R.J.N. Phillips, Phys. Rev. D. **41** (1990), 884; J.F. Gunion *et al.*, *The Higgs Hunter's Guide* (Addison-Wesley, Reading, MA, 1990).

[28] G. Eilam, J.L. Hewett, A. Soni, Phys. Rev. D. **44** (1991), 1473; J.L. Dïaz-Cruz *et al.*, Phys. Rev. D. **41** (1990), 891.

[29] G. Folgi, invited lecture at the Trieste Workshop on the Search for New Elementary Particles: Status and Prospects, Trieste, Italy, May 20–22, 1992.

[30] C. Haber *et al.*, Nucl. Inst. and Meth. **A289** (1990), 388; B. Barnett *et al.*, Nucl. Inst. and Meth. **A315** (1992), 125.

[31] R. Raja, private communication (1992).

CONSTRAINTS ON NEW PHYSICS FROM PRECISION ELECTROWEAK DATA

G. L. FOGLI
Dipartimento di Fisica di Bari, Bari, Italy
Sezione INFN di Bari, Bari, Italy

ABSTRACT

All precision electroweak data are in impressive agreement with the minimal Standard Model, and predictions about the t quark and Higgs boson masses can be derived. Beyond the Standard Model, searching for effects of new physics, a comparison is possible with its simplest extensions, as technicolour theories and the Minimal Supersymmetric Standard Model. A brief review of these topics is given.

1. Introduction

In the Standard Model, radiative corrections introduce a dependence from heavy, unseen virtual particles [1]. The program of estimating radiative correction effects to the different processes within the Standard Model has been carefully pursued in the last ten years [2], with a twofold design

- to verify the consistency of the Standard Model at the 1–loop level with its expected structure of spontaneously broken renormalizable gauge theory;
- to obtain information about its mass spectrum, in particular searching for indications in favour of the still unseen massive particles.

The electroweak data have reached several years ago already the sensitivity to radiative corrections. Starting from the structure of the neutral currents extracted from the low–energy (LE) data (in particular from the ν-q deep–inelastic scattering [3]), and combining them with the first data about the vector boson masses [4], it has been possible first to obtain upper bounds on the top quark mass [5], and then derive indications about its expected value [3], [6]. The advent of the first LEP data allowed to refine the estimate of the top quark mass [7], further restricted by the increasing precision of the line shape measurements at LEP and of the hadronic collider data (UA2 and CDF experiments) [8]. At present, the analysis of the complete data set coming from low- and high-energy experiments leads to an extraordinary consistency check for the Standard Model, with clear indications in favour of a top quark mass around 130 GeV and a Higgs boson relatively light [9].

The Standard Model is very well verified at the 1-loop level, and, from this point of view, there is no reason to go beyond it. However, the Standard Model is unsatisfactory from a theoretical point of view, being unable to give raisonable answers to several fundamental questions. This leads to suppose that the Standard Model cannot be the whole story. On the other hand, the present accuracy of the experimental data not only permits a careful comparison with the standard electroweak theory, but is presumably not too far from being sensitive to possible forms of new physics, expected to appear below 1 TeV. We can then compare the precision electroweak measurements with the effects induced by possible extensions of the Standard Model, by adopting for these effects a parametrization as model-independent as possible. Along this line, a comparison can be performed with "technicolour", the non-perturbative approach in which the gauge symmetry breaking has a dynamical origin related to fermions condensates strongly interacting at a new energy scale. Alternatively, the Standard Model could be the explored sector of a more sophisticated (and satisfactory) theory based on supersymmetry. In this case, the so-called Minimal Supersymmetric Standard Model (MSSM) represents the "minimal" realization of this approach, can be faced perturbatively, and is naturally the first candidate to be tested.

In the present talk the previous points will be concisely reviewed. In Section 2 the comparison with the minimal Standard Model is briefly reported, with the most recent predictions about the masses of the top quark and of the Higgs boson. In Section 3 the problem of reparametrize the radiative correction effects in a model-independent way is considered, with an estimate of the uncertainties related to this kind of schematization. In Section 4 we analyze in detail the two main lines followed in the realization of the simplest schemes of new physics, i.e. technicolour theories and MSSM. In Section 5, an anticipation [10] of the impact within the minimal Standard Model of the new preliminary data of the four LEP experiments [11] is reported. Finally, in Section 6 the conclusions are drawn.

2. Radiative corrections in the minimal Standard Model

A global analysis of the electroweak data within the minimal Standard Model [9] involves many different precise electroweak data, sensitive to different combinations of the so-called "oblique" corrections (the universal gauge boson vacuum polarization effects) and to specific vertex and box contributions.

The LE data come from three different sectors, each characterized by a specific elementary interaction (ν–q, ν–e and e–q) whose structure is described in a model–independent parametrization by well–specific neutral current couplings. The ν–q sector is dominated by the deep–inelastic processes, from which the four squared chiral couplings $u^2_{L,R}$, $d^2_{L,R}$ can be derived: here the couplings are taken from the analysis of the neutral current structure performed in ref. [3]. The ν–e sector is parametrized in terms of the couplings g_L and g_R: here to the old data discussed in ref. [12] the recent precise determination of the CHARM II Collaboration is added [13]. Finally, the e–q interaction is parametrized in terms of the the couplings C_{1u}, C_{1d} and $C_{2ud} = C_{2u} - \frac{1}{2}C_{2d}$ (see for example the analysis of ref. [14]). At present, it is dominated by the atomic physics parity violation experiment of the Boulder group [15], whose experimental result is to be compared with the theoretical predictions by taking into account the recent improvement [16] in the calculations of the atomic matrix elements controlling the parity–violating effect measured in cesium.

The high–energy data imply mainly the LEP experiments. At LEP, the line-shape is measured with very high precision, and, beside the impressive estimate of M_Z, also σ_0^h, the total width Γ_Z and the ratio of the hadronic to the leptonic width $R = \Gamma_h/\Gamma_\ell$ are known at a high level of accuracy. More recently, also the partial widths in heavy quarks $\Gamma_{b\bar{b}}$ and $\Gamma_{c\bar{c}}$ become to be sensitive to radiative corrections. The other source of information from LEP is represented by the asymmetries (ASY) $A^\ell_{\rm FB}$, $A^b_{\rm FB}$ and $A^\tau_{\rm pol}$, in particular the leptonic asymmetry being measured with very good precision. An independent measurement, of considerable importance because it leads to an estimate of the mixing parameter $s^2_{\rm w}$ independent of the top and Higgs mass §, is the measurements of the vector boson mass ratio M_W/M_Z, performed at the hadronic colliders by the UA2 and CDF Collaborations [17].

The accuracy reached in a large part of the experimental measurements quoted above requires a comparable accuracy in the estimate of the radiative correction effects. This has been accomplished by refining the approach to the radiative correction calculations through a certain number of "technical" improvements, as:

- A recursion procedure is introduced in the estimate of $s^2_{\rm w}$, in order to reproduce consistently for each assigned pair of values of m_t and M_H, the 1–loop relation which relates $s^2_{\rm w}$, M_Z and the radiative correction $\Delta r(s^2_{\rm w}\,,\, M_Z\,,\, m_t\,,\, M_H)$ [2], [8].
- A procedure is adopted for resumming higher order in $\alpha_{\rm em}$ [18], [19].
- The $O(\alpha_{\rm em}\alpha_s)$ contributions to the vacuum polarization $(t\,,\, b)$ diagrams at high energies [20] are properly included [9].
- Leading [21] and next–to–leading [22] QCD corrections to the hadronic Z^0 decay widths (up to $O(\alpha_s^3)$) are included in the estimate of Γ_h.

---

§ Here and in the following the mixing parameter is introduced according with the Sirlin's "on–shell" definition preposed in the first of refs. [2].

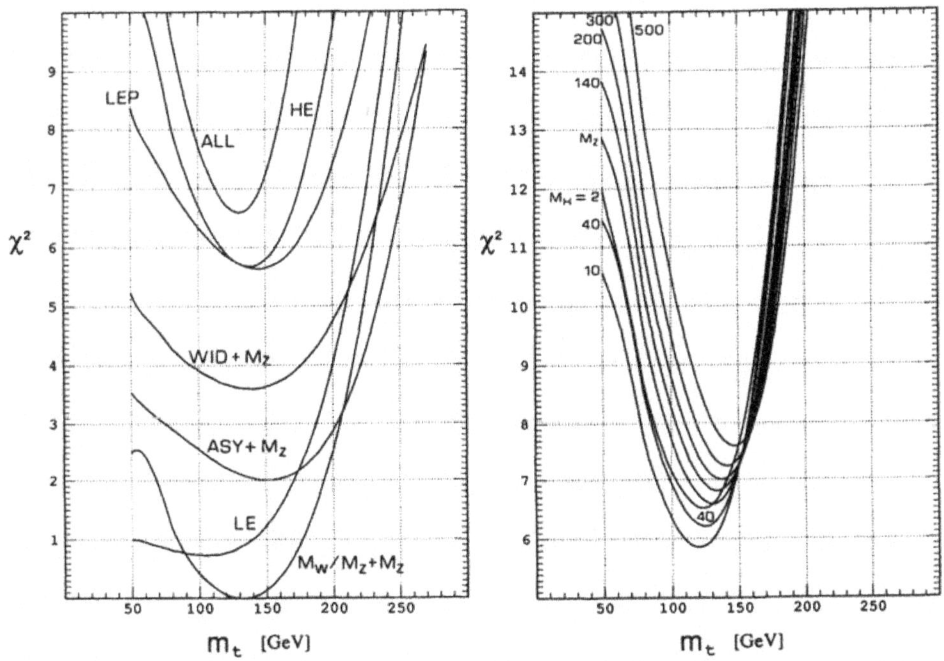

Fig. 1 — (a) The χ^2 curves as functions of m_t with fixed $M_H = M_Z$, for different combinations of the precision electroweak data, including ν–q, e–q and ν–e scattering data (LE), LEP measurements of Z^0 decay asymmetries (ASY) and widths (WID), the ensemble of the LEP data (LEP), LEP and $p\bar{p}$ collider data on M_W/M_Z (HE), and the unconstrained global fit (ALL). (b) The χ^2 curves, as functions of m_t, for unconstrained global fits with different values of M_H.

- The theoretical uncertainties still affecting Δr, due i) to the effect of extrapolating α_{em} from the Thompson limit to M_Z (± 0.0009) [23], ii) to the specific renormalization scheme, [24] and iii) to the choice of the α_s scale [25], are explicitly introduced, through their effects on M_Z. In this way they contribute in a non–negligible way to the present uncertainties affecting m_t and M_H.

When all these aspects are properly taken into account, it becomes possible to constrain severely the parameters of the Standard Model still unknown, through a simultaneous comparison of all these inputs with their theoretical estimates at the 1–loop level (details on the method can be found for example in [6], [8]). In Fig. 1a the χ^2 curves relative to different combinations of the experimental data are reported as function of m_t, for fixed $M_H = M_Z$. This gives an idea of the contributions of the different sectors in constraining m_t. In Fig. 1b all data are taken into account simultaneously, and the χ^2 curves are reported as functions of m_t for different values of the Higgs mass. In this way the following limits on m_t are obtained:

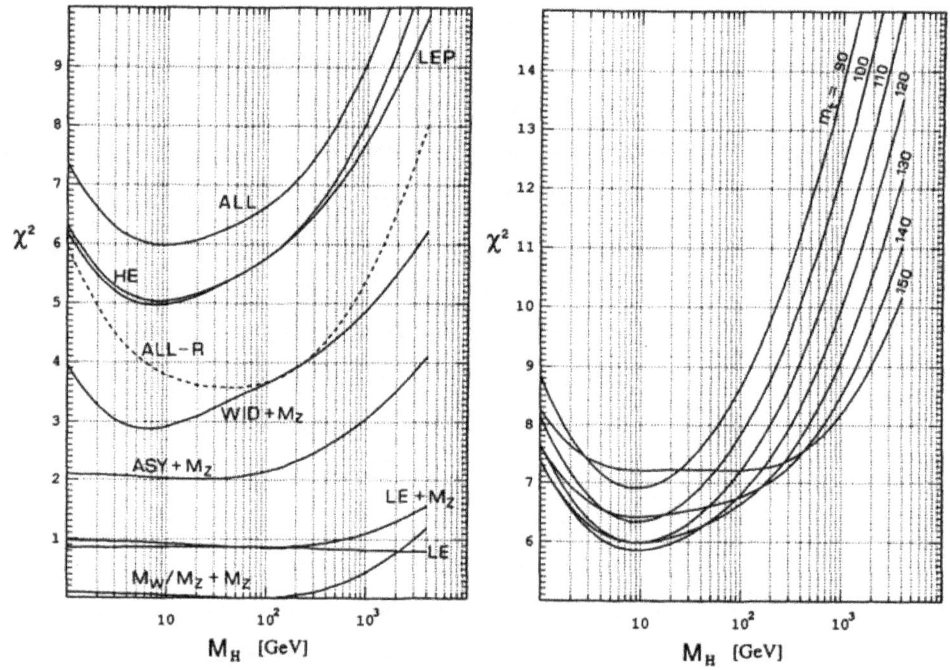

Fig. 2 — (a) The χ^2 curves as functions of M_H with fixed $m_t = 130$ GeV, for different combinations of the precision electroweak data. The notation is as in Fig. 1, with an extra curve (dashed) corresponding to the omission of LEP data on $R = \Gamma_h/\Gamma_\ell$ from the global fit. (b) The χ^2 curves as functions of M_H, for unconstrained global fits with different values of m_t.

$$\text{all data}\Big|_{M_H=M_Z} : \qquad m_t = 130 \begin{matrix} +25 \\ -28 \end{matrix} \text{ GeV}, \tag{1a}$$

$$\text{all data}\Big|_{M_H \text{ unconstrained}} : \qquad m_t = 120 \begin{matrix} +27 \\ -28 \end{matrix} \text{ GeV}, \tag{1b}$$

together with a first indication about the dependence of χ^2 on M_H.

A more specific analysis in terms of M_H is reported in Fig. 2: in (a) for different data sets assuming m_t fixed at 130 GeV, and in (b) including all the data simultaneously but for different values of m_t. From the analysis a precise preference in favour of a rather light Higgs emerges.

The simultaneous dependence on both M_H and m_t is considered in Fig. 3, which reports the two allowed regions corresponding to $\Delta\chi^2 = 1$ and 2.7 (68% and 90% of confidence level on one of the two masses if the other is unconstrained). In Fig. 3 we

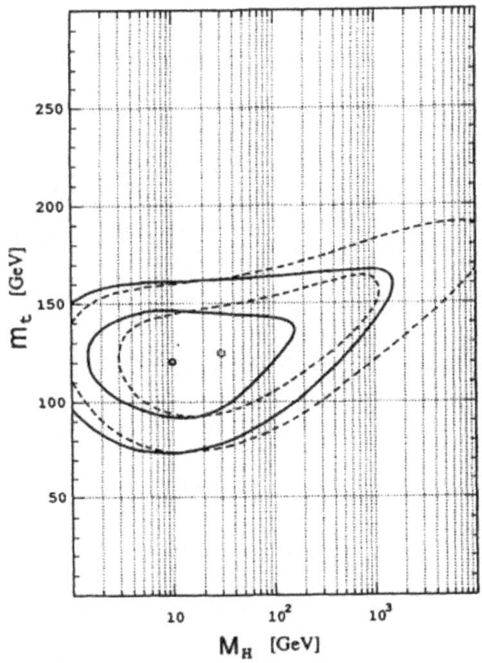

Fig. 3 — The $\Delta\chi^2 = 1$ and 2.7 contours in the (M_H, m_t) plane for unconstrained global fits in the Standard Model. The solid lines are for all data, the dashed lines for all data with $R = \Gamma_h/\Gamma_\ell$ omitted.

distinguish between the case in which all the data are considered (full curves) and the case in which the measurement of the ratio $R = \Gamma_h/\Gamma_\ell$ is not included (dashed curves). The ratio R has a considerable influence on the contours of the allowed regions, and this is to be interpreted as a signal that the results must be taken with some caution until the dependence on specific single measurements will show a less crucial behaviour (see, however, Section 5).

Limits about M_H can be derived from the χ^2 behaviour. For fixed $m_t = 130$ GeV the χ^2 is minimized when $M_H \simeq 40$ GeV, and gives

$$2.6 \text{ GeV} < M_H < 500 \text{ GeV} \qquad 68\% \text{ C.L.} \quad , \tag{2a}$$

$$0.8 \text{ GeV} < M_H < 1600 \text{ GeV} \qquad 90\% \text{ C.L.} \quad . \tag{2b}$$

Taking the envelope of the global χ^2 curves, one obtains for m_t left unconstrained a central value $M_H \simeq 10$ GeV, relatively insensitive to m_t, and

$$1.4 \text{ GeV} < M_H < 160 \text{ GeV} \qquad 68\% \text{ C.L.} \quad , \tag{3a}$$

$$0.5 \text{ GeV} < M_H < 1500 \text{ GeV} \qquad 90\% \text{ C.L.} \quad . \tag{3b}$$

3. Toward a model–independent parametrization

As anticipated in the Introduction, the comparison of the electroweak precision data with physics beyond the Standard Model could be the way to reveal some indications in favour of new physics. A first attempt in this direction is based on the possibility of parametrizing in a very general way those contributions to radiative corrections which are universal, i.e. independent of the specific process under consideration. This is the case of all the effects coming from the vector boson vacuum polarizations ("oblique" corrections) [26], which can appear only in finite number of combinations in the 1–loop contributions to the electroweak processes.

Actually, the vacuum polarization tensors are calculated at the q^2 which characterizes the process under consideration, but, if we expand in q^2 and retain only the first two terms (in other words, "linearizing" the q^2–dependence), after a standard renormalization technique specifying α_{em}, G_F and M_Z (here assumed fixed at its experimental value), we have the possibility of parametrizing all the "oblique" contributions in terms of three parameters alone [27], [28], [29]. In the literature these parameters are indicated as S, T, U [27] or as ε_1, ε_2, ε_3 [29]. They are simply related to the conventional radiative corrections $\Delta\varrho$, Δr_w (which is the more familiar Δr with the $\Delta\alpha$ contribution removed), and $\Delta\kappa$:

$$\alpha T = \varepsilon_1 \; = \; \Delta\varrho \, , \tag{4a}$$

$$\alpha S = \; 4s_w^2\varepsilon_3 = \; 4s_w^2c_w^2(\Delta\varrho + \Delta r_w) + 4s_w^2(c_w^2 - s_w^2)\Delta\kappa \, , \tag{4b}$$

$$\alpha U = -\,4s_w^2\varepsilon_2 = -4s_w^2c_w^2\Delta\varrho + 8s_W^4\Delta\kappa - 4s_W^4\Delta r_w \, , \tag{4c}$$

$\Delta\varrho$, Δr_w and $\Delta\kappa$ being specified in terms of the renormalized 1–loop vacuum polarizations $\hat{\Pi}_i$ $(i = \gamma\, ,\, W\, ,\, Z)$ through the relations

$$\Delta\varrho = \; \hat{\Pi}_W(0) - \hat{\Pi}_Z(0) \, , \tag{5a}$$

$$\Delta r_w = -\,\hat{\Pi}_W(0) + \hat{\Pi}_\gamma(M_Z^2) \, , \tag{5b}$$

$$\Delta\kappa = -\,\frac{c_w}{s_w}\hat{\Pi}_{\gamma Z}(M_Z^2) \, . \tag{5c}$$

From their structure one can infer that

- $S \propto \varepsilon_3$ is the isospin–symmetric measure of the size of the Higgs contribution;
- $T \propto \varepsilon_1$ is the isospin–breaking term containing the leading m_t–dependent contribution $\sim m_t^2/M_Z^2$;
- $U \propto \varepsilon_2$ contains non–leading terms $\sim \ln(m_t^2/M_Z^2)$, expected to be very small.

The procedure introduced above of extracting the "oblique" corrections in terms of the three parameters S, T, U or $\varepsilon_{1,2,3}$ must be taken with some caution, since it involves, however, some more or less evident approximations, related to the following points:

a) Expansion of the vacuum polarization amplitudes ("linearization"). This seems a reasonable approximation, since for new physics arising at a scale Λ, higher orders are suppressed as increasing powers of q^2/Λ^2, and $q^2 \leq M_Z^2$ for physics at LEP and at lower energies).

b) Further non–universal radiative corrections coming from box and vertex diagrams. They are rather small in the Standard Model, and expected still small in possible extensions of it. But they cannot be neglected and a convenient way of introducing them is needed. Two attitudes are possible:

   i) To assume a "oblique correction dominance hypothesis", by estimating non–oblique corrections within the Standard Model [30]. Of course, this introduce the strong hypothesis that new physics involves primarily new contributions to oblique corrections.

   ii) To include some of the non–oblique corrections within the parameters ε_i's (introducing a sort of "hybrid" ε_i's), estimated starting from three specific experimental quantities [29]. It is reasonable to choose those best measured, which do not introduce a dependence on parameters difficult to be controlled, as α_s: a typical choice is Γ_ℓ, $A_{\rm FB}^\ell$ and M_W/M_Z [29]. This approach, of course, becomes more and more approximate as far as new independent measurements are added.

c) A lower accuracy in the estimate of radiative corrections. This cannot be avoided, being connected to the impossibility of controlling the value of $s_{\rm w}^2$ through the recursion relation (valid in the Standard Model) when the point in which the χ^2 is estimated is not a point compatible with the Standard Model itself. The only way out here is to introduce a "reference point", corresponding to assigned values of m_t and M_H, and to estimate $s_{\rm w}^2$ and the non–oblique corrections with reference to this point. Of course, one obtains different results depending on the reference point, which gives an estimate of the uncertainty affecting the model–independent approach.

In Fig. 4, taken from ref. [31], the attitude ii) is assumed. The figure shows the $\Delta\chi^2 = 1$ allowed region in the plane $(\varepsilon_1, \varepsilon_3)$, with the improvement obtained by adding progressively new data in the analysis: in a) only Γ_ℓ, $A_{\rm FB}^\ell$ and M_W/M_Z are included, in b) $A_{\rm pol}^\tau$ is added, in c) $A_{\rm FB}^b$ is also included, and, finally, in d)

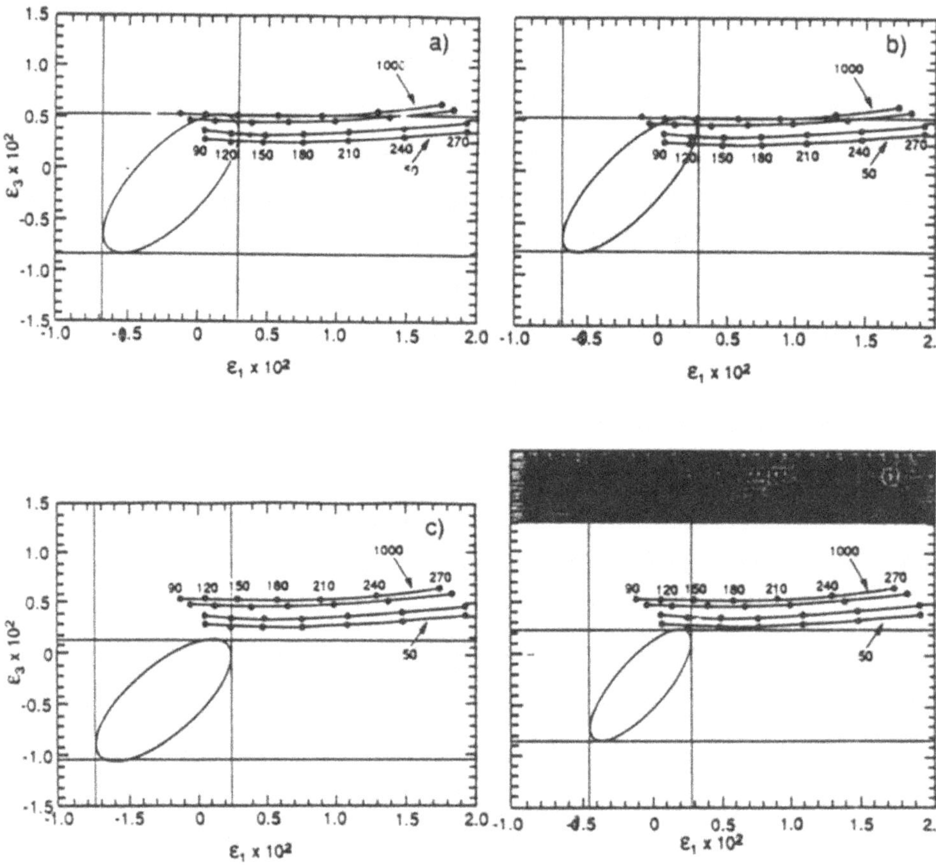

Fig. 4 — Allowed region in the plane $(\varepsilon_1\,,\,\varepsilon_3)$, with the improvement obtained by adding progressively new data in the analysis: in a) only Γ_ℓ, $A^\ell_{\rm FB}$ and M_W/M_Z are included, in b) $A^\tau_{\rm pol}$ is added, in c) $A^b_{\rm FB}$ is also included, and, finally, in d) both HE and LE data are taken (with the exclusion of $\Gamma_{b\bar{b}}$ because of its strong dependence on m_t at the level of non-oblique corrections). The figure is taken from ref. [31].

both HE and LE data are taken (with the exclusion of $\Gamma_{b\bar{b}}$ because of its strong dependence on m_t at the level of non-oblique corrections).

Conversely, in Fig. 5, taken from ref. [30], the "oblique correction dominance hypothesis" is assumed, all data are simultaneously included and $(m_t\,,\,M_H) = (130\ \text{GeV}\,,\,M_Z)$ is taken as reference point. The $\Delta\chi^2 = 1$ allowed regions are reported in the three planes (a) $(\delta\varepsilon_1\,,\,\delta\varepsilon_3)$, (b) $(\delta\varepsilon_1\,,\,\delta\varepsilon_2)$ and (c) $(\delta\varepsilon_2\,,\,\delta\varepsilon_3)$, the $\delta\varepsilon_i$'s being the corrections with respect to the reference point, where $\delta\varepsilon_i \equiv 0$ is

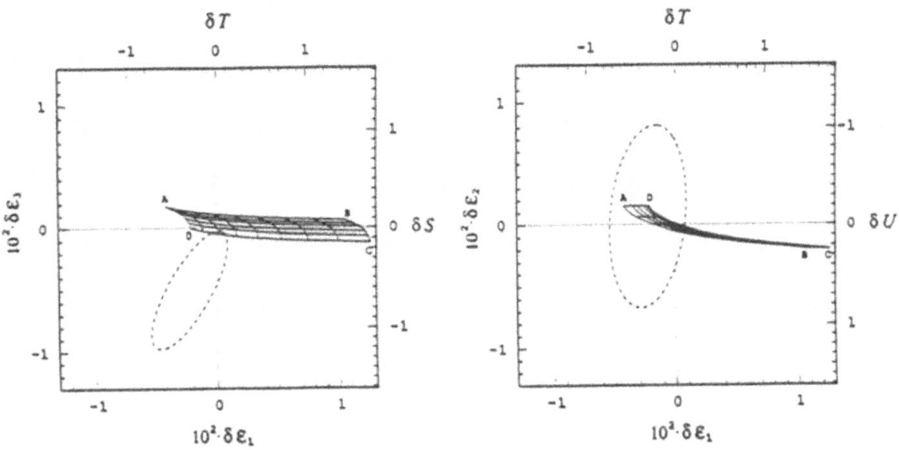

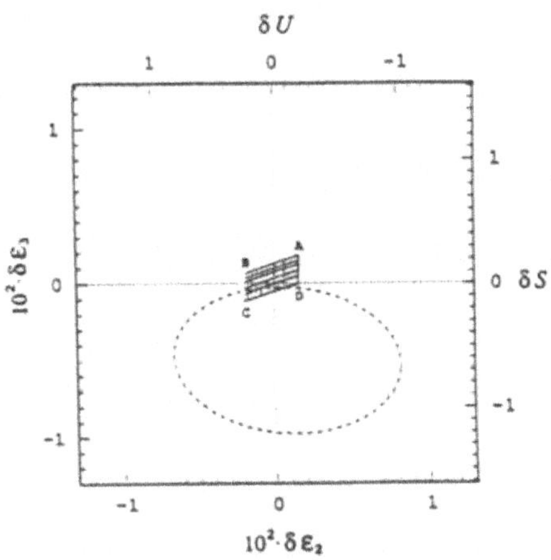

Fig. 5 — Allowed region favoured by electroweak data at the $\Delta\chi^2 = 1$ level in the three planes (a) $(\delta\varepsilon_1\,,\,\delta\varepsilon_3)$, (b) $(\delta\varepsilon_1\,,\,\delta\varepsilon_2)$, and (c) $(\delta\varepsilon_2\,,\,\delta\varepsilon_3)$, when $(m_t\,,\,M_H) = (130\text{ GeV}\,,\,M_Z)$ as reference point is taken. The Standard Model is bounded by the extreme points A, B, C, D, whose reference values in terms of $(m_t\,,\,M_H)$ are (masses in GeV): $A \equiv (90, 1000)$, $B \equiv (250, 1000)$, $C \equiv (250, 50)$, $D \equiv (90, 50)$.

assumed.

In Fig. 6 the allowed region is reconstructed in the space $(\delta\varepsilon_1\,,\,\delta\varepsilon_2\,,\,\delta\varepsilon_3)$. The Standard Model, represented by a surface delimited by the choice 90 GeV $< m_t <$ 250 GeV, 50 GeV $< M_H <$ 1 TeV, has a very small portion within the allowed region, here represented by the ellipsoid. This, in principle, leaves some room to hypothetical extensions of the Standard Model able to realize a better agreement with the data.

Finally, in Fig. 7 different $\Delta\chi^2 = 1$ ellipses, related to different choices of the reference point, are represented in the plane $(\delta\varepsilon_1\,,\,\delta\varepsilon_3)$. The comparison of the previous reference point with those considered in Fig. 7 (i.e. (masses in GeV): $A \equiv (90, 1000)$, $B \equiv (250, 1000)$, $C \equiv (250, 50)$, $D \equiv (90, 50)$) shows that in the model–independent parametrization the choice of the reference point cannot be neglected, and introduces a certain degree of approximation in the approach.

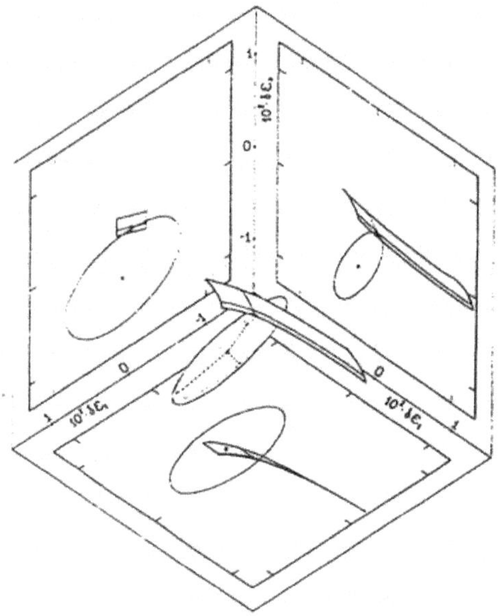

Fig. 6 — The ellipsoid in the 3-parameter $(\delta\varepsilon_1\,,\,\delta\varepsilon_2\,,\,\delta\varepsilon_3)$ space favoured by electroweak data at the $\Delta\chi^2 = 1$ level, together with the two-dimensional manifold calculated in the Standard Model. Also indicated are projections onto the $(\delta\varepsilon_i\,,\,\delta\varepsilon_j)$ planes shown in Fig. 5. As reference values $(m_t\,,\,M_H) = (130\ \mathrm{GeV}\,,\,M_Z)$ have been used.

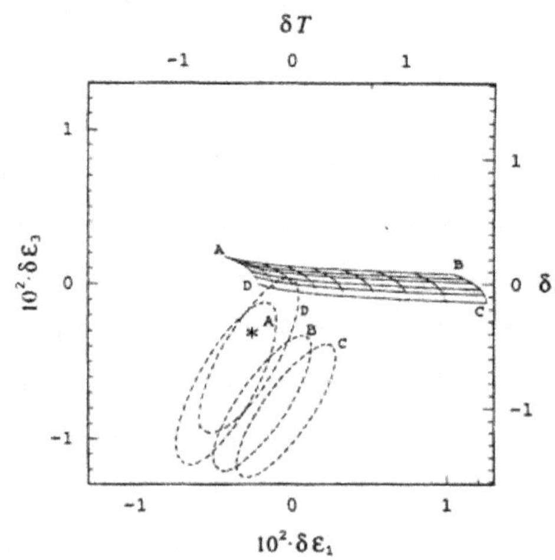

Fig. 7 — As for Fig. 5, but for the different choices of the reference values of $(m_t\,,\,M_H)$ corresponding to the points A, B, C, D of fig. 5. In each case the origin has been shifted to the Standard Model $(m_t\,,\,M_H) = (130\ \mathrm{GeV}\,,\,M_Z)$ point.

4. Constraints on extensions of the Standard Model

In this Section we examine two possible general extensions of the Standard Model, the first related to the non-perturbative ideas leading to the technicolour theories, the other representing the minimal realization of the perturbative extension of the Standard Model induced by supersymmetry, i.e. the Minimal Supersymmetric Standard Model.

4.1 Comparison with technicolour theories

The problem of a comparison with technicolour relies essentially on the estimate of the contributions to the parameters T and S (or ε_1, ε_3) within the technicolour scheme. This has been performed in ref. [32], through a dispersive theoretical approach, by using the strong interaction data and including resonant and non-resonant contributions to the spectral functions. Starting from ref. [32], it can be estimated [30]

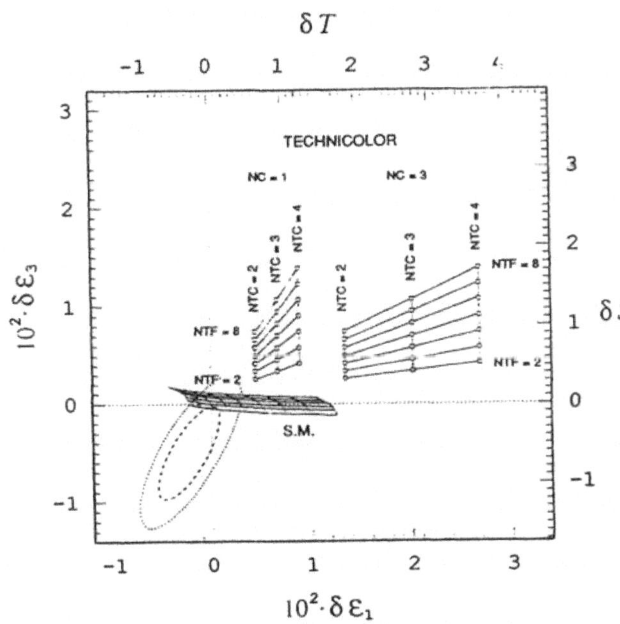

Fig. 8 — Predictions of different models based on technicolour theories (see text), compared with the $\Delta\chi^2 = 1$ and 2.7 ellipses and the Standard Model. As reference point again $(m_t\,,\, M_H) = (130\ \mathrm{GeV}\,,\, M_Z)$ is taken.

$$\delta T_{\rm TC} \simeq 0.3 N_{\rm C} N_{\rm TC} \ , \qquad \delta S_{\rm TC} \simeq 0.05 N_{\rm TF} N_{\rm TC} + 0.127 \ , \tag{5}$$

where $N_{\rm TC}$ is the number of technicolours ($N_{\rm TC} \geq 2$) and $N_{\rm TF}$ the number of technifermions ($N_{\rm TF}$ from 2 to 8). The constants come from simple expressions estimated for $m_t = 130$ GeV. An alternative analysis of the technicolour contributions can be found in ref. [33], with similar conclusions: $\delta T_{\rm TC}$ can be relatively small for small masses of the bosons of extended technicolour, but $\delta S_{\rm TC} \geq 0.38$ for $N_{\rm TC} = 4$. In other words, it is very difficult to find a small value of $\delta S_{\rm TC}$.

In Fig. 8 the predictions of different technicolour realizations are reported in the plane ($\delta\varepsilon_1$, $\delta\varepsilon_3$), and compared with the Standard Model estimate and the $\Delta\chi^2 = 1$ and 2.7 ellipses. It is clearly seen that, if technicolour theories cannot be ruled out, however data do not point in their direction.

4.2 Comparison with the Minimal Supersymmetric Standard Model

The interest for supersymmetry is due to its possibility of solving in a natural way several theoretical problems affecting the Standard Model, in particular the hierarchy problem. Moreover, if the indication emerged in the previous Section in favour of a relatively light Higgs is taken seriously, this is exactly what the MSSM predicts, with, at the tree level, the two CP–even neutral Higgses, h and h', the former lighter and the latter heavier than the neutral vector boson Z^0. First order radiative corrections modify this situation [34], but do not alter the essential message, that the lighter CP–even neutral Higgs is required to have a mass of the order of the Z^0 mass.

There is not room here to discuss the MSSM and its parametrization (for a detailed introduction see for example [35]). We only remind the reader that in the MSSM a new rich structure is present, realized by the supersymmetric partners of the usual particles. As a consequence, there is a lot of new contributions to the radiative corrections, which make the MSSM rather complicate to the 1–loop level. However, these new contributions become smaller and smaller as the masses of the sparticles become larger, because of the "decoupling" typical of the objects whose masses do not arise from a spontaneous symmetry breaking of the theory.

Within the MSSM several authors evaluated radiative corrections to the electroweak parameters [36], but a systematic comparison with the precision electroweak data is still lacking [37]. There exist, however, several attempts of estimating the effects expected from specific sectors of the MSSM. They will be briefly reported in the following.

One of the first attempts of analyzing the effect of MSSM radiative corrections on the electroweak parameters is presented in ref. [38], where the contributions of the supersymmetric partners of the t and b quarks are included in the estimate of m_t from electroweak data, and found to alter of a few GeV the mass of the t quark with respect to the value expected in the Standard Model. In Fig. 9 the estimate

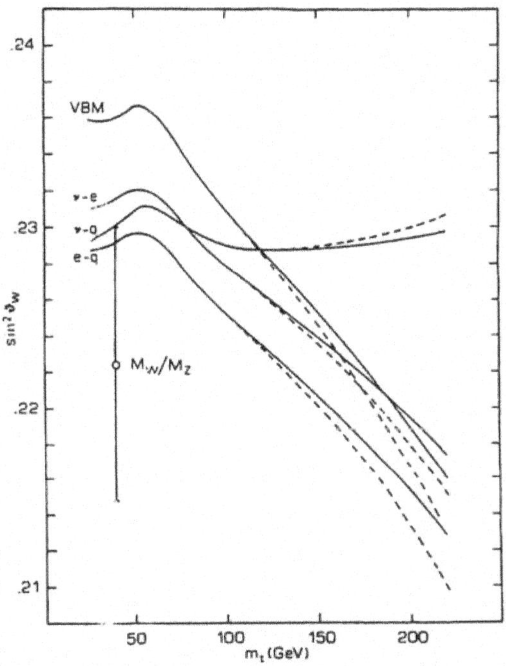

Fig. 9 — Dependence on m_t of the central values of $\sin^2\vartheta_w$ extracted from the different electroweak sector: ν–q, ν–e, e–q and the vector boson masses (VBM). Also shown the the central value and the one–standard deviation range of $\sin^2\vartheta_w$ extracted from M_W/M_Z, independent of m_t. The solid line are calculated in the minimal Styandard Model, , whilst the dashed lines are calculated including the squark loops assuming $m_{\tilde{q}} = 150$ GeV and $\tan\beta = 1.3$

of ref. [38] is reported: the effect of reducing m_t is clearly rather small, at least if the squark masses are above about 200 GeV.

More recently, the potential effect on the t–quark mass coming from the gaugino–higgsino sector of the MSSM has been examined in ref. [39]. By assuming

i) $m_\chi^\pm \geq 50$ GeV (mass of the lightest chargino),

ii) $m_\chi + m_{\chi'} \geq M_Z$ (masses of the lightest and next-to-lightest neutralinos),

iii) $m_s > 50$ GeV (common slepton mass),

in order to satisfy the bounds on the sparticle masses coming from the experiment, it is estimated that if both the lightest chargino and the (common) slepton masses take their minimum values, then one obtains to maximize the effect on the top quark mass, appearing lighter than expected in the Standard Model. The effect can be expressed in terms of the parameters ε_i's and is reported in Fig. 10, taken from ref. [39].

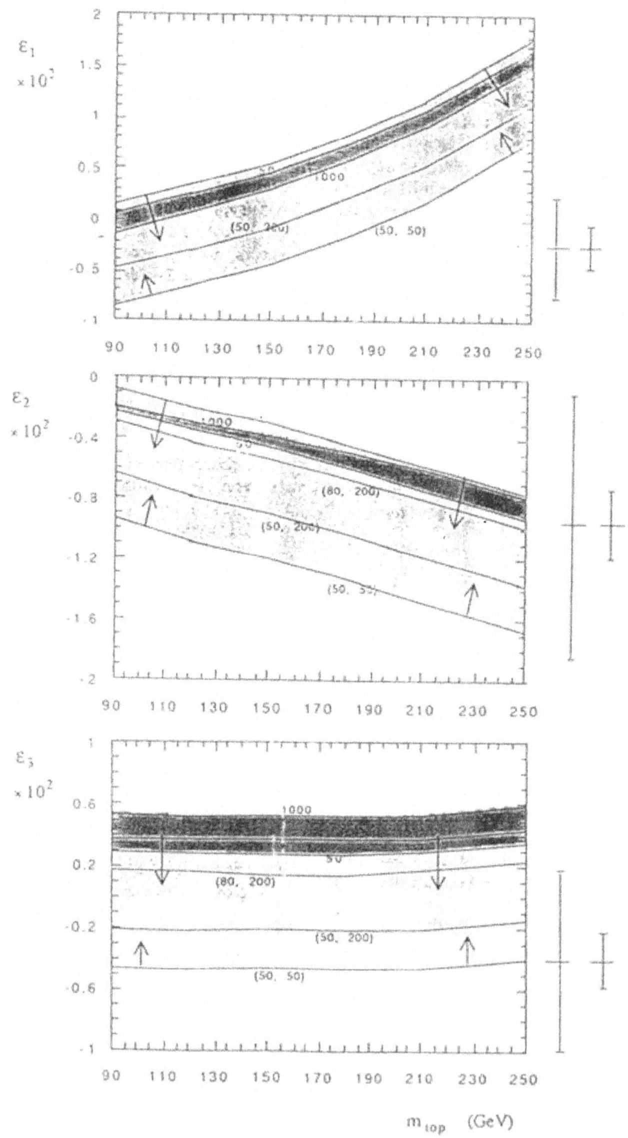

Fig. 10 — Ranges of values of ε_1, ε_2 and ε_3 in the Standard Model (dark grey region) for 50 GeV $< M_H <$1 TeV and in the MSSM (light grey region), as functions of m_t. The minimum values (maximum effects) attainable in the MSSM are indicated for different choices of the lightest chargino mass and of the common slepton mass ($m_\chi^\pm$, m_s). Also shown is the experimental value with 1 σ error (present and projected at the end of the LEP I phase).

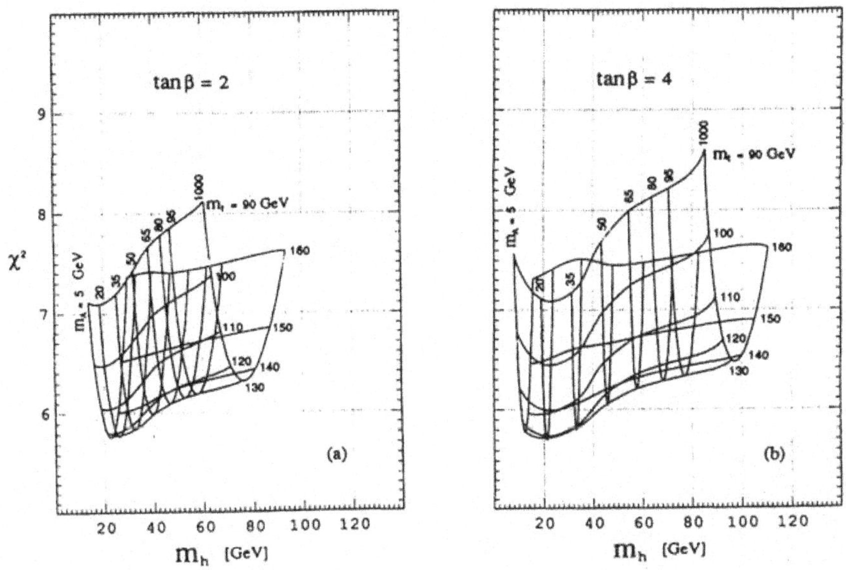

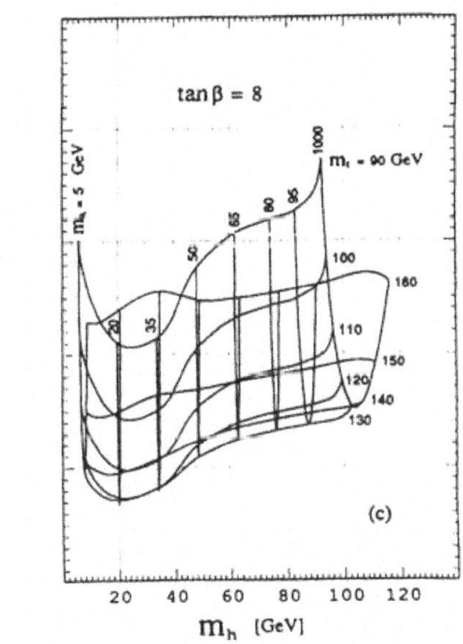

Fig. 11 — The χ^2 curves as functions of m_h for (a) $\tan\beta = 2$, (b) $\tan\beta = 4$, (c) $\tan\beta = 8$. In each figure the curves are shown for different values of m_t in the range 90 to 160 GeV. The values of m_A corresponding to different points on these curves are indicated and joined together. It can be seen that $m_h, m_A \simeq 20$ GeV and $m_t \simeq 130$ GeV are slightly preferred.

Finally, in ref. [40] the effect due to the Higgs sector of the MSSM is examined. Here two kinds of radiative correction contributions must be included: those affecting the different processes at the 1–loop level, and, more relevant, those affecting directly the masses of the supersymmetric Higgses [34]. In [40] the question in

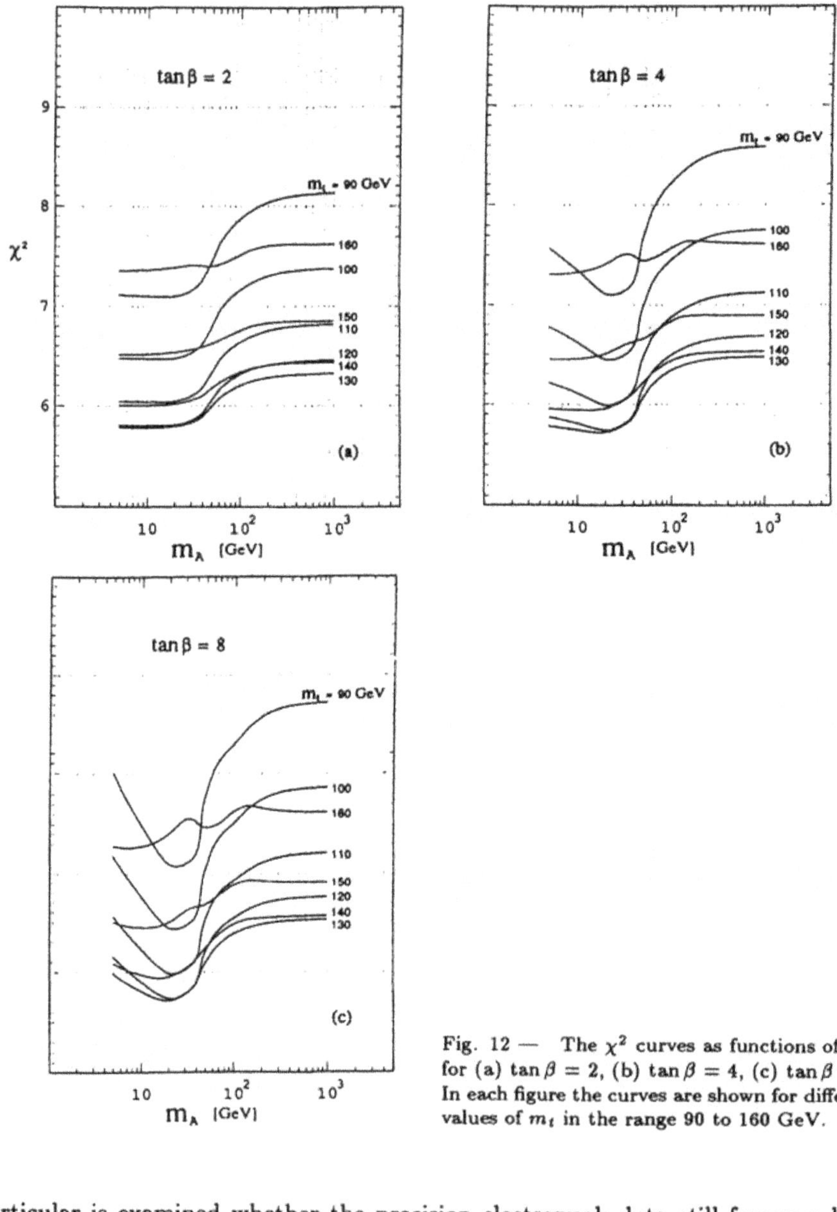

Fig. 12 — The χ^2 curves as functions of m_A for (a) $\tan\beta = 2$, (b) $\tan\beta = 4$, (c) $\tan\beta = 8$. In each figure the curves are shown for different values of m_t in the range 90 to 160 GeV.

particular is examined whether the precision electroweak data still favour a light Higgs boson even in the MSSM, at least in the limit of heavy sparticles. It is found that the preference for a light Higgs is maintained, although the optimal value of the Higgs mass is rather higher, and then closer to its direct experimental limit. In

Fig. 11, taken from ref. [40], we report the χ^2 curves as functions of m_h (the lighter CP-even neutral Higgs of the MSSM) for different values of m_t in the range 90 to 160 GeV and three different values of $\tan\beta$ (the ratio of the two Higgs v.e.v.'s). It is seen that $m_h \simeq 20$ GeV and $m_t \simeq 130$ GeV are slightly preferred. In Fig. 12 the χ^2 curves are plotted as functions of m_A (the mass of the pseudoscalar Higgs expected in the MSSM), for the same values of m_t and $\tan\beta$. It is seen that $m_a \simeq m_h$ is preferred, the effect being more marked in presence of a relatively light top quark.

5. Updating the SM parameters with the last (preliminary) LEP data

In this Section we anticipate, with respect to ref. [37], a re-analysis of the radiative corrections effects within the Standard Model in which the recent updating of the LEP data [11] is taken into account. The relevant point here is the possibility

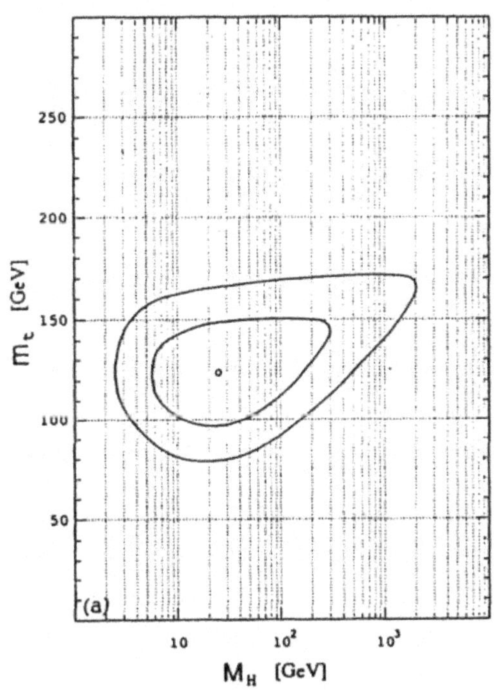

Fig. 13 — As in Fig. 3, the $\Delta\chi^2 = 1$ and 2.7 contours in the (M_H, m_t) plane for for the complete data set, but updating the LEP data according with ref. [11].

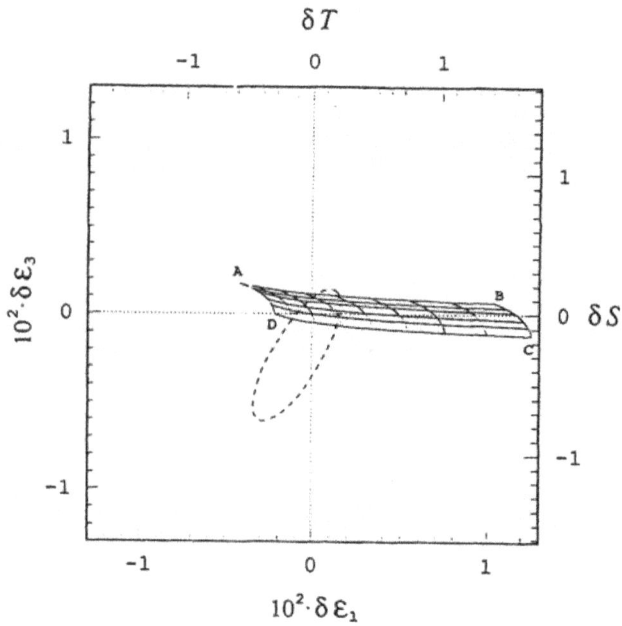

Fig. 14 — Updated model-independent analysis reported in the plane ($\delta\varepsilon_1$, $\delta\varepsilon_3$) (or (δT , δS)). Note the greater overlap of the $\Delta\chi^2 = 1$ experimental ellipse with the Standard Model region, if compared to what is reported in Fig. 5.

of removing some of the slight discrepancies still present in the analysis of the Standard Model performed in Section 2. In particular, the new LEP data realize a better agreement with the Standard Model (a better χ^2), the minimum χ^2 indicates a rather higher value of the Higgs mass (closer to the present experimental limit from LEP [41]), and the new value of $R = \Gamma_h/\Gamma_\ell$ is now not in contradiction with the other precision electroweak measurements. In Fig. 13 the analysis of Fig. 6 is newly reported, with all the data included, without an anomalous dependence on the ratio $R = \Gamma_h/\Gamma_\ell$.

As a consequence, the preferred region in the model -independent parametrization discussed in Section 3 is now in better agreement with the Standard Model, as can be seen in Fig. 14, which shows the projection of the $\Delta\chi^2 = 1$ ellipsoid onto the plane ($\delta\varepsilon_1$, $\delta\varepsilon_3$), compared with the region describing the Standard Model.

6. Conclusions

It is evident, in particular from the content of Section 5, that the Standard Model is getting better and better, and, at least at the concrete level of a comparison with the data, new physics is not required.

However, the present, impressive accuracy of both theory and experiment gives for the first time the possibility of analyzing radiative correction effects beyond the Standard Model.

The analysis seems to disfavour technicolour theories, in particular for large N_{TC} and N_{TF}.

Viceversa, we see good perspectives for supersymmetry, in particular for its minimal realization, the Minimal Supersymmetric Standard Model. In fact

- data agree with a light "lightest Higgs" (with some optimism, slightly better than in the Standard Model);
- if sparticles masses are not too high, m_t is expected slightly lighter than in the Standard Model, which could be a first signal of supersymmetry.

References

[1] M. Veltman, Nucl. Phys. **B123** (1977) 89;
M.S. Chanovitz, M.A. Furman and I. Hinchliffe, Phys. Lett. **B78** (1978) 285.

[2] A few refs. about this subject are:
A. Sirlin, Phys. Rev. **D22** (1980) 971;
W.J. Marciano and A. Sirlin, Phys. Rev. **D22** (1980) 2695, E Phys. Rev. **D31** (1985) 213;
W.F.L. Hollik, Fortschr. Phys. **38** (1990) 165.

[3] G.L. Fogli and D. Haidt, Z. Phys. C – Particles and Fields **40** (1988) 379.

[4] UA1 Collaboration, C. Albajar et al., Phys. Lett. **B185** (1987) 233; **B191** (1987) 462A;
UA2 Collaboration, R. Ansari et al., Phys. Lett. **B186** (1987) 440.

[5] U. Amaldi, A. Böhm, L.S. Durkin, P. Langacker, A.K. Mann, W.J. Marciano, A. Sirlin and H.H. Williams, Phys. Rev. **D36** (1987) 85;
G. Costa, J. Ellis, G.L. Fogli, D.V. Nanopoulos and F. Zwirner, Nucl. Phys. **B297** (1988) 244.

[6] J. Ellis and G.L. Fogli, Phys. Lett. **B213** (1988) 526, Phys. Lett. **B231** (1989) 1896.

[7] J. Ellis and G.L. Fogli, Phys. Lett. **B232** (1989) 139;
P. Langacker, Phys. Rev. Lett. **63** (1989) 1920.

[8] J. Ellis and G.L. Fogli, Phys. Lett. **B249** (1990)543;
P. Langacker and M. Luo, Phys. Rev. **D44** (1991) 817.

[9] J. Ellis, G.L. Fogli and E. Lisi, Phys. Lett. **B274** (1992) 456.

[37] This analysis will be presented in more detail in:
J. Ellis, G.L. Fogli and E. Lisi, preprint BARI-TH.111/92, CERN-TH.6568/92 (1992).

[11] The LEP data we use in the present analysis have been presented in preliminary form by:
D. Schaile, in Proceedings of *Les Rencontres de Physique de la Vallèe d'Aoste* on "Results and Perspectives in Particle Physics" (La Thuile, Aosta Valley, 1992), to appear.

[12] G.L. Fogli, Europhys. Lett. **4** (1987) 527.

[13] CHARM II Collaboration, as presented by R. Beyer, Proceedings of the *International Lepton–Photon Symposium and Europhysics Conference on High Energy Physics*, S. Hegarty, K. Potter and E. Quercigh eds. (Geneva, 1991) Vol. 1, p. 46.

[14] G.L. Fogli, Z. Phys. C – Particles and Fields **43** (1989) 229.

[15] M.C. Noecker, B.P. Masterson and C.E. Wieman, Phys. Rev. Lett. **61** (1988) 310.

[16] S.A. Blundell, W.R. Johnson and J. Sapirstein, Phys. Rev. Lett. **65** (1990) 1411.

[17] UA2 and CDF Collaborations, as presented by H. Plothow–Besch, Proceedings of the *International Lepton–Photon Symposium and Europhysics Conference on High Energy Physics*, S. Hegarty, K. Potter and E. Quercigh eds. (Geneva, 1991) Vol. 1, p. 36.

[18] M. Consoli, W.F.L. Hollik and F. Jagerlehner, Phys. Lett. **B227** (1989) 167.

[19] F. Halzen and B.H. Kniehl, Nucl. Phys. **B353** (1991) 567.

[20] T.H. Chang, K.J.F. Gaemers and W.L. van Neerven, Nucl. Phys. **B202** (1982) 407;
A. Djouadi and C. Verzegnassi, Phys. Lett. **B195** (1987) 265;
B.H. Kniehl and J.H. Kühn, Nucl. Phys. **B329** (1990) 547;
B.H. Kniehl, Nucl. Phys. **B347** (1990).

[21] L.R. Surguladze and M.A. Samuel, Phys. Rev. Lett. **66** (1991) 560;
S.G. Gorishny, A.L. Kataev and S.A. Larin, Phys. Lett. **B259** (1991) 144.

[22] K.G. Chetyrkin and J.H. Kühn, Phys. Lett. **B248** (1990) 359.

[23] H. Burkhardt, F. Jagerlehner, G. Penso and C. Verzegnassi, in *Polarization at LEP*, ed. by G. Alexander, G. Altarelli, A. Blondel, G. Coignet, E. Keil, D.E. Plane and D. Treille, CERN Yellow Report CERN 88–06, Vol. I (1989) p. 145.

[24] F. Jegerlehner, Z. Phys. C – Particles and Fields **C32** (1986) 425;
W.F.L. Hollik and H.J. Timme, Z. Phys. C – Particles and Fields **C33** (1986) 125.

[25] G.J.H. Burgers, F. Jegerlehner, B.A. Kniehl and J.H. Kühn, in *Z Physics at*

LEP, ed. by G. Altarelli, R. Kleiss and C. Verzegnassi, CERN Yellow Report CERN 89-08, Vol. I.

[26] D. Kennedy and B.W. Lynn, Nucl. Phys. **322** (1989) 1;
B.W. Lynn, M.E. Peskin and R.G. Stuart, in *"Physics at LEP"* J. Ellis and R. Peccei eds. (CERN, Geneva, 1986).

[27] M.E. Peskin and T. Takeuchi, Phys. Rev. Lett. **65** (1990) 964.

[28] W.J. Marciano and J.L. Rosner, Phys. Rev. Lett. **65** (1990) 2963.

[29] G. Altarelli and R. Barbieri, Phys. Lett. **B253** (1990) 161.

[30] J. Ellis, G.L. Fogli and E. Lisi, Phys. Lett. **B285** (1992) 238.

[31] G. Altarelli, R. Barbieri and S. Jadach, Nucl. Phys. **B369** (1992) 3;
an updated version of this analysis has been presented by G. Altarelli in the Proceedings of the *International Lepton–Photon Symposium and Europhysics Conference on High Energy Physics*, S. Hegarty, K. Potter and E. Quercigh eds. (Geneva, 1991) Vol. 1, p. 683.

[32] M. E. Peskin and T. Takeuchi, SLAC preprint SLAC-PUB-5618 (1991).

[33] T. Appelquist and G. Triantaphyllou,Yale University preprint YCTP-P49-91 (1991).

[34] Y. Okada, M. Yamaguchi and T. Yanagida, Prog. Theor. Phys. Lett. **85** (1991) 1;
J. Ellis, G. Ridolfi and F. Zwirner, Phys. Lett. **B257** (1991) 83;
H.E. Haber and R. Hempfling, Phys. Rev. Lett. **66** (1991) 1815.

[35] H.P. Nilles, Phys. Rep. **110** (1984) 1;
H.E. Haber and G.L. Kane, Phys. Rep. **117** (1985) 75;
A.B. Lahanas and D.V. Nanopoulos, Phys. Rep. **145** (1987) 1;
R. Barbieri, Riv. N. Cimento **11** (1988) 1.

[36] For a recent review, see:
J. Sóla, Univ. of Barcelona preprint UAB-FT-265 (1991).

[37] A similar analysis is however in preparation:
J. Ellis, G.L. Fogli and E. Lisi, in preparation.

[38] A. Bilal, J. Ellis and G.L. Fogli, Phys. Lett. **B246** (1990) 459.

[39] R. Barbieri, M. Frigeni and F. Caravaglios, Phys. Lett. **B279** (1992) 169.

[40] J. Ellis, G.L. Fogli and E. Lisi, preprint BARI-TH.97/92, CERN-TH.6418/92 (1992).

[41] E. Gross, CERN preprint CERN-PPE/92-91 (1992).

Massive Lepton Searches at LEP/SLC and LEP200

Gregor Herten
Fakultät für Physik, Universität Freiburg,
7800 Freiburg, Germany

ABSTRACT

The LEP and SLC experiments have placed stringent limits on massive leptons. A 4th generation sequential lepton family with a left handed doublet and right-handed singlet is only allowed if $M_{L^0} > 42.8$ GeV and $M_{L^\pm} > 42.8$ GeV. From a measurement of the process $e^+e^- \rightarrow \nu\bar{\nu}\gamma$ the number of light neutrino generations has been measured to be 3.14±0.24±0.12. Unstable heavy neutrinos (with standard couplings to the Z^0) are excluded for $M_{L^0} < 45$ GeV (for Dirac neutrinos). Leptoquarks are excluded for masses below 41 to 44 GeV depending on the charge assignment and the decay branching ratios.

1. Introduction

Searches for massive leptons in e^+e^- collisions have a long tradition. 25 years ago a proposal[1] was presented to search for leptoquarks and heavy leptons at the e^+e^- collider Adone in Frascati. The result of the experiment has been published in 1973 excluding heavy leptons below ≈ 1 GeV.[2] Two years later a positive signal in the $e\mu$ channel was found at higher energies by M. Perl et al. at the e^+e^- collider SPEAR, which was interpreted as the heavy lepton τ.[3] Since then the search for heavy leptons has been one of the first tasks at all new colliders. Especially e^+e^- collisions are well suited for heavy lepton searches, because the signature is distinctive with little background. Sequential heavy leptons have been excluded at DORIS, PETRA/PEP and Tristan for masses below 29.6 GeV.[4] Searches by UA1 at the CERN $p\bar{p}$ collider have excluded charged heavy leptons below 41 GeV.[5] LEP offers the possibility to reach higher masses and to search for new leptons with non-standard couplings.

This article is organized as follows: In section 2 the mass limits are reviewed which can be obtained from the precise measurements of the Z^0 width at LEP. Any production of additional leptons, charged or neutral, results in an increased total Z^0 width. Section 3 presents the direct measurement of Z^0 decays into invisible neutral leptons from the observation of single photon events. Then the strategies for direct searches for heavy leptons and leptoquarks are explained. Finally I present an outlook for massive lepton searches at LEP 200. All limits will be given with 95% C.L. Since the experiments use different methods to obtain limits on new particle production it is not easy to combine the results. I shall therefore present measurements of individual experiments and refer to similar measurements by other groups.

2. Limits from the Z^0 width

The LEP experiments (Aleph, Delphi, L3 and Opal) have accumulated in total about 2 million hadronic Z^0 decays during the 1989, 1990 and 1991 data taking periods. Precise tests of the Standard Model parameters have been performed and the couplings of the Z^0 to charged leptons, neutrinos and quarks have been measured with unprecedented precision. The four LEP experiments have published a combined electroweak analysis[6] of the 1989 and 1990 data. Recently, an update has been presented which also includes the 1991 data.[7] The LEP measurements are in good agreement with the predictions of the Standard Model.

The electroweak data can be used to obtain limits on the production of new particles in Z^0 decays. Any additional Z^0 decay mode $Z^0 \to X$ increases the total width of the Z^0. Depending on the event topology of the final state X the measured hadronic, leptonic or invisible width of the Z^0 will appear to be larger than in the Standard Model. Thus a comparison of the measured Z^0 widths with the Standard Model expectations provides limits on new particle production. Strictly speaking, one assumes that the new physics which is responsible for the $Z^0 \to X$ decay does not modify the Standard Model Z^0 partial widths into known leptons and quarks.

The Standard Model predictions are calculated[8] allowing the top quark and Higgs boson masses to vary between $91 \leq m_t \leq 200$ GeV, $50 \leq m_H \leq 1000$ GeV, and $0.110 \leq \alpha_s \leq 0.126$. For a conservative limit on new particle production the lowest Standard Model prediction $\Gamma_Z = 2468$ MeV has been used (corresponding to $m_Z = 91.175$ GeV, $m_t = 91$ GeV, $m_H = 1000$ GeV, $\alpha_s = 0.110$).

Using the combined LEP data[6] one obtains $\Gamma_Z = 2487 \pm 10$ MeV. Assuming a Gaussian measurement error and dividing the area above the Standard Model prediction in proportion 95:5 the dividing line gives the position of the 95 % C.L. upper bound. From the above values one finds an upper bound of $\Gamma_Z < 2524$ MeV. Thus Z^0 decays into new particles with $\Gamma(Z^0 \to X) > 37$ MeV or $\mathrm{Br}(Z^0 \to X) > 1.5\%$ are excluded independent of the event topology.

Similarly one can use the measured Z^0 invisible width $\Gamma_{inv} = 498 \pm 8$ MeV[6] and finds an upper limit for the production of additional invisible decays of the Z^0 of $\Delta\Gamma_{inv} < 5.7$ MeV.

The expected Z^0 decay widths into charged and neutral leptons, for $Z^0 \to L^+L^-$ and $Z^0 \to L^0\bar{L}^0$ are given by the following formulae:

$$\Gamma_{L^+L^-} = \frac{G_F M_Z^3}{12\pi\sqrt{2}}[\beta(3-\beta^2)g_V^2 + 2\beta^3 g_A^2] \quad (1)$$

$$\text{Dirac neutrinos} \quad \Gamma_{L^0\bar{L}^0} = \frac{G_F M_Z^3}{12\pi\sqrt{2}}\frac{1}{4}\beta(3+\beta^2) \quad (2)$$

$$\text{Majorana neutrinos} \quad \Gamma_{L^0\bar{L}^0} = \frac{G_F M_Z^3}{12\pi\sqrt{2}}\beta^3, \quad (3)$$

where $g_A = I_3 = -\frac{1}{2}$, $g_V = I_3 - 2Q_L sin^2\theta_W$, and $\beta = \sqrt{1 - 4m_L^2/s}$; Q_L is the electric charge and m_L the mass of the corresponding lepton.

From the upper bound on Γ_Z one can exclude charged leptons below 26.2 GeV, Majorana neutrinos below 34.6 GeV and Dirac neutrinos below 42.6 GeV. Similarly, one finds that from the measurement of the invisible width Dirac and Majorana neutrinos are excluded below 40.3 and 45.2 GeV, respectively.

This method can be extended to the hadronic and leptonic widths of the Z^0 and can provide more stringent limits on new particle production with hadronic or leptonic event topologies.[9]

3. Single photon events

A study of events with single photons allows a direct measurement of Z^0 decay rates into neutrinos and other weakly interacting particles (with $m < 0.5\sqrt{s}$) which are not observed in the detector. At c.m.s. energies around the Z^0 mass the dominant contribution to single photon production comes from events with photon radiation in the initial state with a subsequent Z^0 decay into neutrinos. W-exchange terms and their interference with the Z^0 terms contribute less than 3% to the total cross section for three neutrino generations.

The differential cross section can be written as:

$$\frac{d^2\sigma}{dE_\gamma \, dcos\theta_\gamma} = H(E_\gamma, cos\theta_\gamma, s)\ \sigma_0(s'),$$

where H is a radiator function describing the initial state radiation of a photon with energy E_γ at an angle θ_γ at c.m.s. energy $\sqrt{s}$. s' is the reduced center of mass energy squared $s' = s(1 - 2E_\gamma/\sqrt{s})$. The cross section $\sigma_0(s)$ is given by:

$$\sigma_0(s) = 12\pi\ M_Z^2 \frac{s\Gamma_e N_\nu \Gamma_{\nu\bar{\nu}}}{(s - M_Z^2)^2 + s^2\Gamma_Z^2/M_Z^2} + W - \text{terms},$$

where $\Gamma_{\nu\bar{\nu}}$ is the Standard Model decay width of the Z^0 in a pair of neutrinos, and M_Z, Γ_Z and Γ_e are the measured LEP values for the Z^0 mass, the total Z^0 width and the Z^0 partial width into e^+e^- pairs, respectively.

The data at LEP has been taken at center-of-mass energies from 88.56 GeV to 93.75 GeV. Since these energies are close to M_Z typical photon energies from initial state radiation are in the few GeV range. Good trigger efficiency for low energy photons ($E > 1\ GeV$) is therefore required to select single-photon events with high efficiency. A complete detector coverage down to about 1° is needed to reject background from radiative Bhabha scattering.

First experimental results on this process have been obtained by Opal[10] and L3.[11] Recently, the L3 collaboration has performed a new measurement with improved statistics, lower trigger threshold, and a better understanding of the trigger efficiency. The trigger efficiency is 25% at 1 GeV and about 80% for $E_\gamma > 1.5$ GeV. The new L3 measurement is shown in Figure 1, where the cross section $\sigma(e^+e^- \to \nu\bar{\nu}\gamma)$ is shown

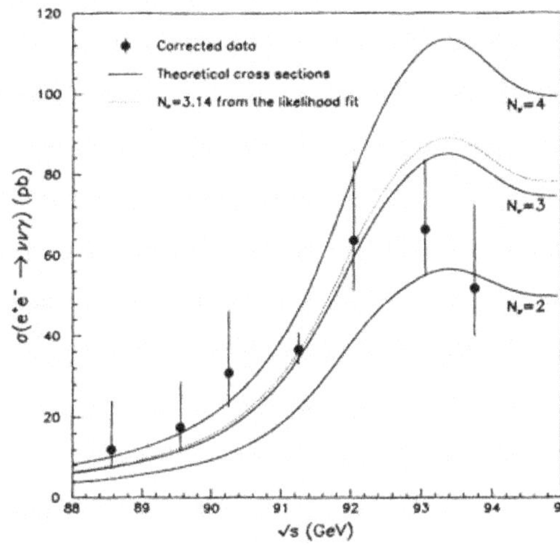

Figure 1: L3 measurement of the cross section $\sigma(e^+e^- \to \nu\bar{\nu}\gamma)$.

as function of $\sqrt{s}$. The measurement is compatible with three neutrino families. From a fit to the cross section one finds for the number of light neutrino generations:

$$N_\nu = 3.14 \pm 0.24(stat) \pm 0.12(sys),$$

which corresponds to a Z^0 invisible width of

$$\Gamma_{inv} = 524 \pm 40(stat) \pm 20(sys)\ MeV.$$

4. Search For Massive Leptons

The mass limits obtained from the Z^0 width can be improved with a direct search for events with massive lepton signature. Assuming a massive sequential lepton, with a left-handed doublet and a right-handed singlet and conserved lepton number, one can distinguish the cases, where the neutrino is lighter or heavier than the charged lepton.

If $M_{L^0} < M_L$ the heavy neutrino is stable and the pair production and decay of massive charged leptons is characterized by events with an $e\mu$ and lepton-hadron signature and missing energy. In the search one requires typically missing energy, missing transverse momentum and isolated leptons. All four LEP experiment and Mark-II at SLC have obtained mass limits with the result that charged lepton masses up to 44.3 GeV are excluded for small neutrino masses.[12–14]

If $M_{L^0} \geq M_L$ and if lepton number is conserved the heavy charged lepton is stable. At LEP one expects therefore to produce pairs of heavy stable charged particles. The same signature is predicted in the SUSY scenario if R-parity is conserved and if the charged slepton is lighter than the lightest neutral SUSY particle. The event

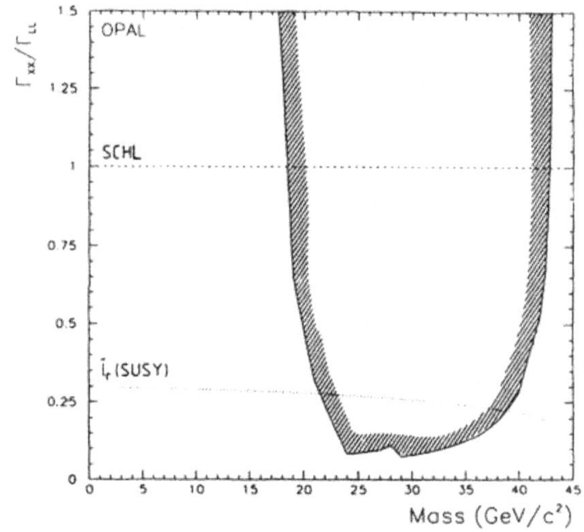

Figure 2: Opal limit on the production of stable charged particles.

signatures in both cases are two back-to-back tracks with low momentum and large time-of-flight. Due to the different spins the cross section is larger for charged sequential leptons than for sleptons (spin 0). Opal has searched for these events using the time-of-flight counters, which are located at the outer radius of the jet chamber. The time difference between the production at the interaction point and the signal in the time-of-flight counters gives a measurement of the velocity. Figure 2 shows the Opal measurement[15]as function of the charged particle mass, expressed as ratio Γ_{XX}/Γ_{ll}, the partial Z^0 decay width of the new stable particle X, Γ_{XX}, divided by the partial width for light leptons, Γ_{ll}. The expected partial widths are indicated for singly-charged heavy leptons (SCHL) and SUSY partners of right-handed leptons. Opal excludes the mass ranges from 18.5 GeV to 42.8 GeV for stable sequential leptons and from 21.5 GeV to 38.8 GeV for stable sleptons.

L3 has performed a similar analysis and has extended the high mass range with a search for low momentum back-to-back tracks in the vertex chamber. No candidate has been found, thus excluding stable sequential leptons below 44.6 GeV.[13]

A combined search by the L3 collaboration[13] for a 4th generation sequential heavy lepton pair with a heavy neutrino L^0 and a charged lepton $L^{\pm}$ is shown in Figure 3. The figure shows the excluded range in the M_{L^0} - $M_{L^{\pm}}$ plane. Contour (a) is derived from the measurement of the total Z^0 width (using $\Delta\Gamma_Z < 71\ MeV$) and contour (b) gives a mass limit for stable neutrinos from the measurement of the invisible width (using $\Delta\Gamma_{inv} < 30\ MeV$). Contour (c) is excluded from the direct search for stable charged leptons and contour (d) from the search for unstable charged leptons. Thus all mass combinations of a 4th generation heavy lepton pair are excluded except if both

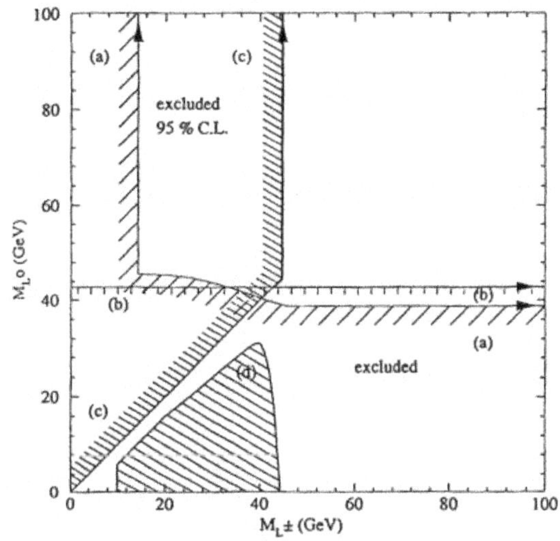

Figure 3: Combined L3 limit on a 4th generation charged lepton and neutrino. Contour (a) is derived from the measurement of Γ_Z; contour (b) from Γ_{inv}; (c) is excluded from a direct search for stable charged leptons and (d) from the search for unstable charged leptons.

masses are larger than 42.8 GeV (for stable Dirac neutrinos).

5. Unstable Neutrinos

So far limits on heavy stable neutrinos have been discussed. One could also imagine that the heavy neutrino mixes with lighter neutrinos, allowing the decay $L^0 \to \ell + W^*$. The decay width of the heavy neutrino is then given by the expression:

$$\Gamma(L^0 \to \ell^- + W^{+*}) = 9 \mid V_{\ell,L^0} \mid^2 \frac{G_F^2 M_L^5}{192\pi^3}, \tag{4}$$

where the factor 9 takes into account the W^* decay modes into $e\nu$, $\mu\nu$, $\tau\nu$, ud and cs (including colour). In case of Majorana neutrinos this factor is increased to 18, because the heavy neutrino is its own antiparticle and therefore the decays $L^0 \to \ell^+ + W^*$ and $L^0 \to \ell^- + W^*$ have equal probability. $|V_{\ell,L^0}|^2$ is the mixing probability between the heavy neutrino L^0 and a light neutrino ν_ℓ, $\ell = e, \mu, \tau$.

The L^0 decay length l_{L^0} depends strongly on the neutrino mass:

$$l_{L^0} = \beta\gamma c\tau_{L^0} \propto \beta \mid V \mid^{-2} M_{L^0}^{-6}, \tag{5}$$

where $\mid V \mid^2 = \mid V_{e,L^0} \mid^2 + \mid V_{\mu,L^0} \mid^2 + \mid V_{\tau,L^0} \mid^2$.

Experimentally three decay length regions are important:

1. Short decay length with $l_{L^0} < 1$ cm, which corresponds to a lifetime less than 60 ps for a neutrino with a mass of 40 GeV. Due to the short decay length the tracks of the decaying particles can be reconstructed with full efficiency in the central tracking chamber.

2. Medium decay length with $l_{L^0} \approx 1$ m. Such a neutrino decays inside the detector volume and the reconstruction efficiency depends critically on the specific cuts used in the event selection.

3. Long decay length with $l_{L^0} > 3$ m. Such decays are not triggered in the LEP detectors and are thus invisible. A limit on long lived neutrinos can only be derived from the measurement of the invisible width and the rate of single photon events as explained in section 3.

As an example, L3 has searched for unstable heavy neutrino decays with short decay length.[13] The search is based on requiring at least one isolated electron or muon with $p > 8$ GeV and missing energy. The total energy measured in the calorimeters should exceed 30 GeV. Since no candidate has been found one excludes unstable neutrinos below 46.4 GeV (Dirac) and 45.1 GeV (Majorana) for the electron, muon and tau decay mode.

Mark-II has performed a special search for the medium decay length.[16] They search for tracks with large impact parameters b and require that the fraction of charged tracks with b/σ_b should exceed 60%, where σ_b is the measurement error on the impact parameter. The detection efficiency for a 35 GeV neutrino is $\approx 45\%$ for an average decay length $< L >= 5$ cm and $\approx 15\%$ for $< L >= 1$ m.

A combined analysis by Aleph of all three regimes is shown in figure 4 as function of the mass and the mixing probability $| U_{\ell L^0} |^2$ $(=| V_{\ell,L^0} |)$ for Dirac neutrinos.[12] The complete range of mixing probabilities is excluded for masses up to ≈ 45 GeV. Contour (a) is obtained from a direct search for L^0; (b) from a search for displaced vertices; (c) from the total hadronic cross section measured at the Z^0 peak. The Delphi limits on $| U_{\ell L^0} |^2$ for Majorana neutrinos are shown in Figure 5 as function of the heavy neutrino mass M_{L^0}.[17] The limits labelled e,μ,τ are obtained from a direct search assuming mixing with one generation only. The limits obtained from Γ_{tot} and Γ_{inv} are also indicated. Similar results have also been obtained by the OPAL experiment.[18]

See-saw model

In the see-saw model every left handed neutrino has a massive right handed partner L^0, which is an isospin singlet. The light neutrinos mix with the heavy neutral singlets and one expects to see events of the type $e^+e^- \rightarrow \bar{\nu}L^0$ and $e^+e^- \rightarrow \nu\bar{L}^0$. The massive neutrinos decay via $L^0 \rightarrow \ell W^*$ and $L^0 \rightarrow \nu_\ell Z^*$. Flavor changing neutral currents are not suppressed since no GIM mechanism operates between left-handed and right-handed particles. Opal has performed a search for the single production of massive neutrinos. The search is based on the missing energy and missing p_T signature of $e^+e^- \rightarrow \nu L^0$ events. The OPAL limits on the mixing probability $| U |^2$ $(=| V_{\ell L} |^2)$

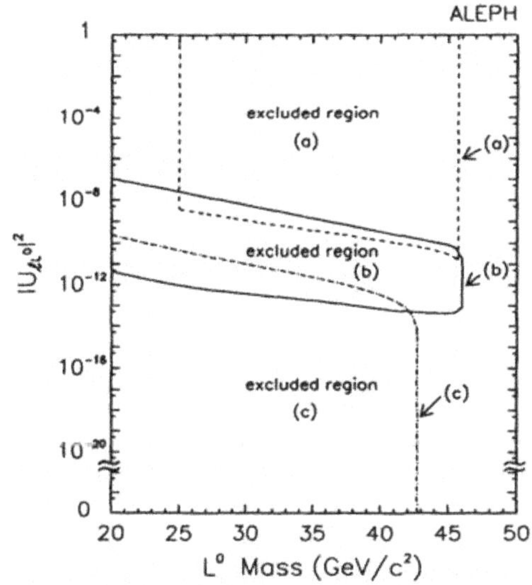

Figure 4: Excluded regions by Aleph for $Z^0 \to L^0\bar{L}^0$ in the $\mid U_{\ell L^0} \mid^2$ versus M_{L^0} plane. Contour (a) is derived from a search for short lived neutral lepton, (b) from a search for displaced vertices and (c) from the measured total hadronic cross section at the Z^0 peak.

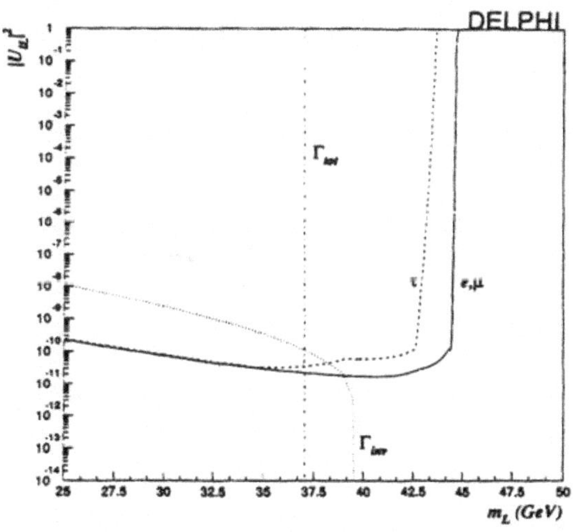

Figure 5: Delphi limits on $\mid U_{\ell L^0} \mid^2$ as a function of m_{L^0} for e, μ and τ mixing for a Majorana neutrino, assuming mixing with one family.

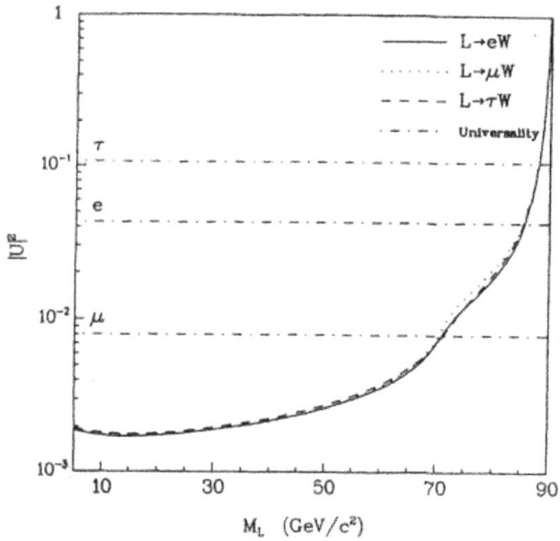

Figure 6: Opal limits on $| U |^2$ as a function of M_L for e, μ, τ mixing in the the process $e^+e^- \rightarrow \nu \bar{L}^0$ or $\bar{\nu} L^0$.

are shown in figure 6. They extend down to $| U |^2 < 3 \times 10^{-3}$ for masses below 50 GeV; $< 8 \times 10^{-3}$ for $M_{L^0} = 70$ GeV; and $< 2 \times 10^{-2}$ for $M_{L^0} = 80$ GeV. Separate curves indicate the case of 100% mixing with electron, muon or tau. The limits assume that 2/3 of the decays proceed via the charged current and 1/3 via the neutral current process. The horizontal lines indicate the weak universality limits.[19]

6. Leptoquarks

Leptoquarks are colour-triplet particles with a Yukawa-type coupling to a lepton-quark pair. They are predicted in several theories beyond the standard model, such as grand unified theories,[20] technicolor,[21] compositeness,[22] and E6 models[23] inspired by superstring theories. The predicted quantum numbers differ from model to model but in most cases they have a distinctive decay into a lepton-quark pair.

As an example the L3 search[24] for leptoquarks will be presented here. The leptoquarks have spin 0, electric charge $-1/3$ or $+2/3$, they are color triplet and SU(2) singlet states. They are pair produced in e^+e^- collisions with the cross section:

$$\frac{d\sigma}{dcos\theta} = \frac{3\pi\alpha^2}{8s} \beta^3 \, sin^2\theta \sum_{j=L,R} | C_j |^2, \tag{6}$$

where

$$C_j = \sum_V g_V^2 Q_V(D) \frac{s}{s - m_V^2 + i m_V \Gamma_V} Q_V^j(e). \tag{7}$$

β is the leptoquark velocity and C_j contains the Z^0 propagator and the leptoquark couplings to the Z^0. In this calculation only the dominant s-channel Z^0 exchange has been included.

Leptoquarks couple to a single generation such that charge $-1/3$ leptoquarks decay into $u_R\ell_R^-$, $u_L\ell_L^-$, $d_L\nu_L$ and the $Q = 2/3$ leptoquarks into $u\bar{\nu}_\ell$, $d\ell^+$ final states. u, d and ℓ are the up-type, down-type quarks and the charged lepton of the same generation. Thus one expects to find events with the signature:

1. $e^+e^- \rightarrow D\bar{D} \rightarrow q\ell + \bar{q}\bar{\ell}$ giving two leptons and two jets.

2. $e^+e^- \rightarrow D\bar{D} \rightarrow q\ell + \bar{q}\bar{\nu}$ giving two leptons, two jets and missing energy.

3. $e^+e^- \rightarrow D\bar{D} \rightarrow q\nu + \bar{q}\bar{\nu}$ giving two jets and missing energy.

A search for scalar leptoquarks from L3 is shown in Figure 7. In this study one has used the following branching ratios: $Br(D \rightarrow q\ell) = 2/3, Br(D \rightarrow q\nu) = 1/3$, and specifically for the third generation leptoquarks $Br(D \rightarrow \nu_\tau b) = 1$, $Br(D \rightarrow \tau b) = 1$. The solid lines correspond to a leptoquark of charge $-1/3$ and the dashed line to a leptoquark of charge 2/3. The lowest mass limit (41.6 GeV) is found for a third-generation leptoquark with charge 2/3. In a further study the mass limits have been obtained as a function of the decay branching ratio $Br(D \rightarrow \ell^- u) = 1 - Br(D \rightarrow \nu d)$. The combined L3 analysis excludes leptoquarks below 41 to 44 GeV depending on the charge assignment and decay branching ratios. Similar results have been obtained by other experiments.[25,26]

7. Massive lepton searches at LEP 200

In this preview for LEP 200 I follow the studies performed at an earlier workshop.[27] At LEP 200 massive charged and neutral lepton masses up to $m_L \leq E_{beam}$ are accessible. Future experiments at the Tevatron will presumably raise the mass limit for charged leptons beyond the LEP 100 limits, but the mass range above 60 GeV is the domain of LEP 200. For the detection of sequential charged leptons one should distinguish two different mass regimes, below and above the W mass. The charged heavy lepton signature for $m_L < m_W$ is similar as at LEP 100. Thus the search strategy is as described in section 4. With sufficient amount of integrated luminosity one should reach mass limits close to the beam energy. The situation changes drastically for lepton masses slightly above the W-mass. There the topology of $e^+e^- \rightarrow L^+L^-$ resembles $e^+e^- \rightarrow W^+W^-$. Since the W-pair production rate is 10 times larger than the lepton pair rate the detection of heavy charged leptons is rather difficult. Small kinematic differences exist but high luminosity is needed to separate lepton-pair and W- pair production.

The search for massive neutrinos is much simpler since the signature is background free. One expects to find final states with three or four hard leptons or two leptons and hadrons with missing energy. These configurations allow a detection of massive neutral leptons up to the highest masses which are kinematically accessible.

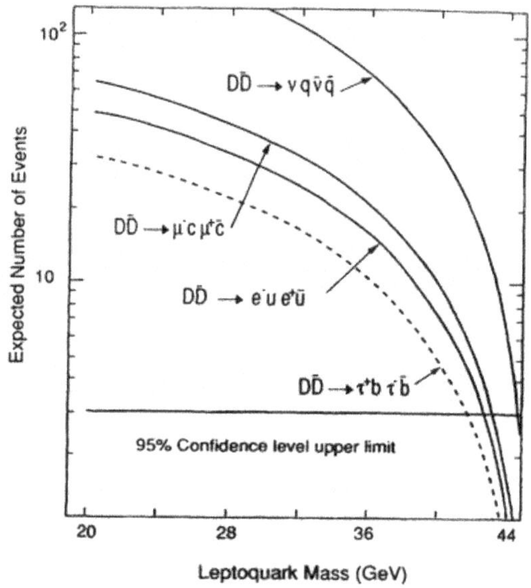

Figure 7: L3 limits on leptoquarks ($Q = -1/3$) in terms of mass and branching ratio belonging to the (a) electron or (b) muon family.

The following table[27] indicates the cross section and the expected number of events for an integrated luminosity of 500 pb^{-1} for charged and neutral leptons at two center-of-mass energies. The cross section is normalized to the point-like QED cross section $\sigma_{point} = 4\pi\alpha^2/3s$.

	$\sqrt{s} = 160GeV$	$\sqrt{s} = 200GeV$
$m_{L^\pm} = 60\ GeV$ $R = \sigma/\sigma_{point}$ #events	 0.92 1565	 1.11 1204
$m_{L^\pm} = 90\ GeV$ $R = \sigma/\sigma_{point}$ #events	 - -	 0.54 588
$m_{L^0} = 60\ GeV$ $R = \sigma/\sigma_{point}$ #events	 0.37 623	 0.36 387
$m_{L^0} = 90\ GeV$ $R = \sigma/\sigma_{point}$ #events	 - -	 0.13 145

Table: Cross section and expected number of events at LEP 200 for 500 pb^{-1}.

The numbers presented in the table show that high luminosity is required at LEP 200 to produce enough heavy lepton events to measure decay properties and to determine the mass precisely.

8. Summary

A 4th generation sequential lepton family with a left handed doublet and right-handed singlet is only allowed if $M_{L^0} > 42.8$ GeV and $M_{L^\pm} > 42.8$ GeV. From a measurement of the process $e^+e^- \to \nu\bar{\nu}\gamma$ the number of light neutrino generations has been measured to be $3.14 \pm 0.24 \pm 0.12$. Unstable heavy neutrinos (with standard couplings to the Z^0) are excluded for $M_{L^0} < 45$ GeV (for Dirac neutrinos). Leptoquarks are excluded for masses below 41 to 44 GeV depending on the charge assignment and the branching ratios.

References

1. M. Bernardini et al., INFN/AE-67/3.
2. M. Bernardini et al., Nouvo Cimento 17A (1973) 383.
3. M. Perl et al., Phys. Rev. Lett. 35 (1975) 1489.
4. Particle Data Group, K. Hikasa et al., Phys. Rev. D 45 (1992) VI.31.
5. C. Albajar et al., Phys. Lett. B185 (1987) 241.
6. LEP Collaborations, Phys. Lett. B 276 (1992) 247.
7. D. Schaile, proceedings, Les Rencontres de Physique de la Vallee D'Aoste, La Thuile (Italy), March 8-14, 1992.
8. ZFITTER program, D. Bardin et al., Nucl. Phys. B 351 (1991) 1; and references therein.

9. Aleph collaboration, D. Decamp et al., CERN-PPE/91-149.
10. Opal collaboration, M.Z. Akrawy et al., Z.Phys. C50 (1991) 373.
11. L3 collaboration, B. Adeva et al., Phys. Lett. B275 (1992) 209.
12. D. Decamp et al., Phys. Lett. B 236 (1990) 511.
13. B. Adeva et al., Phys. Lett. B 251 (1990) 321.
14. M.Z. Akrawy et al., Phys. Lett. B 240 (1990) 250.
15. M.Z. Akrawy et al., Phys. Lett. B 252 (1990) 290.
16. Mark-II Collaboration, C.K. Jung et al., Phys. Rev. Lett. 64 (1990) 1091; P.R. Burchat et al., Phys. Rev. D 41 (1990) 3542.
17. P. Abreu et al., Phys. Lett. B 274 (1992) 230.
18. M.Z. Akrawy et al., Phys. Lett. B 247 (1990) 448.
19. M. Gronau et al., C.N. Leung, J.L. Rosner, Phys.Rev. D29 (1984) 2539.
20. J. Pati, A, Salam, Phys. Rev. D 10 (1974) 275; R. Mohapatra, G. Segré, L. Wolfenstein, Phys. Lett. B 145 (1984) 433.
21. S. Dimopoulos, L. Susskind, Nucl. Phys. B 155 (1979) 237; S. Dimopoulos, Nucl. Phys. B 168 (1980) 69; E. Eichten, K. Lane, Phys. Lett. B 90 (1984) 125.
22. B. Schrempp, F. Schremmp, Nucl. Phys. B 231 (1984) 109; Nucl. Phys. B 242 (1984) 203; Phys. Lett. B 153 (1985) 101.
23. V. Angelopoulos et al., Nucl. Phys. B 292 (1987) 59; D. Schaile, P. Zerwas, *Proceedings of the Workshop on Physics at Future Accelerators*, CERN 87-01 (1988). N. Tracas, S.Vlassopoulos, Phys. Lett. B 220 (1989) 285.
24. L3 collaboration, B. Adeva et al., Phys. Lett. B 261 (1991) 169.
25. DELPHI collaboration, P. Abreu et al.,Phys. Lett. B 276 (1991) 222.
26. OPAL collaboration, G. Alexander et al., Phys. Lett. B 262 (1991) 123.
27. P. Igo-Kemenes et al. in ECFA Workshop on LEP200, Aachen 29.9-1.10.1986, CERN 87-08, Vol.2, p.251.

Compositeness at LEP

Maria Bardadin-Otwinowska

Lab. de Physique Corpusculaire
IN2P3-CNRS, Univ. Blaise Pascal
F - 63 177 Aubière CEDEX, France

Abstract

Searches for compositeness made by four LEP collaborations are reported. Limits are set on excited fermion masses and couplings. A limit on the branching ratio $Z \to \gamma\gamma\gamma$ is determined. Four-fermion and two-fermion, two-boson contact terms are studied in the reactions $e^+e^- \to \ell^+\ell^-$ and $e^+e^- \to \gamma\gamma$ respectively and limits are obtained on the energy scale of a new interaction.

Introduction

The regular pattern of three fermion families is suggestive of an underlying structure. However the old idea of postulating another layer of matter did not result in a convincing theory of a new constituent-binding interaction, The main difficulty is to reconcile the mass scale of presently known particles with a much larger fundamental compositeness scale, Λ, which is of the order of 1 TeV or larger, as suggested by the absence of any evidence of compositeness in the experimental data.

The present strategy consists in looking for possible non-standard phenomena using phenomenological models as a guide [1]. Several scenarios are possible: only the fermions are composite while the W and Z are genuine gauge bosons, or W and Z are composite too. The most spectacular proof of the substructure would be the discovery of excited fermions or bosons. More subtle effects might show up in the form of small deviations from the standard model predictions in standard processes. They could be detectable at energies well below Λ provided that the precision of the data is high.

This paper gives a review of recent results on compositeness from LEP including unpublished updates where available. Limits on excited fermion masses and couplings are set from the Z lineshape measurements, from direct searches and from indirect effects of the e^* exchange in the reaction $e^+e^- \rightarrow \gamma\gamma$. A summary of the best limits on the Zf*f couplings is given for f=e, μ, τ, ν and q. The boson sector is tested by looking for an anomalous $Z\gamma\gamma\gamma$ coupling.

Finally, a new analysis of contact terms is presented. Two-fermion, two boson contact interactions are searched for in the reaction $e^+e^- \rightarrow \gamma\gamma$ using the 1989/90 data of four LEP collaborations. Preliminary results from ALEPH are reported on a search for four-fermion contact terms in the reaction $e^+e^- \rightarrow \ell^+\ell^-$.

Width limits from Z lineshape

Any new particle having a large coupling to the Z boson would increase significantly the Z width. The maximum allowed values of the total and topological widths available for non-standard Z decays, compatible both with the Z lineshape measurements and the standard model predictions, were determined in ref. [2]. and are updated here using the data collected at LEP during 1989-91.

The following lineshape variables are used: the total Z width Γ_{tot}, the peak hadronic cross section σ^0_{had}, hadron to lepton ratio R and invisible to lepton width ratio Γ_{inv}/Γ_l. The experimental values are averages of the measurements reported by four LEP Collaborations [3]. The standard model predictions are calculated [4] for three lepton families and 5 quarks, assuming that new channels mix incoherently with the standard decays. The top and Higgs masses and α_s are allowed to vary within the limits: $91 < m_t < 200\,\mathrm{GeV}/c^2, 51 < m_{\mathrm{Higgs}} < 1000\ \mathrm{GeV}/c^2$, $0.119 <$

$\alpha_s(m_Z) < 0.129$; the range of $\alpha_s(m_Z)$ corresponds to one standard deviation from the average of the LEP measurements using event topology [5]. The most conservative bound is chosen for each variable. The allowed range of each variable is determined and hence limits on topological widths are derived at 95% c.l.. The results are listed in Table 1. The words "hadronic" or "leptonic" do not refer to quantum numbers of a new decay channel but indicate that its topology is similar to that of a $q\bar{q}$ or a lepton pair event, respectively. Other decays, in particular the neutral ones, belong to the "invisible" category.

Topology	Source	Width limit (MeV)	Br.ratio limit (%)
any	Γ_t	37	1.5
non-hadronic	σ_h^0	21	0.7
purely leptonic	R	2.5	0.1
invisible	Γ_{inv}/Γ_l	19	0.7

Table 1: Maximum allowed values of widths and branching ratios for non-standard Z decays, at 95 % c.l., derived from Z lineshape parameters measured at LEP.

In spite of smaller experimental errors, no improvement is found with respect to previous results [2] since the measured value of the Z width has increased. The sensitivity of the method is limited anyway to branching ratios of the order of 10^{-3} as long as the top mass remains unknown.

For comparison, a direct search performed by a single LEP experiment in the most favourable conditions (full efficiency and zero background) allows the possibility of reaching branching ratios down to $5 \cdot 10^{-6}$.

Excited fermion mass limits

Excited leptons are expected to couple to the Z boson like ordinary fermions and therefore should be copiously produced in pairs at LEP1, provided they are not too heavy. Limits on excited fermion masses are set from the above width limits using the relation [1] between the $Z \rightarrow f^*\bar{f}^*$ decay width and the f^* mass. For excited neutrinos three types of couplings are considered: Dirac, Majorana and homodoublet [6].

The maximum allowed total width, 37 MeV, excludes the existence of excited fermions with a mass smaller than the value given in Table 2, independent of the f^* decay mode. Better limits are achieved for excited leptons under the assumption (which could be false [7]) of a purely radiative deexcitation $f^* \rightarrow f\gamma$.

Excited fermion	any decay	radiative decay
u-type quark	41.7	-
d-type quark	44.6	-
charged lepton	29.7	35.5
Dirac neutrino	43.6	45.0
Majorana neutrino	36.3	39.9
homodoublet neutrino	45.4	45.5

Table 2: Limits on excited fermion masses in GeV/c^2. Masses between 0 and the value given in the table are excluded at 95% confidence level. The first limit is valid for any decay mode, the second one assumes radiative decay of excited leptons.

Direct searches for the reaction $e^+e^- \to f^*\bar{f}^*$

A direct search for excited fermions pairs allows the exploration of a small mass region not excluded from the Z width. Excited electrons, muons, taus and neutrinos are searched for assuming the radiative decay mode $f^* \to f\gamma$. No signal is found and f^* masses up to about $m_Z/2$ are excluded under the assumption that the radiative branching ratio is 100%.

Excited neutrino data are interpreted in ref. [8] in terms of a new mechanism with a $Z\nu^*\nu^*$ coupling k times smaller than the standard model coupling of the Dirac-type neutrino. The following limit is set [9]:

$$k \times Br(\nu^* \to \nu\gamma) < 0.02$$

for ν^* masses up to 37 GeV.

Excited quarks are expected to decay into a photon or a gluon, with branching ratios of 8% and 92%, respectively [10]. No q^* signal is found. The limits on the two branching ratios [8] exclude the values of the q^* mass larger than 45 GeV independent of the branching ratios.

In conclusion, LEP 1 has almost reached the kinematical limit $m_Z/2$. Not much further progress on excited lepton mass limits is expected before LEP2 .

Gauge boson-f^*-f couplings

Searches for excited fermions in the reaction $e^+e^- \to f^*f$ at LEP are based on a model [6] which assumes the existence of left-handed and right-handed excited fermion doublets. The transition between an excited fermion, an ordinary fermion and a gauge boson V = Z, W or γ is magnetic. The Vf^*f couplings, c_{Vf^*f}, are related to each other by $SU(2) \times U(1)$ invariance and depend on only two unknown parameters f and f'. The interaction Lagrangian is proportional to c_{Vf^*f}/Λ, where Λ is the compositeness scale.

Limits on the $\gamma f^* f$ couplings were set from searches for the reaction $e^+e^- \to f^*f \to f^+f^-\gamma$ at lower energy e^+e^- colliders. The Zf^*f couplings became accessible at LEP 1 where s-channel Z exchange is a dominant process. The $\gamma e^* e$ coupling remains present in low-angle scattering $e^+e^- \to e^*e$. Events produced via the Ze^*e and $\gamma e^* e$ couplings can be easily separated since they have two distinct topologies: in the first case the final electron and positron are visible in the detector, whereas in the second case one of them is emitted at a low angle and escapes detection.

The decays of excited fermions can be radiative, $f^* \to f\gamma$, or weak, $f^* \to fZ$ or $f^* \to fW$, where the gauge boson can be real or virtual. The branching ratios depend on the Vf^*f couplings and are unknown, unless a relation between f and f' is assumed. The simplest choice $f = f'$ implies $c_{\gamma\nu^*\nu} = 0$, i.e. it excludes radiative ν^* decays. In this paper f and f' are considered unrelated. The expected number of events of the reaction $e^+e^- \to f^*f$ is proportional to $(c_{Vf^*f}/\Lambda)^2 \times$ B($f^* \to$ X), where X is the f^* decay channel studied. A negative search result is expressed in terms of a limit on the square root of this quantity as a function of the f^* mass.

$Z\ell^*\ell$

Single ℓ^* production has been sought in the reactions $e^+e^- \to \ell^+\ell^-\gamma$, where $\ell = e, \mu$ or τ by all LEP collaborations [8,9,11,12,13]. After a suitable renormalisation of the published limits [2], the $Z\ell^*\ell$ couplings are similar. The most stringent results for an excited lepton mass $m^* = 50\,\text{GeV}$ are:

$$
\begin{aligned}
(c_{Ze^*e}/\Lambda) \;\; B^{1/2}(e^* \to e\gamma) &< (\;8\,\text{TeV})^{-1},\\
(c_{Z\mu^*\mu}/\Lambda) \;\; B^{1/2}(\mu^* \to \mu\gamma) &< (10\,\text{TeV})^{-1},\\
(c_{Z\tau^*\tau}/\Lambda) \;\; B^{1/2}(\tau^* \to \tau\gamma) &< (\;7\,\text{TeV})^{-1}.
\end{aligned}
$$

No results have been reported on the weak decays modes $\ell^* \to Z^*\ell$ and $\ell^* \to W^*\nu$.

$Z\nu^*\nu$

Two decay channels of the ν^* were studied: $\nu^* \to \nu\gamma$ [8,9,14] and $\nu^* \to$ We [14]. No ν^* candidates are found. Fig.1 shows the limits on the $Z\nu^*\nu$ coupling plotted as a function of the ν^* mass. The radiative mode limits from ALEPH and L3 are combined according to the formula $L^{-1} = L_1^{-1} + L_2^{-1}$ valid for a "zero-event limit".

The present limits for $m^* = 50$ GeV are:

$$
\begin{aligned}
(c_{Z\nu^*\nu}/\Lambda) \;\; B^{1/2}(\nu^* \to \nu\gamma) &< (10\,\text{TeV})^{-1},\\
(c_{Z\nu^*\nu}/\Lambda) \;\; B^{1/2}(\nu^* \to \text{eW}) &< (\;3\,\text{TeV})^{-1}.
\end{aligned}
$$

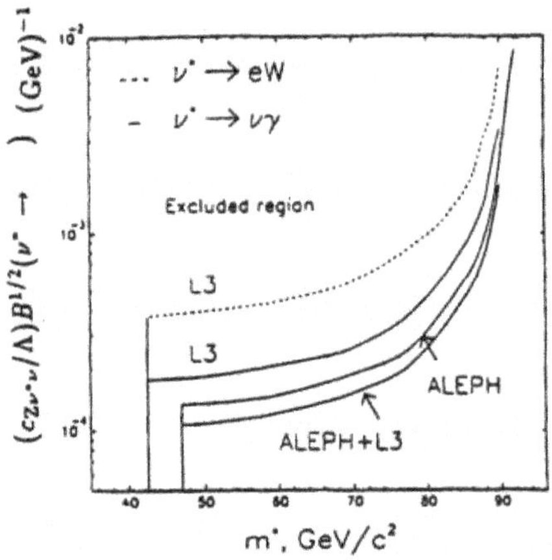

Figure 1: Limits on the excited neutrino coupling

An interesting possibility opens at LEP2: ν^* masses larger than the total energy available and the ν^*eW coupling can be explored in the reaction $e^+e^- \rightarrow W^+W^-$.

Zq*q

Direct searches for single q* production have been performed by the LEP collaborations [8,9,15,16,17] through the decays $q^* \rightarrow q\gamma$ and $q^* \rightarrow qg$. No excited quark signal is found. Although the formalism of q* production in e^+e^- interactions is the same as for ℓ^* production, the limits are reported in terms of the quantity f/λ following the formulae of ref [18].

A summary of the results on the $q^* \rightarrow q\gamma$ decay mode, taken from ref [19], is shown in fig.2a. Results on the $q^* \rightarrow$qg mode are given in Fig.2b.

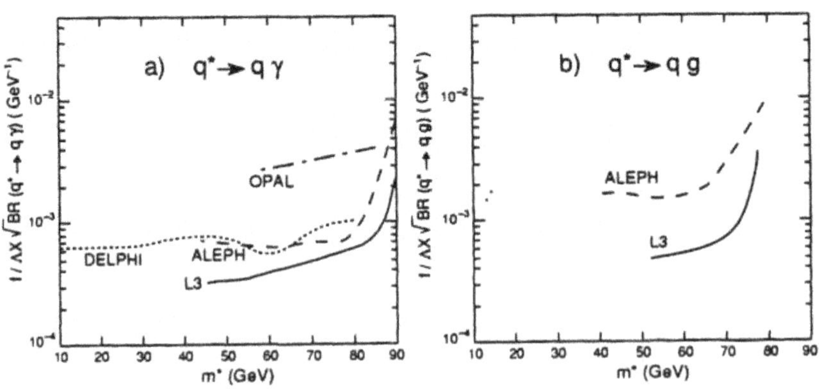

Figure 2: Limits on the excited quark couplings

The best limits on the Zq*q coupling for $m^* = 50\,\mathrm{GeV}$ are:

$$\begin{array}{lll} (f/\lambda) \;\; B^{1/2}(\mathrm{q}^* \to q\gamma) & < & (\,3\,\mathrm{TeV})^{-1}, \\ (f/\lambda) \;\; B^{1/2}(\mathrm{q}^* \to \mathrm{q}g) & < & (\,5\,\mathrm{TeV})^{-1}. \end{array}$$

γe*e

The γe*e coupling is studied at LEP in the t-channel process $e^+e^- \to e^*e \to e(e)\gamma$. Limits are set on the coupling $\lambda/m^* = \sqrt{2}c_{\gamma e^* e}/\Lambda$ used by lower energy experiments. The results obtained by four LEP expreriments [8,9,11,12,13] are similar. The limit for an e^* mass of about 50 GeV is typically

$$(\lambda/m^*)B^{1/2}(\mathrm{e}^* \to e\gamma) < (\;2\;\;\mathrm{TeV})^{-1}.$$

A direct search for e^* production is limited to the mass region below the kinematic limit $m^* < m_Z$. Higher masses can be reached indirectly by looking for possible deviations from QED due to e^* exchange in the reaction $e^+e^- \to \gamma\gamma$. The limit derived in ref. [2] from the combined 1989/90 angular distribution data of four LEP experiments [8,20,21,22] excludes a region in the λ/m^* versus m^* plane extending up to m^*=114 GeV. The forbidden values of λ/m^* are of the order of $(100\ \mathrm{GeV})^{-1}$ or larger.

Search for anomalous gauge boson couplings

The standard model calculation [23] predicts Z→ $\gamma\gamma\gamma$, Z→ $gg\gamma$ and Z→ ggg branching ratios well below the present sensitivity of the LEP experiments. However non-standard effects due to a composite Z boson or to its vector or scalar partners might enhance the branching ratios up to 10^{-5} [24,1,25]. Experimental searches for the decays Z→ $\gamma\gamma\gamma$ [8,9,20,26,22] and Z→ $gg\gamma$ [8,9,17] gave negative results. The branching ratio limits reported by individual experiments are summarized in table 3. The limit from ALEPH is derived assuming a contact interaction.

decay mode	standard model prediction	experimental upper limit (10^{-5})			
		ALEPH	DELPHI	L3	OPAL
Z→ $\gamma\gamma\gamma$	$1\cdot10^{-9}$	0.9	3.6	3.3	3
Z→ $gg\gamma$	$7\cdot10^{-7}$	9.2			330

Table 3: The Z→ $\gamma\gamma\gamma$ and Z→ $gg\gamma$ branching ratios.

Figure 3 shows the energy variation of the ratio of the observed to the predicted number of events for the combined sample from four LEP experiments including the 1991 statistics. Altogether 50 events are found while 59 are expected from QED.

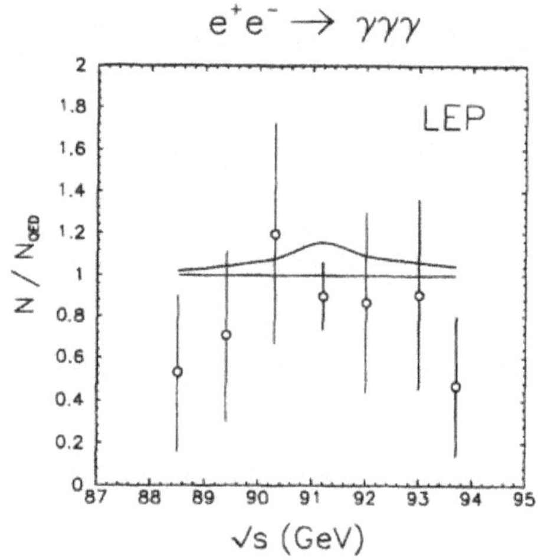

Figure 3: Number of observed $\gamma\gamma\gamma$ events divided by the QED prediction for four LEP experiments. The curve is the maximum $Z \to \gamma\gamma\gamma$ contribution

A fit is performed to the number of observed events versus energy by a maximum likelihood method with a Poisson probability distribution. The fit gives an unphysical negative value of the $Z \to \gamma\gamma\gamma$ branching ratio. A conservative limit is taken as the 95% c.l. deviation from $B = 0$:

$$B(Z \to \gamma\gamma\gamma) < 1.4 \cdot 10^{-5}.$$

The curve in fig.3 shows the corresponding maximum allowed resonant contribution.

It has been argued [27] that the anomalous coupling affects the Z peak more strongly than the wings. There is no evidence for such an effect, the observed number of events at 91.22 GeV being 30 while 33.5 are predicted by QED.

Contact terms

The contact term approach [28,29] assumes that a new interaction at a large energy scale Λ can be considered at present energies as a perturbation to the standard model. The Lagrangian is written as a sum

$$\mathcal{L} = \mathcal{L}_{s.m.} + \mathcal{L}_{contact}.$$

where $\mathcal{L}_{contact}$ is an effective Lagrangian of the new interaction. The contact Lagrangian is built from the fields of presently known particles and it is proportional to a power of Λ depending on the dimensions of the fields involved. Chiral invariance of the fermion currents is required in order to keep the fermion masses much smaller than Λ.

In the context of compositeness the contact terms are remnants of an underlying constituent-binding force. However they can also be generated by various other mechanisms such as heavy boson exchange, anomalous couplings etc. In any case, the formalism of contact terms is a useful way of parametrising possible deviations from the standard model. The parameter Λ sets the scale of new physics beyond the standard model.

Examples of contact terms studied at LEP are represented in fig. 4 and described briefly below. Results of searches for the $ee\gamma\gamma$ and $ee\ell\ell$ contact terms are reported in the following two sections. More details of the analysis can be found in [30].

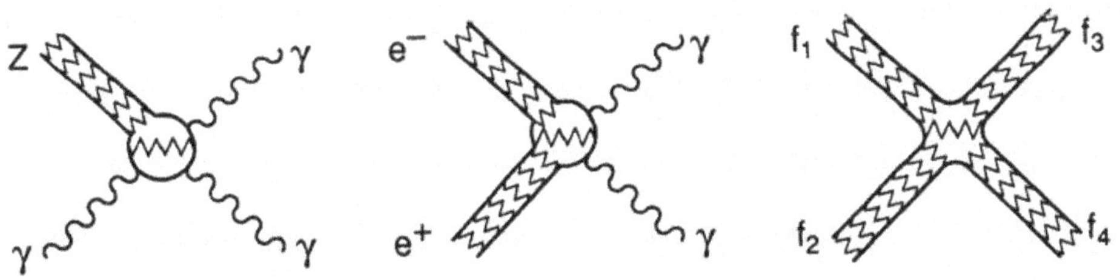

Figure 4: Diagrams of contact terms generated by compositeness.

$Z\gamma\gamma\gamma$

The $Z\gamma\gamma\gamma$ vertex is an example of the four-boson contact interactions. The effective Lagrangian has the form

$$\mathcal{L}_{Z\gamma\gamma\gamma} = \frac{e^3}{\Lambda^4}\left(Z^{\mu\nu}F_{\mu\nu}\right)\left(F^{\alpha\beta}F_{\alpha\beta}\right) + \text{perm.}$$

The width of the decay Z→ $\gamma\gamma\gamma$ calculated from the contact term model [24] is proportional to Λ^{-8}. The limit on the branching ratio Z→ $\gamma\gamma\gamma$ determined from the combined LEP data implies a very weak constraint $\Lambda > 25.4$ GeV.

$ee\gamma\gamma$

This two-boson, two-fermion vertex is described by the following Lagrangian:

$$\mathcal{L}_{ff\gamma\gamma} = \frac{2ie^2}{\Lambda^4}F^{\mu\sigma}F^{\nu}_{\sigma}\left(\eta_L\overline{\psi}_L\gamma_\mu\partial_\nu\psi_L + \eta_R\overline{\psi}_R\gamma_\mu\partial_\nu\psi_R\right)$$

where η_L and η_R define the chirality of the fermion current.

The $ee\gamma\gamma$ contact term is searched for through the reaction $e^+e^- \rightarrow \gamma\gamma$. The leading term in the cross section is proportional to Λ^{-4} and comes from the interference of the contact term with the standard QED process.

$ee\ell\ell$

The Lagrangian of a four-fermion contact interaction is built from two chirality conserving. flavour diagonal fermion currents:

$$\mathcal{L}_{ffff} = \frac{g^2}{2\Lambda^2}\Big[\eta_{LL}\overline{\psi_L}\gamma_\mu\psi_L\overline{\psi_L}\gamma^\mu\psi_L + \eta_{RR}\overline{\psi_R}\gamma_\mu\psi_R\overline{\psi_R}\gamma^\mu\psi_R + \eta_{RL}\overline{\psi_R}\gamma_\mu\psi_R\overline{\psi_L}\gamma^\mu\psi_L + \eta_{LR}\overline{\psi_L}\gamma_\mu\psi_L\overline{\psi_R}\gamma^\mu\psi_R\Big]$$

By convention, it is normalised to the strong coupling constant, $g^2/4\pi = 1$.

Previous searches for the $eeff$ contact terms were performed at PEP and PETRA using the data of the reactions $e^+e^- \rightarrow f^+f^-$ [31]. At these energies fermion pair production is governed by QED and the reaction amplitude is real. The contact term amplitude is also real. The interference term, proportional to Λ^{-2}, is much larger than the contact term squared, which is of order Λ^{-4}. At the Z resonance the situation is very different: the standard amplitude is almost purely imaginary and the leading interference term is strongly suppressed. Therefore the sensitivity to contact terms is strongly reduced at LEP 1. Maximal interference is restored off the Z peak, in particular at LEP 2 [32].

Search for $ee\gamma\gamma$ contact interaction

The two-boson, two-fermion contact interactions are studied in the reaction $e^+e^- \rightarrow \gamma\gamma$ using the combined data of four LEP experiments. Four different contact models corresponding to different helicities of the electron current are considered. In each model both signs of the contact term amplitude are possible and thus two different Λ parameters are defined, denoted Λ^+ and Λ^-. The models are defined in table 4.

Model	η_R	η_L
L	0	±1
R	±1	0
$L+R$	±1	±1
$L-R$	±1	∓1

Table 4: Definition of the $ee\gamma\gamma$ contact models.

The differential cross section for the reaction $e^+e^- \rightarrow \gamma\gamma$ is the sum of two contributions: one from QED and the other from the contact interaction (including its interference with QED). The "cut-off" model is also studied; it assumes a breakdown of QED at small distances and is parametrised by a form factor Λ. At the tree level the prediction is:

$$\frac{d\sigma}{d\Omega} = \frac{\alpha^2}{s}\frac{1+\cos^2\theta}{1-\cos^2\theta} + \left(\frac{d\sigma}{d\Omega}\right)^{contact}$$

where the second term depends on the model and is given by the formulae:

$$\left(\frac{d\sigma}{d\Omega}\right)^{L,R} = \pm\frac{\alpha^2 s}{4\Lambda^4}(1+\cos^2\theta) + \frac{\alpha^2 s^3}{32\Lambda^8}(1-\cos^4\theta)$$

$$\left(\frac{d\sigma}{d\Omega}\right)^{L+R} = \pm\frac{\alpha^2 s}{2\Lambda^4}(1+\cos^2\theta) + \frac{\alpha^2 s^3}{16\Lambda^8}(1-\cos^4\theta)$$

$$\left(\frac{d\sigma}{d\Omega}\right)^{L-R} = \frac{\alpha^2 s^3}{16\Lambda^8}(1-\cos^4\theta)$$

$$\left(\frac{d\sigma}{d\Omega}\right)^{cut-off} = \pm\frac{\alpha^2 s}{2\Lambda^4}(1+\cos^2\theta)$$

The R and L models are identical. The "cut-off" and $L+R$ models are very similar. The model $L-R$ is insensitive to the sign of the contact term amplitude and is expected to impose the weakest costraints since it depends only on Λ^{-8}.

The analysis is performed using the combined LEP data collected in 1989 and 1990 [8,20,21,22]. A maximum likelihood fit is performed to the experimental angular distribution, taking into account radiative corrections [33] and experimental efficiencies. Instead of Λ, the parameter $\varepsilon = (1/\Lambda)^4$ is fitted since it is expected to have a gaussian error. Figure 5 shows the results for the four models studied. The curves represent the model predictions for the one-sided, 95% c.l. limits on ε, ε_{min} and ε_{max}.

The limits on the energy scale Λ are

$$\Lambda^+ = (1/\varepsilon_{max})^{0.25}$$
$$\Lambda^- = (-1/\varepsilon_{min})^{0.25}$$

The resolving power of the experiment is defined as $\lambda = (1/\sigma)^{0.25}$, where σ is the one-sided, 95% c.l. error on ε. (For the model $L-R$, $\varepsilon = (1/\Lambda)^8$ and the power 0.25 is replaced by 0.125). When the calculated value of Λ^+ or Λ^- exceeds λ, then the limit on Λ is set equal to λ. Table 5 summarises the values of ε (with 95% c.l. errors), λ, Λ^+ and Λ^-. Similar limits have been reported from TRISTAN [34,35,36].

Future prospects for an improvement of these limits are not optimistic: the limit increases in proportion to the luminosity raised to the power 1/8.

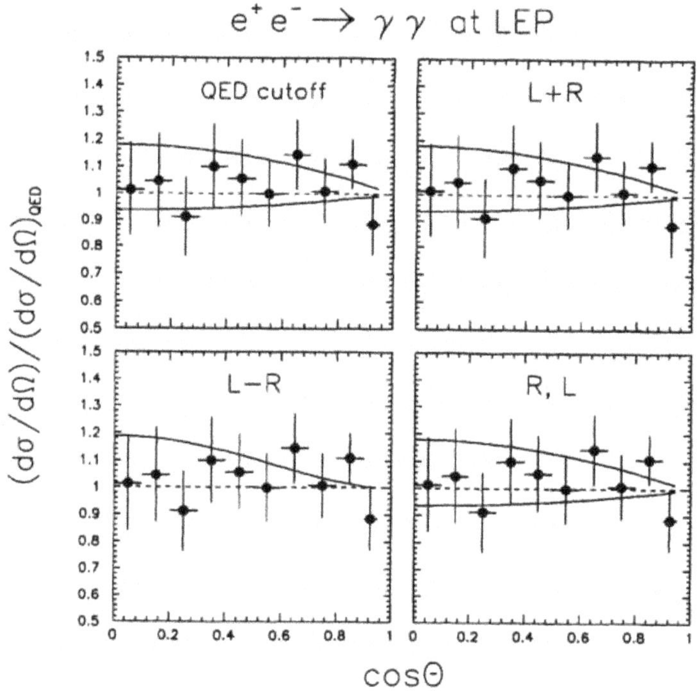

Figure 5: Results of the fits of contact models for the reaction $e^+e^- \to \gamma\gamma$. The data points show the measured differential cross section divided by the standard model predictions. The curves show the maximum allowed deviations at 95% c.l..

Model	ε	λ (GeV)	Λ^+ (GeV)	Λ^- (GeV)
L, R	$(0.29^{+0.66}_{-0.67})10^{-8}$ GeV^{-4}	111	101	111
L + R	$(0.15^{+0.35}_{-0.33})10^{-8}$ GeV^{-4}	131	119	131
L - R	$(0.14^{+0.49}_{-0.45})10^{-15}$ GeV^{-8}	82	79	-
cut-off	$(0.15^{+0.36}_{-0.34})10^{-8}$ GeV^{-4}	130	118	130

Table 5: Results of fits of the $ee\gamma\gamma$ contact terms to the data. The errors on ε are at 95% c.l. . λ is the resolving power and the limits on Λ satisfy $\Lambda^\pm < \lambda$.

Four-fermion contact interactions

Four-fermion contact interactions are searched for in the reactions

$$\begin{aligned} e^+e^- &\to e^+e^- \\ e^+e^- &\to \mu^+\mu^- \\ e^+e^- &\to \tau^+\tau^- \\ e^+e^- &\to \ell^+\ell^-. \end{aligned}$$

where ℓ denotes the any charged lepton. The data [37] are taken at 14 energies around the Z resonance. The integrated luminosity is 19.7 pb^{-1}, with 60% of it from the peak.

The differential cross sections for all lepton flavours agree well with the standard model predictions calculated for $\sin^2\theta_{Wl}(m_Z^2) = 0.233$ and $\rho_\ell(m_Z^2) = 1.0026$. These values correspond to the best estimates of the top and Higgs masses which were obtained from a standard model fit to various electroweak data not including the measurements of $\sin^2\theta_W$ at LEP.

Four-fermion contact terms are now included in order to set limits on the energy scale Λ. The models are named after the chiralities of the fermion currents and are listed in table 6. Two energy scales, Λ^+ and Λ^-, correspond to the two possible signs of the contact amplitude.

Model	η_{LL}	η_{RR}	η_{RL}	η_{RL}
LL	± 1	0	0	0
RR	0	± 1	0	0
VV	± 1	± 1	± 1	± 1
AA	± 1	± 1	∓ 1	∓ 1
RL	0	0	± 1	0
LR	0	0	0	± 1

Table 6: Definition of the parameters of the four-fermion contact models.

The differential cross section for the reaction $e^+e^- \to \ell^+\ell^-$ can be written symbolically as

$$\frac{d\sigma}{d\Omega} \sim |\gamma_s + Z_s + \gamma_t + Z_t + C|^2$$

After development of this expression, the following terms appear:

$|\gamma_s + Z_s|^2$: standard model s-channel exchange amplitude squared,

$|C|^2$: contact term amplitude squared, proportional to $1/\Lambda^4$,

$2\,\Re e\,[(\gamma_s + Z_s)C]$: interference between the s-channel standard model amplitude and the contact term, proportional to $1/\Lambda^2$.

$2\,\Re e\,[(\gamma_s + Z_s)C]$: interference between the s-channel standard model amplitude and the contact term, proportional to $1/\Lambda^2$.

The following terms are present only for the reaction $e^+e^- \to e^+e^-$:

$|\gamma_t + Z_t|^2 + 2\,\Re e\,[(\gamma_t + Z_t)(\gamma_s + Z_s)]$: standard model contributions due to the t-channel and (s,t) interference. This term will be called 't-channel' for short,

$2\,\Re e\,[(\gamma_t + Z_t)C]$: interference between the standard model t-channel exchange amplitude and the contact term. The contribution of Z_t will be neglected since $Z_t << \gamma_t$.

In the improved Born approximation the differential cross section including all terms except the 't-channel' has the form :

$$\left(\frac{d\sigma}{d\Omega}\right)^{Born} = a(s,\Lambda)(1+\cos^2\theta) + b(s,\Lambda)\cos\theta + c(s,\Lambda) + d(\cos\theta,\Lambda).$$

The coefficients $a(s,\Lambda)$, $b(s,\Lambda)$, $c(s,\Lambda)$ and $d(\cos\theta,\Lambda)$ are known functions of s and Λ. They also depend on the coefficients $\eta_{i,j}$. Initial state radiative corrections are taken into account by convoluting the Born cross section with the radiator function [38]. Then the 't-channel' term [39,40] is added to the model predictions.

Five models (the RL and LR models are indistinguishable) are fitted to the set of 14 differential cross sections for each reaction using the maximum likelihood method with a Poisson probability distribution. Systematic errors on the luminosity and on the detection and selection efficiencies are taken into account. The parameter $\varepsilon = 1/\Lambda^2$ is determined by the fit. The results of the fit change very little (compared to the errors from the fit) when the standard model parameters $\sin^2\theta_{Wl}(m_Z^2)$ and $\rho_l(m_Z^2)$ vary within an acceptable range.

Limits on the energy scale Λ are deduced in a similar way as for the $ee\gamma\gamma$ interactions and the condition $\Lambda < \lambda$ is also imposed. The results are given in table 7. Examples of the fits are plotted in figs 6 and 7.

The limits on Λ are of the order of TeV. For the s-channel processes of μ or τ production the limits are comparable with or weaker than those at lower energies. The values obtained for the electrons and all leptons are considerably higher due to the interference between the t-channel photon exchange and the contact term amplitudes. The fitted values of ε with one standard deviation errors are compared in fig. 8-11 with the results obtained at PEP, PETRA and TRISTAN [31]. For electrons the results of ALEPH are more stringent than those from any individual model except LR, and they exceed the combined PEP/PETRA limit for the LL and RR models.

Electrons	ε (TeV^{-2})	λ (TeV)	Λ^+ (TeV)	Λ^- (TeV)
LL	$0.136^{+0.151}_{-0.150}$	2.0	1.6	2.0
RR	$0.147^{+0.176}_{-0.173}$	1.8	1.5	1.8
VV	$0.057^{+0.052}_{-0.053}$	3.4	2.6	3.4
AA	$0.007^{+0.071}_{-0.072}$	2.9	2.8	2.9
LR	$0.155^{+0.181}_{-0.180}$	1.8	1.4	1.8

Electrons	ε (TeV^{-2})	λ (TeV)	Λ^+ (TeV)	Λ^- (TeV)
LL	$0.144^{+0.250}_{-0.248}$	1.5	1.3	1.5
RR	$0.173^{+0.328}_{-0.326}$	1.3	1.1	1.3
VV	$0.206^{+0.112}_{-0.111}$	2.3	1.6	2.3
AA	$-0.067^{+0.087}_{-0.088}$	2.6	2.6	2.1
LR	$0.443^{+0.260}_{-0.262}$	1.5	1.0	1.5

Electrons	ε (TeV^{-2})	λ (TeV)	Λ^+ (TeV)	Λ^- (TeV)
LL	$0.351^{+0.273}_{-0.271}$	1.4	1.1	1.4
RR	$0.564^{+0.382}_{-0.379}$	1.2	0.9	1.2
VV	$-0.053^{+0.113}_{-0.113}$	2.3	2.3	2.0
AA	$0.176^{+0.090}_{-0.090}$	2.6	1.7	2.6
LR	$-0.410^{+0.253}_{-0.255}$	1.5	1.5	1.1

Electrons	ε (TeV^{-2})	λ (TeV)	Λ^+ (TeV)	Λ^- (TeV)
LL	$-0.072^{+0.068}_{-0.069}$	2.9	2.9	2.3
RR	$-0.088^{+0.072}_{-0.072}$	2.9	2.9	2.1
VV	$-0.011^{+0.027}_{-0.028}$	4.7	4.7	4.2
AA	$-0.071^{+0.038}_{-0.038}$	4.0	4.0	2.7
LR	$0.165^{+0.134}_{-0.135}$	2.1	1.6	2.1

Table 7: Results of the contact term fits to the ALEPH data for reactions $e^+e^- \to \ell^+\ell^-$. The parameter ϵ is given with one standard deviation errors. The requirement $\Lambda^\pm < \lambda$ is imposed.

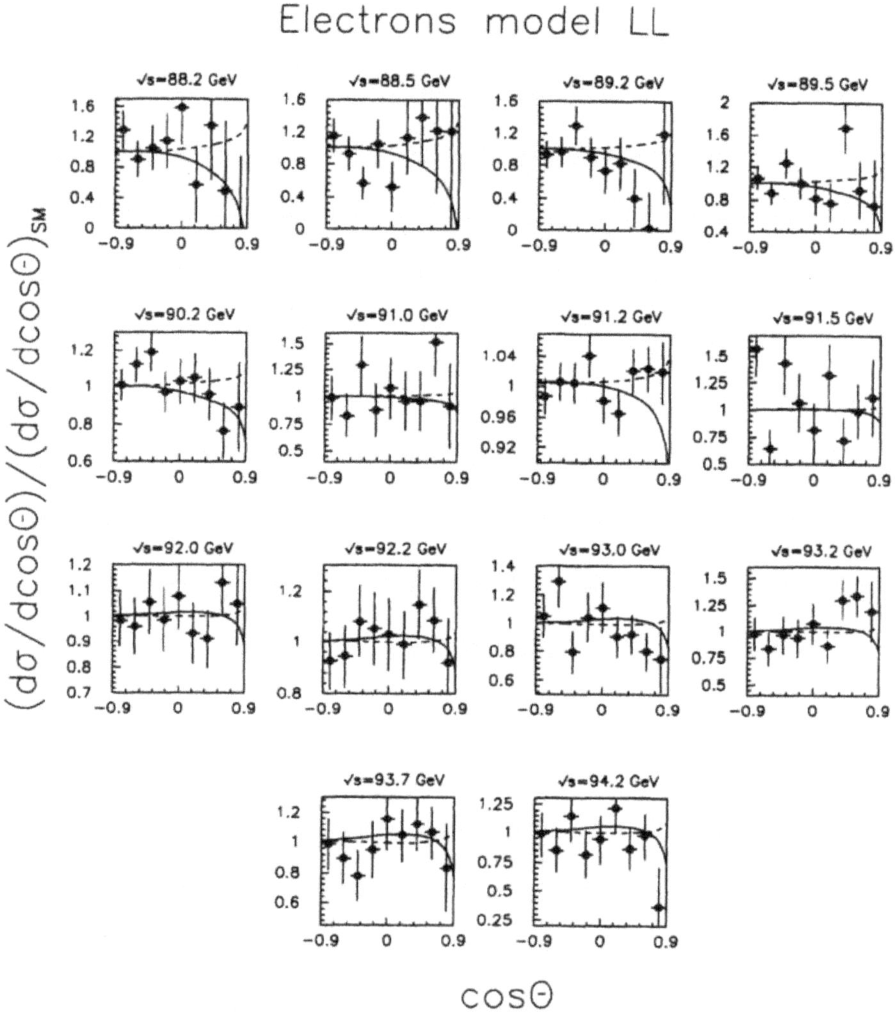

Figure 6: Results of the fit of contact model LL for the reaction $e^+e^- \rightarrow e^+e^-$. The data points are the measured differential cross sections divided by the standard model predictions. The curves are the maximum allowed deviations at 95% c.l. The t-channel contribution is subtracted.

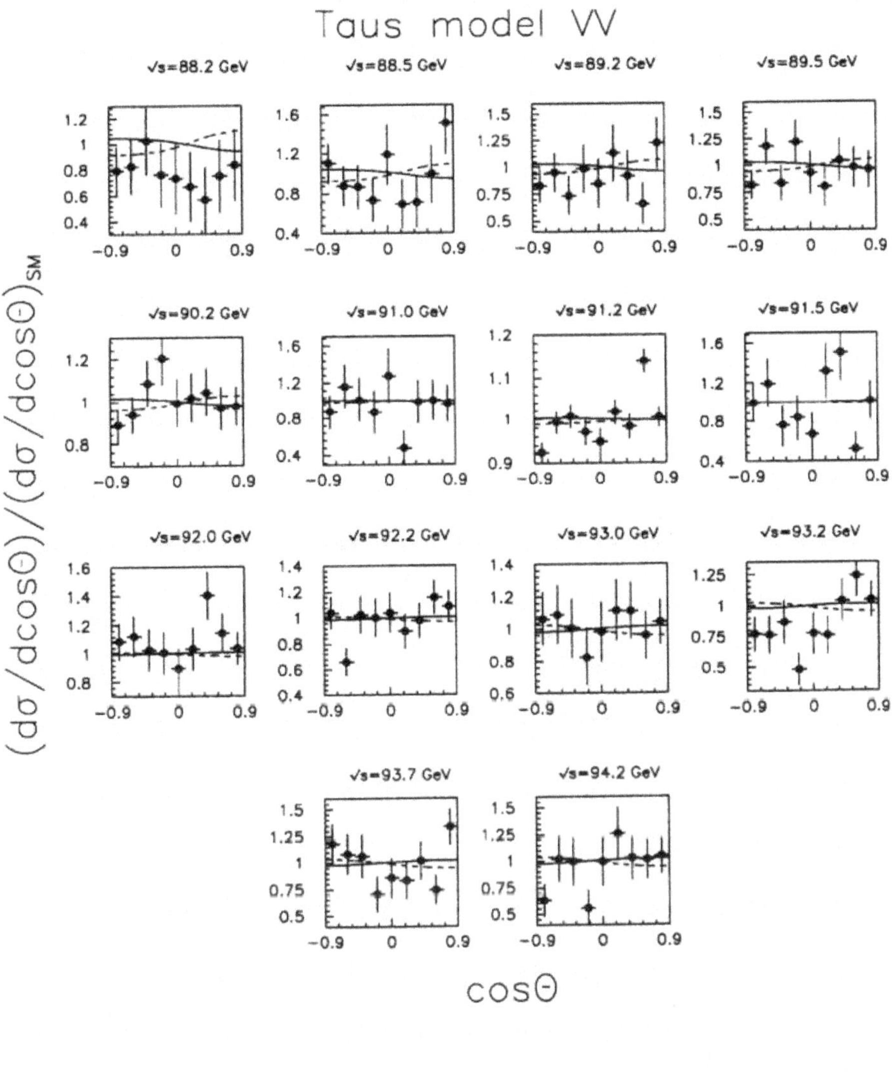

Figure 7: Results of the fit of contact model VV for the reaction $e^+e^- \rightarrow \tau^+\tau^-$. The data points are the measured differential cross sections divided by the standard model predictions. The curves are the maximum allowed deviations at 95% c.l.

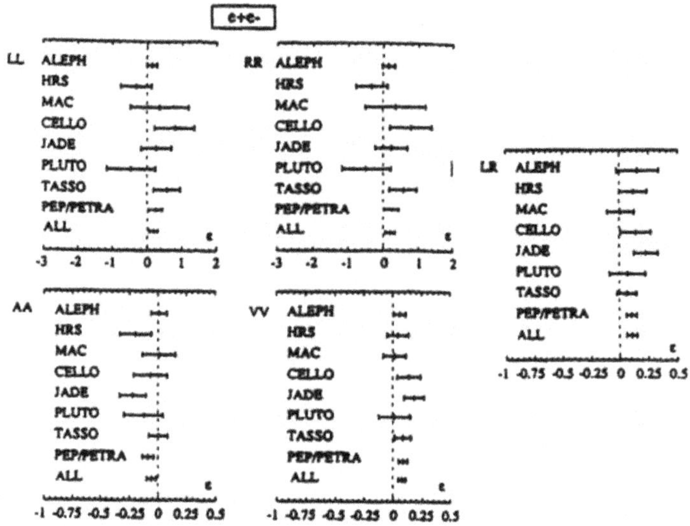

Figure 8: Values of ε (in TeV^{-2}) from the reaction $e^+e^- \to e^+e^-$ plotted with one standard deviation errors. 'ALL' is the weighted average of ALEPH and PEP/PETRA.

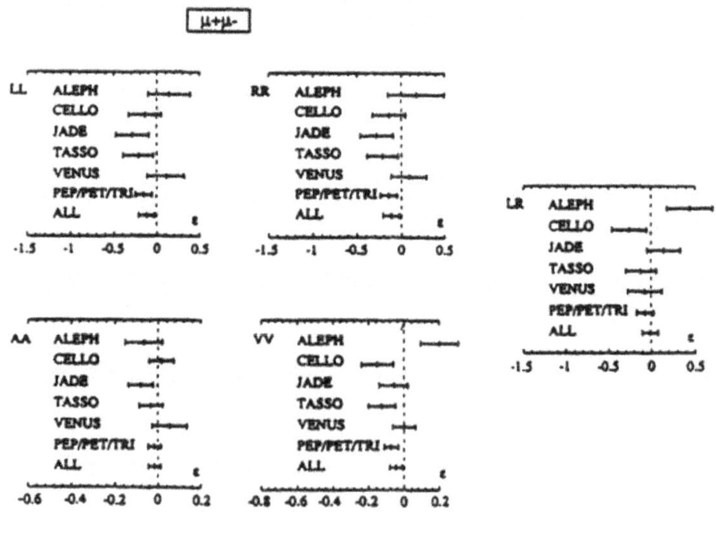

Figure 9: Values of ε (in TeV^{-2}) from the reaction $e^+e^- \to \mu^+\mu^-$ plotted with one standard deviation errors. 'ALL' is the weighted average of ALEPH and PEP/PETRA/TRISTAN.

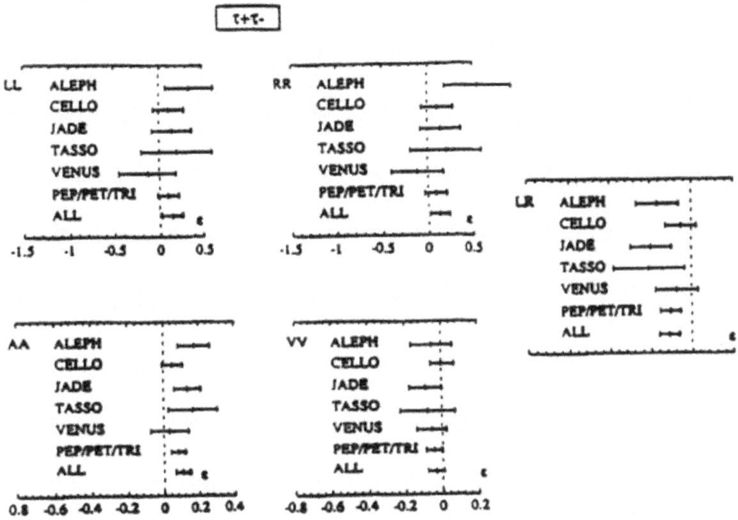

Figure 10: Values of ε (in TeV^{-2}) from the reaction $e^+e^- \rightarrow \tau^+\tau^-$ plotted with one standard deviation errors. 'ALL' is the weighted average of ALEPH and PEP/PETRA/TRISTAN.

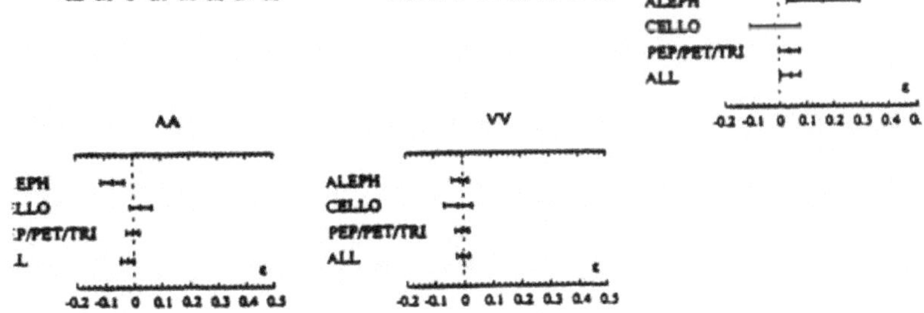

Figure 11: Values of ε (in TeV^{-2}) from the reaction $e^+e^- \rightarrow \ell^+\ell^-$ plotted with one standard deviation errors. 'ALL' is the weighted average of ALEPH and PEP/PETRA/TRISTAN.

The best currently available limits on four-lepton contact interactions are determined from the combined data of ALEPH and of lower energy experiments. Each limit is calculated from the weighted average value of ε and ~~it~~ is required to be smaller than the corresponding resolving power. The results, given in table 8, vary from 3 to 6 TeV.

	Λ^+_{LL}	Λ^-_{LL}	Λ^+_{RR}	Λ^-_{RR}	Λ^+_{VV}	Λ^-_{VV}	Λ^+_{AA}	Λ^-_{AA}	Λ^+_{RL}	Λ^-_{RL}
electrons	1.6	2.2	1.5	2.1	2.8	4.5	3.9	2.8	2.3	3.6
muons	2.5	1.9	2.5	1.9	4.1	3.0	4.5	3.9	2.5	2.4
taus	1.8	2.3	1.8	2.3	3.9	2.9	2.5	4.0	2.3	1.4
leptons	3.5	2.9	3.4	2.7	5.9	5.3	5.6	4.5	3.1	4.1

Table 8: Limits on $\Lambda^{\pm}$ in TeV at 95% c.l. from the combined data of PEP, PETRA, TRISTAN and ALEPH.

Conclusions

Searches for substructure at LEP gave negative results: no excited fermions are discovered, no effects predicted by composite models in standard processes are found.

The Z lineshape measurements exclude a new Z decay (coherent with the standard decays) having a width > 37 MeV or an "invisible" width > 19 GeV, at 95% c.l. Hence limits on excited fermion masses are 43.6 GeV/c^2 for the ν^*, 29.71 GeV/c^2 for the charged lepton, 41.7 and 44.6 GeV/c^2 for the u- and d-type quarks, independent of the f^* decay mode. Topology-dependent limits are more stringent, e.g. 45.0 GeV/c^2 for an "invisible" Dirac neutrino. The windows between these limits and $m_Z/2$ can be covered by direct searches for excited fermion pair production. Present results are incomplete since some of the f^* decay modes are not yet explored. In the the mass region $m_Z/2 < m^* < m_Z$ limits are derived from the searches for single f^* production. Results are presented in the 2-dimensional plane of $c\sqrt{B}/\Lambda$ versus m^*, where c is the Zf^*f or the γf^*f coupling, B is the f^* decay branching ratio and Λ is the compositeness scale. Typical limits on this quantity for excited fermions of $m^* =$ 50 GeV are in the range from (3 TeV)$^{-1}$ to (10 TeV)$^{-1}$, depending on the flavour.

The reaction $e^+e^- \rightarrow \gamma\gamma\gamma$ is compatible with QED and shows no indication for an anomalous $Z\gamma\gamma\gamma$ coupling. The branching ratio limit is $B(Z \rightarrow \gamma\gamma\gamma) < 1.4 \cdot 10^{-5}$.

Possible deviations from the standard model are looked for in the framework of contact terms models which contain one free parameter: the energy scale of a new interaction, Λ. The following vertices are studied: four-bosons. two-bosons and two-fermions, and four fermions. Typical values of the limits on Λ are 25 GeV, 100 GeV and a few TeV, respectively.

The $ee\gamma\gamma$ contact terms are studied in the reaction $e^+e^- \to \gamma\gamma$ assuming various chiralities of the electron current: L, R, $L+R$ and $L-R$. The results are similar to those reported at TRISTAN. Limits on Λ from four-fermion contact terms are set by ALEPH in the range 1.4-3.4 GeV for electrons, 1.1-2.6 TeV for muons and taus, and 1.6-4.7 TeV for common leptons, depending on the chiralities of the fermion currents. The most stringent limits are obtained for the reactions $e^+e^- \to e^+e^-$ and $e^+e^- \to \ell^+\ell^-$ due to the interference of the t-channel photon exchange with the real contact term amplitude. For some models the results from combined lower energy data are improved.

The limits are expected to increase in proportion to the number of events to the power 1/4. The statistics can be easily multiplied by 4 when the data of other experiments are analysed. Limits of the order of 10 TeV should be achieved at LEP 1 in the future.

The best currently available limits on Λ from the combined common lepton data of ALEPH, PEP, PETRA and TRISTAN are about 3 TeV for LL and RR, 3.5-4 GeV for RL and LR, and between 5 and 6 TeV for VV and AA models.

Acknowledgements

It is a pleasute to thank the organisers for a very stimulating Workshop. I am also indebted to Dr Qian Jian for communicating unpublished L3 data.

References

[1] F. Boudjema and F. Renard, "Compositeness", in "Z Physics at LEP1", (G. Altarelli, R. Kleiss and C. Verzegnassi, eds), CERN 89-08, vol 2 (1989) 115.

[2] M. Bardadin-Otwinowska, Z.Phys. C55 (1992) 163.

[3] J. Nash, presented at the XXVIIth Rencontre de Moriond (1992).

[4] G. Burgers, "The shape and size of the Z resonance" in Polarisation at LEP, (G.Alexander et al.,ed.), CERN 88-06 (1988) 121.
Program GAMMAZ courtesy of G. Burgers and W. Hollik.

[5] Frank L. Linde, "Measurement of α_s at LEP", XXVIIth Rencontre de Moriond, March 1992.

[6] K. Hagiwara, S. Komamiya and D. Zeppenfeld, Z.Phys. C29 (1985) 215.

[7] F. Boudjema and A. Djouadi, Phys.Lett. B240 (1990) 485.

[8] ALEPH Collab. - D. Decamp et al., "New Particle Searches in Z Decays with the ALEPH detector", preprint CERN-PPE/91-149 (1991), to be published in Physics Reports.

[9] ALEPH Collab., update including the 1991 data, unpublished.

[10] F.M. Renard, Phys.Lett.132B (1983) 449.

[11] DELPHI Collab., P. Abreu et al., "Search for Excited Charged Leptons in Z Decays", preprint CERN-PPE/91-100 (1991).

[12] L3 Collab., B. Adeva et al., Phys.Lett. B247 (1990) 41,
L3 Collab., B. Adeva et al., Phys.Lett. B250 (1990) 205,

[13] OPAL Collab., M. Akrawy et al., Phys.Lett. B244 (1990) 135.

[14] L3 Collab., B. Adeva et al., Phys.Lett. B252 (1990) 525.

[15] DELPHI Collab., P. Abreu et al., Z.Phys. G53 (1992) 555.

[16] L3 Collab., contribution to to Geneva EPS-LP Conference (1991).

[17] OPAL Collab., M. Akrawy et al., Phys.Lett. B246 (1990) 285.

[18] F.M. Renard, Nuovo Cimento 77A (1983) 1.

[19] M. Begalli, A. De Min, "Limits on new physics from the study of isolated photons in hadronic Z^0 decays at LEP", to be published in CERN Yellow Book "Photon Radiation from Quarks"

[20] DELPHI Collab., P. Abreu et al., Phys.Lett B268 (1991) 296.

[21] L3 Collab., B. Adeva et al., Phys.Lett. B250 (1990) 183.

[22] OPAL Collab., M. Akrawy et al., Phys.Lett. B257 (1991) 531.

[23] M. Laursen at al., Phys.Rev. D23 (1981) 2795.

[24] F.M. Renard, Phys.Lett.116B (1982) 269.

[25] E.W.N. Glover and J.J. van der Bij, "Rare Z decays", in "Z Physics at LEP1", (G. Altarelli, R. Kleiss and C. Verzegnassi, eds), CERN 89-08, vol 2 (1989) 1.

[26] L3 Collab., "A Test of Quantum Electrodynamics in the Reaction $e^+e^- \rightarrow \gamma\gamma(\gamma)$", preprint CERN-PPE/92-87 (1992).

[27] M. Baillargeon and F. Boudjema, "New Physics Through Final State Photons", preprint ENSLAPP-A-366/92 (1992), to be published in CERN Yellow Book "Photon Radiation from Quarks"

[28] E. Eichten, K. Lane and M. Peskin, Phys. Rev. Lett. 50 (1983) 811.

[29] B. Schrempp, F. Schrempp, N. Wermes and D. Zeppenfeld, Nucl. Phys. B296 (1988) 1.
F. Schrempp, DESY 89-047, (1989).

[30] F. Badaud 'Study of compositeness at LEP with ALEPH', thesis, Blaise Pascal University, Clermont-Ferrand, 1992 (unpublished).

[31] H. Kroha, University of Rochester, preprint UR1241.

[32] D. Treille et al., proc. of ECFA Workshop on LEP 200, A. Bohm and W. Hoogland eds, vol.2, Report CERN 87-08 (1987) 414.

[33] F. Berends and R. Kleiss, Nucl.Phys. B186 (1981)22.

[34] AMY Collaboration, H.J. Kim et al., KEK preprint 89-52 (1989).

[35] VENUS Collaboration, K. Abe at al., Z.Phys. C45 (1989) 175.

[36] TOPAZ Collaboration, K. Shimozawa et al., Phys.Lett B284 (1992) 144.

[37] ALEPH Collab., D. Decamp et al., Z. Phys. C53 (1992) 1.
'Update of Electroweak Parameters from Z Decays', to be published.

[38] F.A. Berends and W.L. Van Neerven, Nucl. Phys. B297 (1988) 429.

[39] W. Beenakker, F. Berends, S.C. van der Marck, Nucl.Phys. B349 (1991) 323.

[40] W. Beenakker, M. Martinez, R. Miquel: Program ALISTAR (unpublished).

The Hunt for the Higgs Boson(s) at LEP: Present and Future

M. Felcini

CERN, PPE Division
CH-1211 Geneva 23, Switzerland

Abstract

The Higgs searches at LEP 100 are reviewed and the prospects for the discovery of Higgs bosons at LEP 200 are discussed.

1 Introduction

With the large amount of data collected at LEP 100 on the Z resonance, the Standard Model (SM) of the electroweak interactions has been subject to very precise tests and up to now it has been found remarkably in agreement with the experimental data. An essential component of the model is the weak isospin doublet of complex scalar fields which is responsible for the spontaneous breakdown of the gauge symmetry. Spontaneous symmetry breaking is needed to generate masses for the bosons and the fermions. After electroweak symmetry breaking, there remains a single neutral scalar particle, the Higgs boson. The search for the Higgs boson is one of the most important efforts for LEP as for other present and future accelerators. In the SM, the Higgs boson mass is unspecified but it must be below 1 TeV to maintain valid the perturbative approach of the theory. In the absence of more stringent theoretical predictions for the Higgs boson mass it is necessary that the whole allowed mass region be explored for its detection. The LEP search for the SM Higgs boson is described in Section 2. The experiments at LEP 100 running on the Z resonance have thoroughly explored the mass range between 0 and 60 GeV without finding any signal. Thus, a lower limit of 60 GeV has been set on the SM Higgs mass.

Additional doublets are necessary in theories beyond the SM, like Supersymmetry. In the simplest version of a supersymmetric model just two Higgs doublets are needed to give mass to the quarks and the charged leptons. The Higgs boson spectrum of this model consists of five physical states, three neutral and a pair of charged ones. Comprehensive searches for these Higgs bosons have been made at LEP 100, covering a large number of decay channels. The results of these searches are independent of the specific Higgs boson decay modes and therefore valid in a general two doublet model. They can be interpreted in the framework of the Minimal Supersymmetric Standard Model (MSSM). They have been used to set limits on the masses of the lightest scalar h and of the pseudoscalar A. The searches for non-minimal Higgs bosons and the interpretation of the results in the MSSM framework are reviewed in Section 3.

Since the Higgs mass range of sensitivity at LEP 100 has been almost fully explored, LEP 200 is needed to proceed further in the Higgs searches. LEP 200 searches are the subject of Section 4. The limits of detectability of the SM Higgs boson at LEP 200 as a function of the center-of-mass (c.o.m.) energy and luminosity are discussed. The prospects for discovery of one or more of the five Higgs bosons of the MSSM are outlined. The regions of the parameter space for which a measurable signal would be predicted are delineated. They depend on the c.o.m. energy, the luminosity and, in addition, on the t-quark mass, which affects the Higgs boson masses and couplings through one-loop radiative corrections.

A summary of the results of present Higgs searches at LEP 100 and the prospects for discoveries at LEP 200 is given in Section 5.

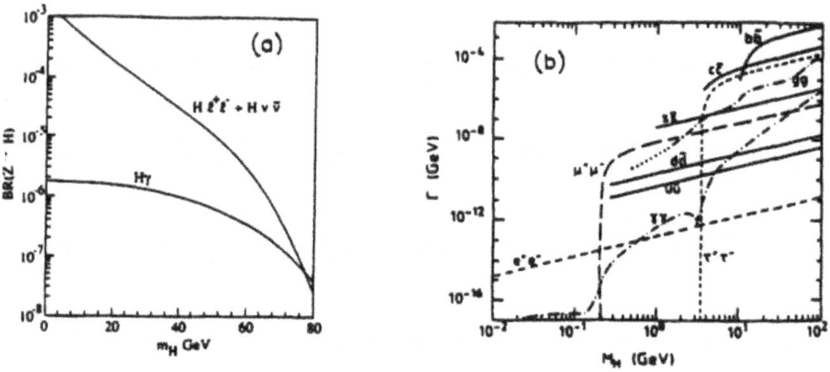

Figure 1: (a) Rate of $Z \to Z^*H_{SM}$ events, in the charged lepton and neutrino channels compared to the rate of $Z \to \gamma H_{SM}$ events. (b) Approximate widths for all two-body partonic decay modes of the SM Higgs boson.

2 Search for the Higgs boson of the Standard Model

At LEP 100 running on the Z resonance, the SM Higgs boson H_{SM} can be produced via the two processes

$$e^+e^- \to Z \to Z^*H_{SM}, \tag{1}$$

$$e^+e^- \to Z \to \gamma H_{SM}. \tag{2}$$

In process (1) the Higgs boson is radiated off the Z. This process is referred to as the Bjorken process. Its rate, determined by the coupling ZZH_{SM}, is well known within the context of the SM. Process (2) proceeds via a loop involving the W boson and fermions, including the top quark. This process is theoretically very interesting because it involves couplings specified by the theory but never measured. So, if detected, it might be used to test the predictions for the Higgs and gauge coupling. However, within the range of Higgs mass $m_H \lesssim 70$ GeV of interest at LEP 100, its rate is expected to be smaller than that of the Bjorken process, as shown in Figure 1. In addition, its experimental search is very difficult because of the large background from hadronic radiative Z decays $Z \to \gamma q\bar{q}$. Thus, the Bjorken process is the only one which can be practically used to search for the SM Higgs boson at LEP 100.

The signatures of Higgs events are determined by the final states of the virtual Z and the decays of the Higgs. These latter depend on the Higgs mass. In the mass region below the $2m_\mu$ threshold the only two possible decays are $H_{SM} \to \gamma\gamma$ and $H_{SM} \to e^+e^-$. In this region, the Higgs is long-lived, while for $m_H > 2m_\mu$ the Higgs decay occurs too

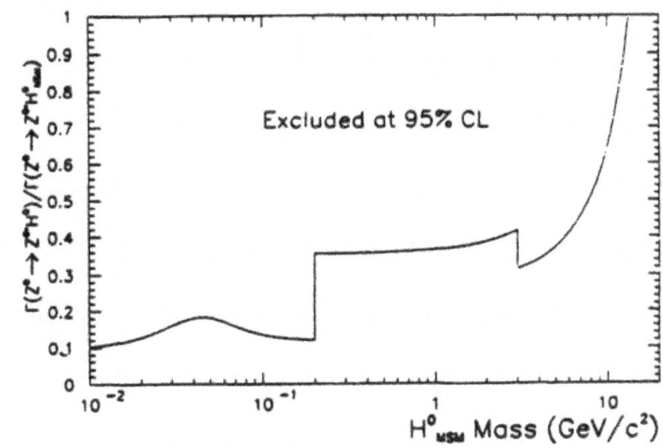

Figure 2: Upper limit on the production rate of a light Higgs boson in Z decays, independent of the Higgs decays, given as a fraction of the SM Higgs production rate in Z decays, in function of the SM Higgs mass, from the OPAL search [5].

fast to be measured directly. In the mass region $2m_\mu < m_H < 2m_\tau$ the Higgs decay is dominantly into a muon pair or gluons. The QCD calculation of the Higgs decay branching ratios is affected by significant uncertainty and it is model dependent. In the mass region $2m_\tau \leq m_H < 2m_b$, the Higgs decays dominantly into a tau pair or a c-quark pair. Above the $2m_b$ threshold, the dominant Higgs decays are into a $b\bar{b}$ quark pair ($\approx 87\%$), into a $c\bar{c}$ quark pair ($\approx 7\%$) and a $\tau^+\tau^-$ pair ($\approx 6\%$).

In the mass region $m_H < 2m_\mu$, if the Higgs decays outside the detector, the cleanest signature is that with a pair of acoplanar leptons from the Z^* decay. If the Higgs decays within the detector, then a clean signature is given by events with the Z^* decaying into $\nu\bar{\nu}$ and the Higgs into a pair of close-by tracks originating from a secondary vertex.

In the mass region $2m_\mu < m_H < 2m_b$, the Higgs has been searched for in events with the $Z^* \to \nu\bar{\nu}$ and a pair of charged tracks or a monojet from the Higgs decay. Though the rates of the Higgs decays are uncertain in this mass region, the analysis is made independent of the Higgs decay modes by assuming as detection efficiency the lowest among those for the different Higgs decays. Being the number of expected events very large, all the four experiments, ALEPH [1], DELPHI [2], L3 [3] and OPAL [4,5] have excluded, with a confidence level larger than 95%, a SM Higgs in the mass region below the $2m_b$ threshold. As an example of the result of the search for the Higgs lighter than $2m_b$, Figure 2 (from [5]) shows the upper limit on the branching ratio of the Z decay into a light Higgs boson, independent of the Higgs decays. This limit was derived considering the two channels [4] $Z \to \nu\bar{\nu}H$, H $\to$electromagnetic final state, and

Exp.	N_h (10^3)	$H_{SM}\nu\bar{\nu}$ ϵ (%)	N_H	$H_{SM}e^+e^-$ ϵ (%)	N_H	m_{rec} (GeV)	$H_{SM}\mu^+\mu^-$ ϵ (%)	N_H	m_{rec} (GeV)	Limit $m_H >$ (GeV)
ALEPH	488	59	0	63	0		68	0		53.0
DELPHI	330	46	0	36	1	35.0 ± 5.0	46	0		47.0
L3	408	59	0	52	1	31.4 ± 1.5	62	1	70.4 ± 0.7	52.0
OPAL	494	57	0	45	0		45	0		52.6

Table 1: Summary of the results of the SM Higgs boson search performed by the four LEP experiments. N_h is the number of hadronic Z decays used, ϵ is the efficiency for a 50 GeV Higgs in the specified channel, N_H is the number of data events surviving all cuts ("Higgs candidates"), m_{rec} is the missing mass recoiling against the dilepton system.

$Z \to e^+e^-, \mu^+\mu^-H$, $H \to$non-electromagnetic final state, thus covering all the possible SM Higgs decays. It can be seen that a SM Higgs boson with mass below 13.3 GeV is excluded at 95% CL, independently of the Higgs decays.

In the mass range $m_H > 2m_b$, the channels used for the SM Higgs search are: the neutrino channel $Z \to \nu\bar{\nu}H_{SM}$, $H_{SM} \to$ hadrons, and the charged lepton channel $Z \to (e^+e^-$ or $\mu^+\mu^-)H_{SM}$, $H_{SM} \to$ hadrons, with relative rates 18.9% and 6.4% respectively. Though the most abundant channel is the hadronic channel $Z \to q\bar{q}H_{SM}$, with 65.5% relative rate, it has not been used by the LEP experiments yet, because of the large QCD background from processes like $e^+e^- \to q\bar{q}gg$ or $e^+e^- \to q\bar{q}g, g \to q\bar{q}$.

The events in the charged leptonic channel have a clear signature characterized by two energetic and isolated leptons, from the decay of the virtual Z, and localized hadronic activity (jets) from the Higgs decay. The event topology depends upon the Higgs mass. For m_H below about 15 GeV, the hadrons from the Higgs decay are close together to form a single jet (monojet) recoiling against an energetic pair of leptons, which carry a large fraction of the event energy. For larger Higgs masses, the hadrons from the Higgs decay are spread over a large portion of solid angle, originating 2 or more hadronic jets, with a large opening angle, recoiling against a lepton pair with a small invariant mass. Practically the only, but irreducible, background in this channel is the four fermion production $e^+e^- \to q\bar{q}\ell\ell$ ($\ell = e$ or μ). Though this process has a small cross section, because it proceeds through higher order diagrams, its rate is comparable to that of the Higgs events in the charged lepton channel. It is interesting to note that all the Higgs candidates, in the higher mass region, have been found in this channel, up to now. They are listed in Table 1. The L3 candidate found in the muon channel [6], shown in Figure 3, with 70.4 ± 0.7 GeV missing mass recoiling against the muon pair, is particularly 'beautiful'. It has in fact a third energetic muon in one of the hadronic jets, as one would expect from the semileptonic decay of a b-quark. Furthermore the

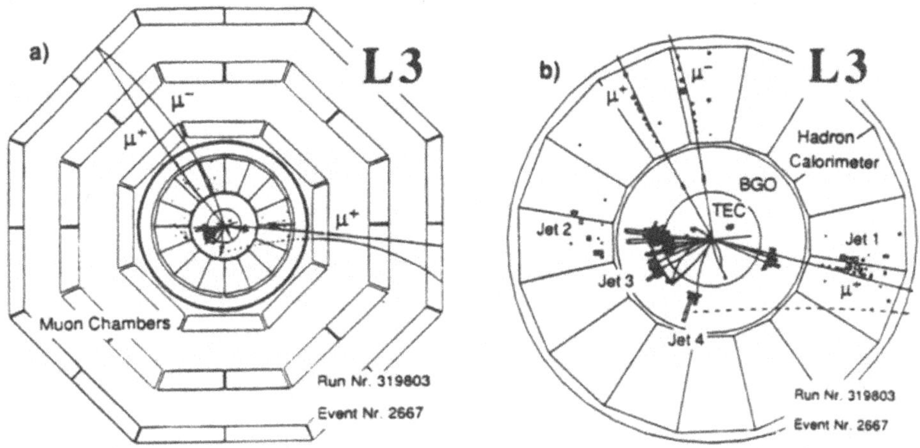

Figure 3: The L3 Higgs candidate found in the $H\mu^+\mu^-$ channel [6] shown in the plane perpendicular to the beam line. Figure a) shows the whole detector, including the muon chambers. Figure b) shows the central region, with the tracking chamber (TEC), the electromagnetic (BGO) and the hadronic calorimeters. The missing mass recoiling against the isolated muon pair is 70.4 ± 0.4 GeV. A 70 GeV Higgs? Only more data will tell.

invariant mass of the the muon pair is small, 6.5 ± 0.2 GeV, as expected for the mass of the virtual Z when the Higgs is heavy. Is this event from a Higgs of 70 GeV? Only more data will tell. For the moment it seems to be compatible with the four fermion background, as the other two candidates at lower masses.

The Higgs events in the neutrino channel are about three times more abundant than in the charged lepton channel. Again the event topology depends upon the Higgs mass. For a relatively light Higgs, below 15 GeV, the signature of a Higgs event is a monojet recoiling against large missing momentum. For a heavier Higgs, the missing momentum carried away by the neutrinos is reduced and the jets from the Higgs tend to have larger opening angles. Thus, as the Higgs mass increases, it is more difficult to distinguish the signal from the background. Main sources of background are hadronic events in which missing momentum results from detector effects, like jet mismeasurement or incomplete containment of the event, or from physics processes like semileptonic decays of heavy quarks with neutrino production. All these background have been rejected by using topological variables like the acoplanarity of the jets and the missing momentum direction as well as the missing momentum isolation. No Higgs candidates have been found in this channel yet.

The efficiencies in the different channels for ALEPH [7], DELPHI [8], L3 [6] and OPAL [9] are reported in Table 1 together with the 95% CL limits derived by each ex-

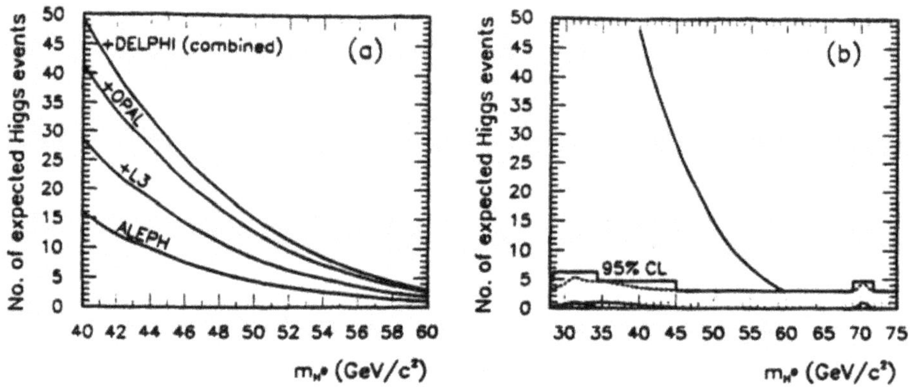

Figure 4: The combined number of Higgs events expected by the four LEP experiments (a) and the combined lower limit on the SM Higgs mass (b). The four experiments exclude a SM Higgs boson lighter than 60 GeV. Conservatively the three events observed in the data have been considered as Higgs candidate when deriving the limit [10].

periment. It is possible to combine the results of the four searches. Given the individual efficiencies, the expected number of Higgs events for the four experiments can be plotted as function of the Higgs mass, as in Figure 4(a) and a lower limit of 60 GeV for the Higgs mass can be derived, as shown in Figure 4(b) (from [10]).

With an increase in statistics of about an order of magnitude, as expected by the end of 1993, one could naively expect to extend the search up to 70 GeV, where the Z→H branching ratio (see Figure 1) is $\lesssim 10^{-6}$, *i.e.* an order of magnitude lower than in the 55–60 GeV range, presently probed. However, with the increased statistics, rare background events from higher order processes, as those from four fermion production, will appear, severely limiting the sensitivity to the Higgs signal in the high mass range. Thus, it is expected that the ultimate Higgs mass limit at LEP 100 will not be above 65 GeV. To explore a higher Higgs mass range LEP 200 is needed!

3 Searches for Higgs bosons beyond the Standard Model

The scalar sector of the Standard Model is not yet constrained by experiment and it is possible that the Higgs structure is more complicated. A strong motivation for extending the Higgs sector comes from Supersymmetry where at least two Higgs doublets are necessary to provide masses for the charged fermions. The neutral members of the doublet acquire vacuum expectation values v_2 and v_1: the former gives masses to up-quarks and the latter to down-quarks and charged leptons.

The physical spectrum of Higgs bosons in a two doublet Higgs model consists of two neutral scalars, h and H, a neutral pseudoscalar A, and a pair of charged scalars $H^{\pm}$. In a general two doublet model, the Higgs sector is described by 6 independent parameters: the 3 masses of the neutral bosons, the mass of the charged boson and 2 angles, the mixing angle α between the scalar bosons h and H and the angle β defined by $\tan\beta = v_2/v_1$. These two angles determine the couplings of the Higgs bosons to the vector gauge bosons and to the fermions. Some couplings to gauge bosons are not allowed. For example, the A boson, being a pseudoscalar, cannot be produced via the Bjorken process and the decays $Z \to hh$ and $Z \to AA$ are forbidden by Bose statistics. The relevant production processes for two doublet Higgs bosons at the Z resonance are

$$e^+e^- \to Z \to Z^*h, \tag{3}$$

$$e^+e^- \to Z \to hA, \tag{4}$$

$$e^+e^- \to Z \to H^+H^-. \tag{5}$$

The possible decays of the neutral Higgs bosons into fermions are the same as for the SM Higgs boson. In addition, if $m_h \geq 2m_A$, the decay $h \to AA$ is also possible. The decay branching ratios of the h boson depend on both the angles α and β. Those of the A boson depend only on $\tan\beta$: for $\tan\beta > 1$ (< 1), the $b\bar{b}$ and $\tau^+\tau^-$ decay rates are enhanced (depleted) while the $c\bar{c}$ decay rate is depleted (enhanced), as compared to the SM Higgs decays. The most likely decays of the charged Higgs boson are into $c\bar{s}$ or into $\tau\nu$. The decay rates depend on $\tan\beta$: for $\tan\beta \gtrsim 3$ the $\tau\nu$ mode is dominant ($> 90\%$), while for $\tan\beta \lesssim 0.4$ the $c\bar{s}$ mode is dominant.

The LEP searches for the non-minimal Higgs bosons, their results and their interpretation in the MSSM are reviewed below.

3.1 Searches for charged Higgs bosons

The Z partial width for the decay into a charged Higgs pair depends only on the mass of the charged Higgs [11]:

$$\Gamma(Z^0 \to H^+H^-) = \frac{G_F m_Z^3}{6\sqrt{2}\pi}\left(\frac{1}{2} - \sin^2\theta_W\right)^2 \beta_{H^\pm}^3, \qquad \text{where } \beta_{H^\pm} = \sqrt{1 - \frac{4m_{H^\pm}^2}{m_Z^2}}. \tag{6}$$

Thus, the number of charged Higgs events expected per 400,000 hadronic Z decays is about 400 for a charged Higgs mass of 40 GeV and it reduces to 30 for a charged Higgs mass of 45 GeV. Searches have been done considering all the three possible final states for the H^+H^- pair:

$$Z^0 \to H^+H^- \to \tau^+\nu\tau^-\bar{\nu}, \qquad \tau\nu cs, \qquad c\bar{s}\bar{c}s. \tag{7}$$

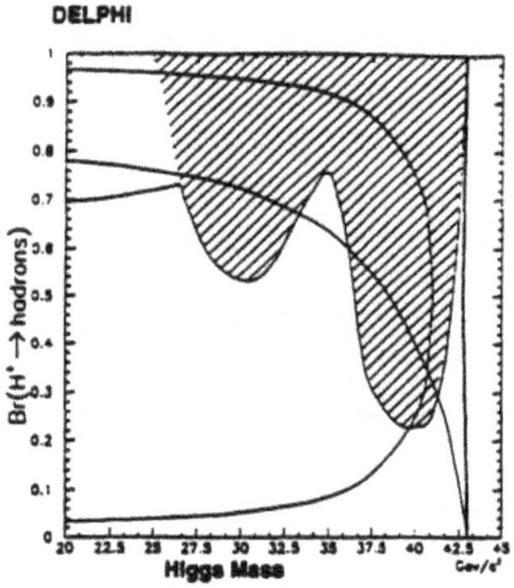

Figure 5: Lower limit on the charged Higgs boson mass from the DELPHI search [12] in the channels $c\bar{s}\bar{c}s$, $cs\tau\nu$ and $\tau^+\nu\tau^-\bar{\nu}$. The vertical line on the right-hand side defines the combined excluded region. Independent of the decay mode, a charged Higgs lighter than 43 GeV is excluded at 95% CL.

Updated results have been recently reported by DELPHI [12] and L3 [3], based on the analysis of 390K and 408K hadronic Z decays, respectively. DELPHI excludes a charged Higgs lighter than 43 GeV, independently of the charged Higgs decay mode. The equivalent limit for L3 is 41 GeV. If Br($H^\pm \to \tau\nu$)=1, L3 excludes a charged Higgs lighter than 44.5 GeV. The result of the updated DELPHI analysis is shown in Figure 5.

3.2 Searches for neutral Higgs bosons

The Z partial widths for the Zh and hA production processes are given by

$$\Gamma(\mathrm{Z} \to \mathrm{Z}^*\mathrm{h}) = \Gamma(\mathrm{Z} \to \mathrm{Z}^*\mathrm{H}_{\mathrm{SM}}) \sin^2(\beta - \alpha), \quad (8)$$

$$\Gamma(\mathrm{Z} \to \mathrm{hA}) = \frac{\lambda^{3/2}}{2}\Gamma(Z \to \nu\bar{\nu}) \cos^2(\beta - \alpha), \quad (9)$$

where λ is the phase-space factor $\lambda = (1-a-b)^2 - 4ab$, with $a = m_{\mathrm{h}}^2/m_{\mathrm{Z}}^2$ and $b = m_{\mathrm{A}}^2/m_{\mathrm{Z}}^2$. The searches in the channels $\mathrm{Z} \to \mathrm{Z}^*\mathrm{h}$ and $\mathrm{Z} \to \mathrm{hA}$ are complementary, in that when the Bjorken process is suppressed the associated hA production becames large. To search for the $\mathrm{Z} \to \mathrm{Z}^*\mathrm{h}$ process, the analysis developed for the SM Higgs boson search is used, except in the case where the h→AA decay can take place, for which additional possible h final states have to be considered, depending on the A mass. ALEPH [13], DELPHI [14]

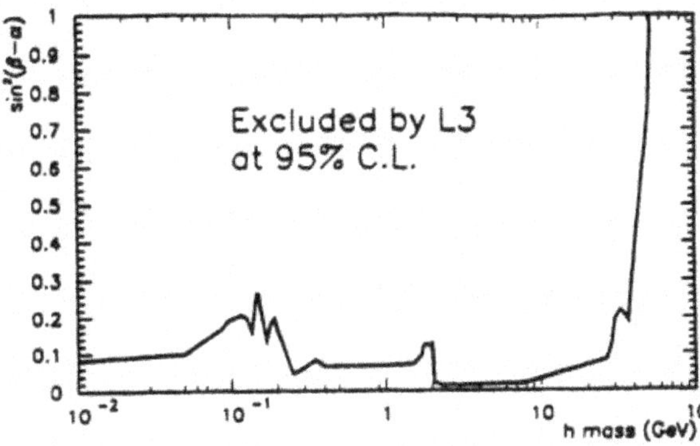

Figure 6: Limit on $\sin^2(\beta - \alpha)$ of the two-doublet Higgs model obtained from the L3 search [3] for $Z \to Z^*h$ process. The limit extends to $m_h = 0$ GeV.

and L3 [3] have recently presented the results of very thorough analyses considering all possible h and A decays, also for very small A masses. These analyses allow to set limits on the suppression factor $\sin^2(\beta - \alpha)$ for h masses between 0 and about 50 GeV. As an example, the result of the L3 search [3] for the $Z \to Z^*h$ process, in a general two-doublet Higgs model, is shown in Figure 6, as a 95% CL upper limit on $\sin^2(\beta - \alpha)$.

For the $Z \to hA$ process, an upper limit on $\cos^2(\beta-\alpha)$ can be derived from the upper bound on the contribution from decays into new particles to the total Z width. This limit, $Br(Z \to hA) < 1\%$, is independent of the h and A decay modes. It is possible to combine the limits on $\sin^2(\beta - \alpha)$ and $\cos^2(\beta - \alpha)$ to derive an excluded region in the m_h-m_A plane. In fact, from the upper limit on $\cos^2(\beta - \alpha)$ one derives a lower limit on $\sin^2(\beta - \alpha)$. The excluded region in the m_h-m_A plane is then the set of m_h-m_A pairs for which the upper limit on $\sin^2(\beta - \alpha)$, from the $Z \to Z^*h$ search, is smaller than the lower limit on $\sin^2(\beta - \alpha)$, derived from the upper limit on $\cos^2(\beta - \alpha)$, from the Z width measurement. The constraint set in this way on the m_h and m_A masses is independent of the h and A decays, thus it holds in a general two doublet model. The region excluded in this way by DELPHI [14] is shown in Figure 7(a). Similar results are obtained by ALEPH [1], L3 [3] and OPAL [15]. With the present data, the method just described is sensitive for h masses not larger than $\simeq$ 40 GeV and A masses not larger than $\simeq$ 50 GeV. The direct search for the $Z \to hA$ decay can provide sensitivity for higher masses. This search has been done in the 4-jets, $\tau^+\tau^-$ + jets and AAA final states. Sophisticated analyses have been developed to take into account the possibility of a very light A and of the $Z \to AAA$ decay. The region excluded, combining the results

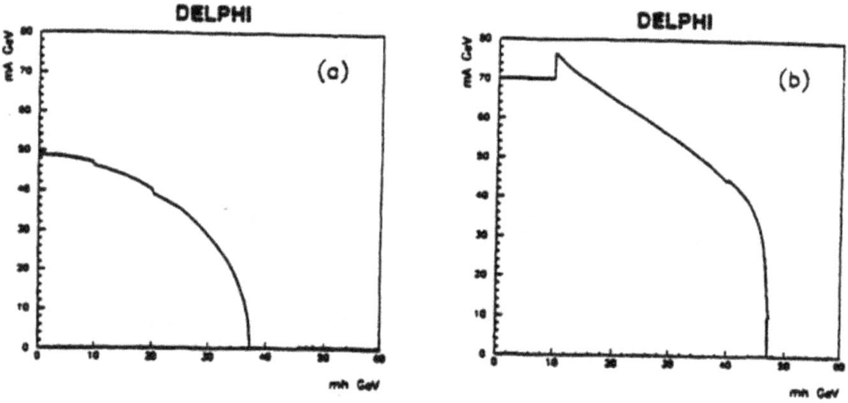

Figure 7: Regions in the m_h-m_A plane excluded by DELPHI [14] (a) in a general two doublet model, from the combination of the results of the Z $\rightarrow$ Z*h (upper limit on $\sin^2(\beta-\alpha)$) and the Z width measurement (upper limit on $\cos^2(\beta-\alpha)$) and (b) including also the results of the Z $\rightarrow$ hA direct search. The exclusion of region (b) is valid only for $\tan\beta > 1$.

of the direct search for Z $\rightarrow$ hA decay together with the limits from the total Z widths and the search for Z $\rightarrow$ Z*h decay, is shown in Figure 7(b). This region is sensibly larger than that obtained from the Z width limit and the Z $\rightarrow$ Z*h search alone. However, the limit including the result of the Z $\rightarrow$ hA direct search is valid only for $\tan\beta > 1$, because in the hA direct search explicit use is made of the $b\bar{b}$ and $\tau^+\tau^-$ decay modes, which for $\tan\beta < 1$ are suppressed.

3.3 Interpretation of the results in the MSSM framework

In the Minimal Supersymmetric Model (MSSM), the conditions on the potential imposed by supersymmetry reduce the number of free parameters of a generic two doublet model from six to two. The choice of m_A and $\tan\beta$ for the two free parameters is theoretically more convenient. Experimentally it is more appropriate to choose m_h and m_A [1]. At tree-level, the angles and the masses of the MSSM Higgs bosons are determined by these

[1]The choice of m_h and m_A , as free parameters of the model, is affected by an ambiguity in $\tan\beta$, in that, for a given pair m_h-m_A , $\tan\beta$ can have two different values. Usually, when deriving an excluded region of the m_h- m_A plane, if the ambiguity occurs, the $\tan\beta$ value which gives the lowest number of expected Higgs events is conservatively taken.

two parameters. In particular, the Higgs masses satisfy the bounds

$$m_h^2 < m_Z^2|\cos^2 2\beta| \leq m_Z^2, \; m_h < m_A, \; m_{H^\pm} > m_W. \tag{10}$$

Due to the large mass of the t-quark, radiative corrections on the Higgs masses cannot be neglected. They modify substantially the tree-level relationships [16–19] Radiative corrections depend on a number of parameters, like the masses of the squarks and their mixing angles. However, assuming that all the squarks have the same mass and that they do not mix, the main effects of the radiative corrections to the Higgs boson masses can be expressed in terms of m_t and a common squark mass scale, $m_{\tilde{q}}$. The upper bound on the mass of the lighter scalar boson h is then given by

$$m_h^2 < m_Z^2|\cos^2 2\beta| + \epsilon < m_Z^2 + \epsilon, \tag{11}$$

where ϵ depends [18] on the t-quark mass m_t and the t-squark mass $m_{\tilde{t}} = m_{\tilde{q}}$

$$\epsilon = \frac{6G_F}{\pi^2\sqrt{2}} m_t{}^4 \ln\left(\frac{m_{\tilde{t}}}{m_t}\right). \tag{12}$$

Two important consequences of the inclusions of radiative corrections are: a) m_h can be heavier than $m_Z^2|\cos^2 2\beta|$, thus no lower limit on $\tan\beta$ can be extracted from the present lower limit on m_h; b) m_h can be heavier than m_A and, if heavier than $2m_A$, the channel h→AA is dominant [20]. Furthermore, the limits become dependent on m_t . However, it turns out that the limit on m_h is almost insensitive to variations of m_t while the limit on m_A decreases significantly as m_t increases. As an example, the region of the m_h-m_A plane exclude at 95% CL by ALEPH [21] is shown in Figure 8 . Similar results have been reported by DELPHI [14], L3 [3] and OPAL [15]. In the derivation of the excluded region in the m_h-m_A plane, ALEPH, L3 and OPAL have chosen the method of varying the amount of radiative corrections ϵ and $\tan\beta$ in theoretically favored ranges, to exclude the m_h-m_A points which are excluded for any combination of ϵ and $\tan\beta$ in the allowed ranges. DELPHI, instead, has chosen to present the excluded region [14] for a given t-quark mass, m_t=135 GeV, and to show the variation of the m_A limit with m_t.

In summary, the present result of the search for the MSSM neutral Higgs bosons h and A at LEP is: at 95% CL, h must be heavier than 43 GeV, irrespective of the A mass, and A must be heavier than 21 GeV, irrespective of the h mass.

4 Higgs searches at LEP 200

We have seen that LEP 100 is approaching its intrinsec limit of sensitivity in the search for the SM Higgs boson, produced at the Z resonance via the Bjorken process. Thus an increase of the c.o.m. energy of the LEP collider is needed to extend the search into a range of higher Higgs masses. How far the range of sensitivity can be extended,

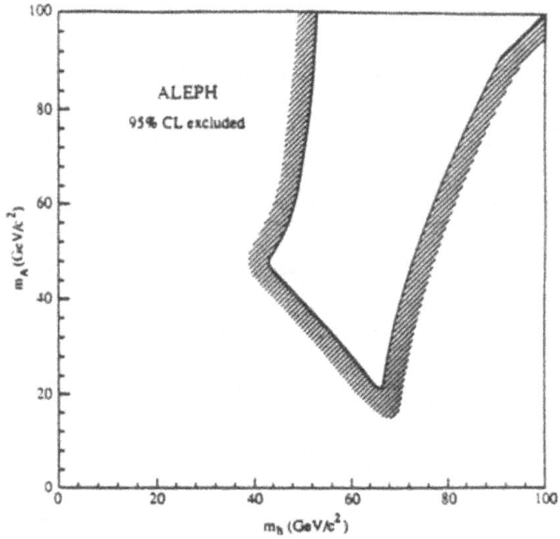

Figure 8: Regions in the m_h-m_A plane of the MSSM model excluded at 95% CL by the ALEPH search [21].

either for the SM Higgs or for the supersymmetric Higgs bosons, will depend both on the luminosity and the c.o.m. energy of the machine. The c.o.m. energy depends on the number of superconducting cavities which will instrument the LEP tunnel. With 192 cavities delivering an accelerating gradient of 5 MV/m a c.o.m. energy of 175 GeV can be obtained. Increasing the number of cavities to 256 will allow to reach a c.o.m. e nergy of 190 GeV [22]. The magnetic limit of the LEP ring is about 240 GeV which might be reached by adding more and/or better performing cavities [23]. An additional demand on the performance of the machine is the integrated luminosity, which, as we will see, should be at least 500 pb^{-1}, to fully exploit the larger kinematical range opened by the higher c.o.m. energy.

Several studies have been done to understand the discovery reach of LEP 200 as a function of the c.o.m energy and the delivered luminosity. The most recent developments concerning the detectability of the SM Higgs as well as the MSSM Higgs bosons are summarized here (details about the most recent studies can be found in [24] for the ALEPH detector and [22] for the comparison among the four LEP detectors).

4.1 Prospects for the SM Higgs boson search

The production mechanism for the SM Higgs at LEP 200 is the Bjorken process with a real Z in the final state

$$e^+e^- \rightarrow Z^* \rightarrow ZH_{SM} \,. \tag{13}$$

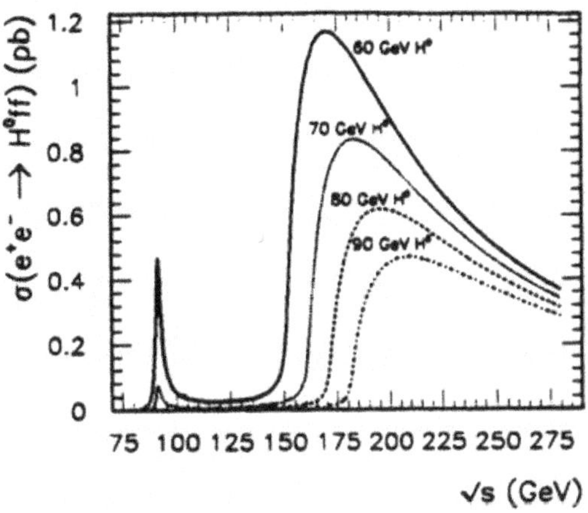

Figure 9: The cross section of the $e^+e^- \to Z^* \to ZH_{SM}$ process, as function of the c.o.m. energy, for different Higgs masses.

The cross section for this process is shown in Figure 9, where it is plotted as function of the c.o.m. energy for several values of the Higgs mass. It can be seen that, for a given Higgs mass, the production cross section reaches its maximum when $\sqrt{s}$ is a few GeV above $m_H + m_Z$. Furthermore, for a given c.o.m. energy, the cross section decreases as the Higgs mass increases. Thus one can naively say that the SM Higgs discovery reach is given by $m_H \lesssim \sqrt{s} - 100$ GeV. To be more precise, the actual sensitivity depends also on the amount of background, which varies with the c.o.m. energy.

The most dangerous backgrounds, at $\sqrt{s} = 190$ GeV and above, are the vector boson pair production processes, $e^+e^- \to WW$ and $e^+e^- \to ZZ$. As at LEP 100, the $H\nu\bar{\nu}$ and He^+e^- ,$\mu^+\mu^-$ channels are used for the search. In addition, due to the much lower QCD background at LEP 200 as compared to LEP 100, the hadronic channel can be used. In all the channels, the general strategy, to reduce especially the WW background, is to require that there is a Z in the event. In the lepton channel, this is done by demanding that the invariant mass of the two leptons is compatible with the Z mass. In the neutrino channel this requirement is applied to the missing mass. In the hadronic channel, after clustering the hadrons in exactly 4 jets, the event is kept only if there is a pair of jets with invariant mass compatible with the Z mass. After this requirement, the WW background is reduced, but the ZZ background is practically left untouched. To reduce it, the most efficient way is to take advantage of the higher decay rate of the Higgs into b-quarks (85%) as compared to that of the Z (15%). Thus, identifying b-quarks in the event, by means of a micro-vertex detector, the signal-to-background ratio is substantially improved. The use of b-tagging is always helpful in reducing the

WW and ZZ background and it is practically mandatory to detect a Higgs boson with mass close to the Z mass. Assuming a b-tagging efficiency of 70%, efficiencies [24] for a Higgs boson of 90 GeV are: $\sim 50\%$ in the lepton channel, $\sim 30\%$ in the neutrino channel and $\sim 15-20\%$ in the hadronic channel. Assuming one can identify b-quarks with high efficiency and purity , the background can be reduced comparable to or lower than the signal (~30-40 events for the analysis in [24]).

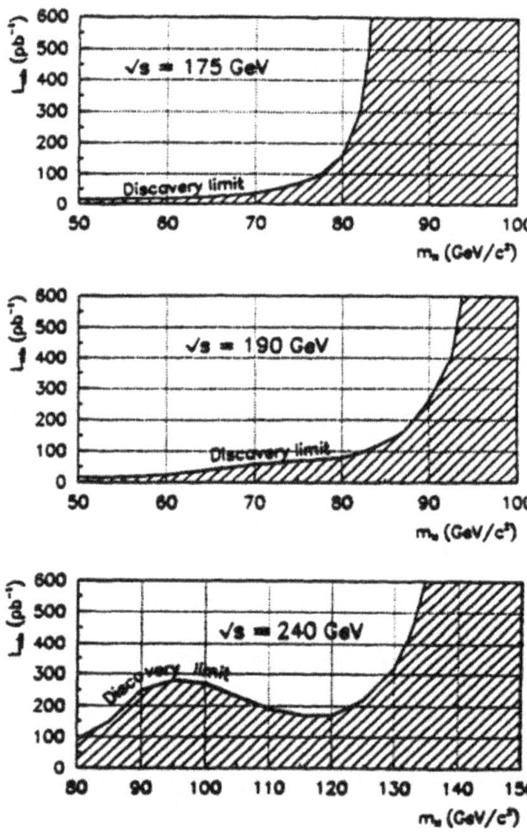

Figure 10: Minimum luminosity needed for Higgs discovery at c.o.m. energies of 175 GeV, 190 GeV and 240 GeV [24] .

Once signal and background are known, one can determine the discovery limit of LEP 200 at a given c.o.m. energy, as the minimum luminosity needed to have a signal of at least five events and which is at least five standard deviations above the expected background. The results are shown in Figure 10 where the discovery limit is plotted as function of the Higgs mass for $\sqrt{s} = 175$ GeV, 190 GeV and 240 GeV. It is seen here that at 175 GeV, 150 pb^{-1} would be enough to discover a Higgs of 80 GeV. A much larger luminosity at this energy would not extend the discovery reach than for a few GeV. At 190 GeV, a luminosity of 500 pb^{-1} would extend the discovery reach up to about 93 GeV. At $\sqrt{s} = 240$ GeV, 500 pb^{-1} would cover the range up to 133 GeV. It should be noted that these results have been derived assuming a highly performing b-tagging device. A worse b-tagging capability would result in a larger amount of luminosity needed to provide the required signal significance [22].

In conclusion, the possibility of running LEP 200 at the highest possible c.o.m energy and luminosity should be seriously considered, since, as it has been shown, this would allow to cover the Higgs mass range between 80 and 130 GeV, which is the most demanding region for the experiments at LHC [25] and SSC [26].

4.2 Prospects for the search of MSSM Higgs bosons

An additional motivation to push the energy and luminosity of the LEP collider up to its ultimate limits comes from the MSSM. In fact, as it will be shown in the following, LEP 200 running at $\sqrt{s} = 240$ GeV would find, with 500 pb^{-1}, at least the lightest Higgs boson, or dismiss the MSSM, if the mass of the t-quark is not too large, *i.e.* $m_t \lesssim 170$ GeV. It is worth noting, at this point, that from electroweak precision measurements, a 90% CL upper limit on the mass of the t-quark of 165 GeV [27], which also holds in the MSSM framework [28], has been set.

The relevant production processes for a Higgs boson in the MSSM at LEP 200 are

$$e^+e^- \to Z^* \to Zh,\ ZH, \tag{14}$$

$$e^+e^- \to Z^* \to hA,\ HA. \tag{15}$$

The $e^+e^- \to Z^*/\gamma \to H^+H^-$ process is also possible, in principle. Within the framework of the MSSM, however, detection of the charged Higgs at $\sqrt{s} = 190$ GeV is excluded, since in the region for $m_A > 50$ GeV, of interest at LEP 200, the charged Higgs is predicted to be heavier than 100 GeV. Within a small range of m_A , for $m_A \lesssim 80$ GeV, the charged Higgs can be lighter than 120 GeV, therefore kinematically accessible at $\sqrt{s} = 240$ GeV, but the production cross section is expected to be very small ($\lesssim 0.05$ pb) and the irreducible WW background much larger, thus making the MSSM charged Higgs detection quite unlikely at LEP 200. For this reason, we will concentrate in the following on the detection of the neutral Higgs bosons.

The Zh production cross section contains an overall factor $\sin^2(\beta - \alpha)$ which suppresses it in certain parameter regions (with $m_A \lesssim 100$ GeV and large $\tan\beta$). The hA and ZH cross sections contain the complementary factor $\cos^2(\beta - \alpha)$. In Figure 11 contours of $\sin^2(\beta - \alpha)$ are plotted in the $\tan\beta$-m_A plane [29]. In the large $\tan\beta$ region, as m_A increases, $\sin^2(\beta - \alpha)$ approches unity, which means h is produced with the same rate as the SM Higgs boson. For large m_A values, m_H and m_{H^+} are also very large, while

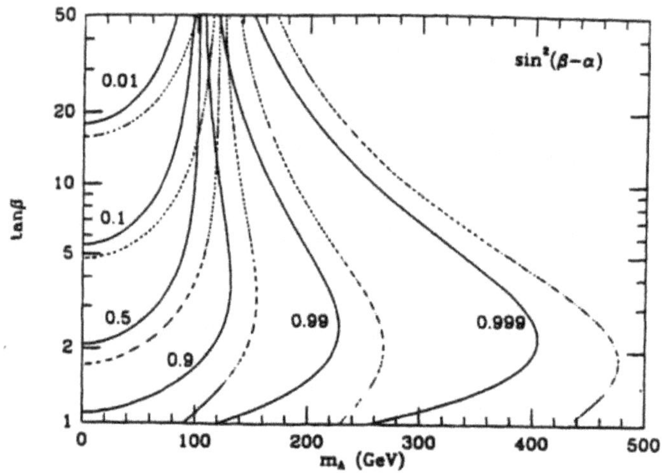

Figure 11: Contours of $\sin^2(\beta-\alpha)$ in the $\tan\beta$-m_A plane [29] . The solid lines correspond to m_t= 160 GeV, the dashed ones to m_t= 120 GeV.

m_h approaches its upper bound (approximated by eq. 11) and only h can be produced. The capability of LEP 200 to detect the lightest Higgs boson depends, for given $\sqrt{s}$ and luminosity, on the mass of the t-quark, which determines the maximum value of m_h . In Figure 12 the contours of m_h^{max} (maximum value of m_h , reached for $m_A \to \infty$) in the $\tan\beta$-m_t plane [29] are shown (for $m_{\tilde{q}}$ = 1 TeV). We see here that, even for large values of m_t ($m_t \lesssim 170$ GeV) and of $\tan\beta$, the lightest Higgs boson is predicted to be lighter than 130 GeV. Thus, considering that the h decay rates into $b\bar{b}$ and $\tau^+\tau^-$ pairs are very close to the SM values over all the parameter space, the lightest Higgs boson should be in the reach of LEP 200, with $\sqrt{s}$ = 240 GeV and 500 pb^{-1} (see Section 4.1), at least for not too large values of m_t ($m_t \lesssim 170$ GeV) and $m_{\tilde{t}}$ ($m_{\tilde{t}} \lesssim 1$ TeV).

In the region $m_A \lesssim 100$ GeV and $\tan\beta \gtrsim 3$, where the Zh cross section is suppressed, if A and H are light, the hA and ZH production processes are kinematically possible and they may have large cross sections. In addition, the A and H decay rates into $b\bar{b}$ and $\tau^+\tau^-$ pairs are very close to the values of the SM Higgs decay rates. Again, the detectability of MSSM Higgs bosons depends on m_t because, larger m_t values determine larger values of the H and h masses. In fact, for very large values of the t-quark mass, the H mass could be so large that the ZH channel might be kinematically very suppressed or not open at all.

From the experimental point of view, the search strategy developed to find the SM Higgs can be efficiently applied for the search of the h and H, produced via the Bjorken process. At $\sqrt{s}$ = 190 GeV, with a global efficiency of ~ 20% for the Higgs signal and 30-40 background events, left after the cuts mentioned in Section 4.1, about 150

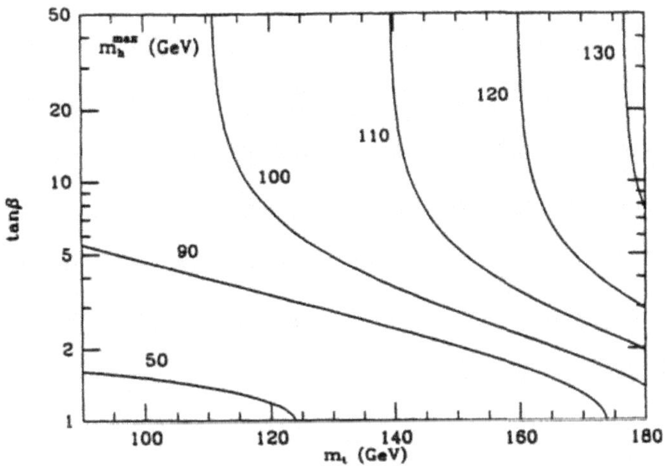

Figure 12: Contours of m_h^{max} in the $\tan\beta$-m_t plane [29] .

produced Higgs events are needed for a signal significance of 5 standard deviation over the background. In Figure 13(a), the contours of the number of Higgs events produced in the Zh channel, for $\sqrt{s}$ = 190 GeV and 500 pb^{-1}, are shown in the $\tan\beta$-m_A plane. Thus, the region for $\tan\beta \lesssim 3$, corresponding to about 150 Zh events produced, would be covered by the Higgs search in the Zh channel.

To search for the hA associate production, the 4-jets and the $\tau\tau$-jets analyses, including b-tagging, can be used, given the fact that h and A decay 85% in $b\bar{b}$ and 8% into $\tau^+\tau^-$ pairs. At $\sqrt{s}$ = 190 GeV and with 500 pb^{-1}, the search in the hA channel, combined with the search in the Zh channel, can cover the region shown in Figure 13(b).

At $\sqrt{s}$ = 240 GeV, all the region for $m_A \gtrsim 100$ GeV can be explored by the search in the Zh channel. The region for $m_A \lesssim 100$ GeV can be covered by the associated searches in the Zh(H) and hA channels. In this region, to have a large enough Higgs production rate, it is important that all three channels, Zh, hA and ZH, are open and not kinematically suppressed. This is possible if m_t is not too large, *i.e.* $m_t \lesssim 170$ GeV. It has been remarked [30,29,25] that in the region $m_A \sim 110-190$ GeV and $\tan\beta \sim 5-15$ the discovery of a MSSM Higgs boson would be difficult, even combining LEP 200 at $\sqrt{s}$ = 190 GeV with LHC and SSC. LEP 200 running at $\sqrt{s}$ = 240 GeV would allow to close this hole (if $m_t \lesssim 170$ GeV).

5 Conclusions

Experiments at LEP 100 exclude a SM Higgs boson with mass lower than 60 GeV. Future searches at LEP 200 are expected to explore the mass range up to $\simeq$ 90 GeV

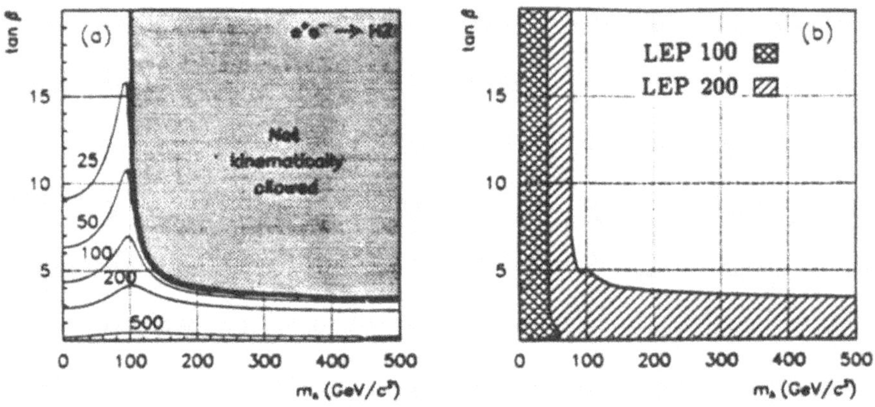

Figure 13: (a) Numbers of Zh events expected at LEP 200 with c.o.m. energy of 190 GeV and 500 pb^{-1} and (b) the domains explored at LEP 100 and at LEP 200, with c.o.m. energy of 190 GeV and with 500 pb^{-1}, by the combined searches in the Zh and hA channels [24]. With LEP 200 running at c.o.m. energy of 240 GeV, the whole parameter space of the MSSM could be explored.

($\simeq$ 130 GeV) for $\sqrt{s}$ = 190 GeV ($\sqrt{s}$ = 240 GeV) and 500 pb^{-1}.

In the context of the MSSM, the searches for non-minimal Higgs bosons at LEP 100 exclude a light scalar h with a mass lower than 43 GeV, irrespective of m_A and a pseudoscalar A with mass lower than 21 GeV, irrespective of m_h. LEP 200, with $\sqrt{s}$ = 190 GeV and 500 pb^{-1}, will cover the region for $m_A \lesssim 80$ GeV, irrespective of $\tan\beta$, and $\tan\beta \lesssim 3$, irrespective of m_A. LEP 200 running at $\sqrt{s}$ = 240 GeV could explore the whole $\tan\beta$-m_A plane. In a region of the parameter space, for $m_A \lesssim 100$ GeV and $\tan\beta \gtrsim 3$, more than one Higgs bosons might be discovered. In most of the $\tan\beta$-m_A space just one Higgs would be discovered, with properties compatible with the SM Higgs boson, within the experimental errors. Thus, there is the possibility to catch at least one Higgs boson at LEP 200, though, likely, it will not be possible to know if it belongs to a minimal sector or to an extended one. Nevertheless, LEP 200 will provide a unique ground of information to develop the most effective strategies for further investigation of the nature of the Higgs sector at LHC and SSC.

Acknowledgments

I am indebted to the organizers of the Workshop, Prof. G. Herten and Prof. M. Perl, for their kind hospitality. I wish to thank all the Higgs hunters of L3, in particular, A. Kunin, M. Pieri and A. Sopczak, for discussions and constructive criticism. I am also grateful to M. Green, J.-F. Grivaz, P. Janot (ALEPH), E. Gross, F. Richard (DELPHI), E. Duchovni and P. Teixeira-Dias (OPAL) for kindly providing their results.

References

[1] ALEPH Collaboration, D. Decamp *et al.*, *Phys. Rep.* **216** (1992) 253.

[2] DELPHI Collaboration, P. Abreu *et al.*, *Z. Phys.* **C 51** (1991) 25.

[3] L3 Collaboration, B. Adeva *et al.*, Preprint CERN, PPE/92-163 (1992) .

[4] OPAL Collaboration, P. D. Acton *et al.*, *Phys. Lett.* **B 268** (1991) 122.

[5] OPAL Collaboration, P. D. Acton *et al.*, OPAL PN048-91, 1991.

[6] L3 Collaboration, B. Adeva *et al.*, *Phys. Lett.* **B 283** (1992) 454.

[7] ALEPH Collaboration, D. Decamp *et al.*, ALEPH 92-37 PHYSIC 92-34, Paper submitted to the XXVI ICHEP Conference, Dallas, August 5–12, 1992 .

[8] DELPHI Collaboration, P. Abreu *et al.*, DELPHI 92-101 PHYS 209, Paper submitted to the XXVI ICHEP Conference, Dallas, August 5–12, 1992 .

[9] OPAL Collaboration, P. D. Acton *et al.*, OPAL PN077-92, Paper submitted to the XXVI ICHEP Conference, Dallas, August 5–12, 1992 .

[10] E. Gross and P. Yepes, Preprint CERN-PPE, 92-153 (1992) .

[11] S. Dawson, J.F. Gunion, H.E. Haber and G.L. Kane,"The Physics of the Higgs bosons: Higgs Hunter's Guide" (Addison Wesley, Menlo Park, 1989).

[12] DELPHI Collaboration, P. Abreu *et al.*, DELPHI 92-75 PHYS 186, Paper submitted to the XXVI ICHEP Conference, Dallas, August 5–12, 1992 .

[13] ALEPH Collaboration, D. Decamp *et al.*, Preprint CERN PPE/, 92-59 (1992) .

[14] DELPHI Collaboration, P. Abreu *et al.*, DELPHI 92-80 PHYS 191, Paper submitted to the XXVI ICHEP Conference, Dallas, August 5–12, 1992 .

[15] OPAL Collaboration, P. D. Acton *et al.*, OPAL PN057-92, Paper submitted to the XXVI ICHEP Conference, Dallas, August 5–12, 1992 .

[16] R. Barbieri and M. Frigeni, *Phys. Lett.* **B 258** (1991) 395.

[17] J. Ellis, G. Ridolfi and F. Zwirner, *Phys. Lett.* **B 257** (1991) 83.

[18] H.E. Haber and R. Hempfling, *Phys. Rev. Lett.* **66** (1991) 1815.

[19] Y. Okada, M. Yamaguchi and T. Yanagida, *Prog. Theor. Phys. Lett.* **85** (1991) 1.

[20] A. Brignole, J. Ellis, G. Ridolfi and F. Zwirner, *Phys. Lett.* **B 271** (1991) 123.

[21] ALEPH Collaboration, D. Decamp *et al.*, ALEPH 92-24 PHYSIC 92-21, Paper submitted to the XXVI ICHEP Conference, Dallas, August 5–12, 1992 .

[22] D. Treille, Talk given at the LEPC Meeting, CERN, November 3, 1992.

[23] C. Rubbia, Proceedings of the LP-HEP 91 Conference, p.441 (1991).

[24] P. Janot, Preprint LAL, 92-20 (1992) , Talk given at the 27th Rencontres de Moriond, March 15–22, 1992, to be published in the Proceedings.

[25] ATLAS Collaboration, CERN/LHCC/92-4, 1992,
CMS Collaboration, CERN/LHCC/92-3, 1992,
L3P Collaboration, CERN/LHCC/92-5. 1992.

[26] GEM Collaboration, SSCL-SR-1184, 1991,
Solenoidal Detector Collaboration, SDC-92-201, 1992.

[27] J. Ellis, G.L. Fogli and E. Lisi, *Phys. Lett.* **B 286** (1992) 85.

[28] J. Ellis, G.L. Fogli and E. Lisi, *Phys. Lett.* **B 274** (1992) 456,
M. Drees, K. Hagiwara and A. Yamada, *Phys. Rev.* **D 45** (1992) 1725 .

[29] Z. Kunszt and F. Zwirner, Preprint CERN-TH., 6150/91 (1991) .

[30] F. Pauss, *Large Hadron Collider: Physics beyond the Standard Model*, Lectures given in the CERN Academic Training Program, Dec. 1991.

HIGGS PHYSICS — MODELS, RECENT RESULTS

G.L. Kane
Randall Lab of Physics
University of Michigan
Ann Arbor, MI 48109-1120

Abstract

This talk is an overview of our present understanding of how to get experimental data that will teach us how nature has implemented the Higgs mechanism. Emphasis is on recent results. Some discussion is given of what can be learned from precision measurements.

INTRODUCTION

In the mid-1980's I and others began to argue that no matter what form nature chooses for Higgs physics, data from LEP + SSC will lead us to the answer. Although there have been some recent challenges, I will continue to argue here that this is probably still basically correct even with our greatly increased knowledge today.

But it is not yet shown that at least one supersymmetric Higgs boson can be detected, as will be discussed below; more work is needed here, and there may be a small region of parameter space that is not accessible to LEP + SSC. We may not know whether SSC can cover that region until data is taken there. If there is a small gap, it can be covered by an e^+e^- (presumably linear) collider with $\sqrt{s}$ only a little above that of LEP (assuming sufficiently large luminosity). Also, if Higgs bosons do exist it is probably necessary to have data from the next linear collider in order to fully untangle their properties and understand the underlying physics.

What about LHC? It could happen that the Higgs bosons have properties that allow LHC to fully detect them. However, LHC has two weaknesses. First, if Higgs bosons are not detected it will be considerably more difficult to really exclude them, because the production cross section at LHC is about 1/3 that at SSC and backgrounds are a little worse. Since excluding them at SSC is thought to be barely possible, at LHC it may not work. If Higgs bosons are not detected it is very important to exclude them in order to know they are not nature's solution. Second, the cross section for producing a pair of W's with $M_{WW} \simeq 1$ TeV is about an order of magnitude smaller at LHC than at SSC, about two orders of magnitude smaller at $M_{WW} \simeq 2$ TeV. Thus it is impossible to make a general study of WW interactions in the TeV region at LHC. Of course, one can invent ad hoc models where W's interact so strongly that their TeV interaction is detectable even at LHC, but there is insufficient motivation for such models to expect the world to be that way, and if an effect is not seen at LHC no physics ideas of significance are excluded, while at SSC it is directly possible to extract the WW cross sections above a TeV. If light Higgs bosons are found, these weaknesses of LHC are not of qualitative significance.

Finally, it should be remarked that all discussions of the physics capabilities of various facilities assume that appropriate detectors are built. It is certainly possible to build detectors that could not detect Higgs bosons even if they were present.

The outline of the rest of the talk is as follows:

- If Higgs bosons exist:
  - — Can a Standard Model h (SM) be observed?
    - △ Note that $M_h \lesssim 2M_Z$ if perturbative unification is valid
    - △ Yes, h can be observed by a combination of methods
    - △ h can be observed even if its decays are invisible
  - — Can the supersymmetric Higgs bosons be observed?
    - △ Maybe.
    - △ Upper limit on M_{h° in any supersymmetric theory?
  - — Can we learn about the existence and mass of a Standard Model h or a supersymmetric h° from the one-loop analysis of precise measurements?
  - — What kind of Higgs boson is it?
- If Higgs bosons are not detected it is necessary to study TeV WW interactions.
  - — Although TeV WW interactions must deviate from their perturbative predictions with a light Higgs boson, how strongly should they interact? Recent papers by H. Veltman and M. Veltman,[1)] and by R. Willey,[2)] argue that the WW interactions will be non-resonant unless new physics exists that is otherwise phenomenologically unmotivated. In particular, arguments relating the behavior of W's to that of pions should not apply unless new physics enters.
  - — Chivukula and Golden[3)] have suggested that WW interaction could produce a large number of inelastic channels, leading to a suppression of elastic WW scattering; perhaps this would make the elastic WW scattering undetectable.
    - △ While their argument could[4)] prevent observing a strong WW interaction in the central region, it will allow observation of "beam pipe W's".[5)]
    - △ The unitarity effects of inelastic channels can lead to large effects that are very different in different charge channels (W^+W^-, W^+Z, W^+W^+).[6)]
  - — If WW interact strongly, perturbative calculations cannot in principle be used; the effective W method is the only alternative. Corrections to this have been studied by Cortese and Petronzio.[7)] Their analysis strengthens the validity of the method in the TeV region, though large corrections are needed for smaller M_{WW}.

— Studies of TeV W's must measure the W polarization rather than assume it, because the new physics should show up in longitudinal W interactions. Probably this requires looking at one $W \to jj$ and the other $W \to \ell\nu$. Probably the techniques to do that are known,[8] though they have not yet been studied in simulations. In particular, they require using the knowledge that all candidate WW events in the TeV region are either perturbative Standard Model physics or new physics.

SUPPOSE HIGGS BOSONS EXIST — CAN A STANDARD MODEL h BE OBSERVED?

Constraints on M_h?

Precision measurements in the past two years, mainly at LEP, have led to increasingly precise values for the electroweak and strong couplings at the electroweak scale. As has been widely discussed, using renormalization group equations one can compute the way the couplings behave as the scale increases, and it is found that they meet at least approximately at a mass scale $M \gtrsim 10^{16}$ GeV in a supersymmetric grand unified theory and in some other theories. Equivalently, one can write $\sin^2 \theta_W$ at the high scale in several unified theories and calculate it at the electroweak scale, obtaining good agreement with the precisely measured value. It would be astonishing if such results were a meaningless accident. Without any commitment to a particular grand unified theory, or even to the existence of a grand unified theory at all (as in some string-motivated approaches), it should now be required of any set of assumptions that is viewed as being likely to describe nature, that approximate coming together of the strength of the forces should be consistent with the assumptions. In particular, the low energy parameters of the theory should be constrained to be in ranges that allow them to have values consistent with the use of perturbation theory up to the high scale, since the coming together of the couplings was based on a perturbative calculation.

It was shown over a decade ago[10] that for a Standard Model Higgs boson the effect of such a requirement gave an upper limit on M_h that was dependent on the value of M_t (because if $M_t \gtrsim M_W$ then the top quark Yukawa coupling affects the equation for the Higgs self-coupling λ, and $M_h \sim \sqrt{\lambda}$). One can summarize the results very approximately in mnemonic form by saying *either* $2M_Z > M_h > 2(M_t - M_Z)$ *or the apparent unification of forces is a meaningless accident.* Any planning for future detectors or colliders that uses a Standard Model Higgs boson with mass outside of this range should provide strong justification for doing so.

Yes, h can be observed

After years of study by a number of people, it now seems to be accepted[11] that it is possible to either detect or fully exclude a simple Standard Model Higgs boson at hadron colliders. The mass range $M_h \lesssim M_Z$ will surely be covered at LEP even if CERN does not choose to raise the energy of LEP to 240 GeV in order to search for M_h up to about 130 GeV, near or at the upper limit in any supersymmetric theory. Combining $h \to \gamma\gamma$ with h

either produced inclusively or in association with W or $t\bar{t}$, with $h \to ZZ^* \to \ell^+\ell^-\ell^+\ell^-$, the hadron colliders can cover the region down to M_Z. To do so is not easy — the detector requirements are severe. But it is of course not necessary to do it all at once. Detectors can be upgraded depending on what is or is not detected, and collider energies and luminosities can be increased.

<u>What if h is invisible?</u>

Several approaches to Higgs physics can give a Higgs boson that decays to undetectable particles, so that it is effectively invisible. Examples are decays of h to Majarons,[12)] to LSP's,[13)] or to Goldstone bosons.[14)] Could such a Higgs boson escape detection?

Probably not.[15)] Assuming that Higgs boson is mainly the one associated with the Higgs mechanism by which W, Z get mass (if it is not there is another one to detect), then its coupling to $W^\pm, Z$ is full strength. In particular, at the SSC or LHC it is produced through the diagram

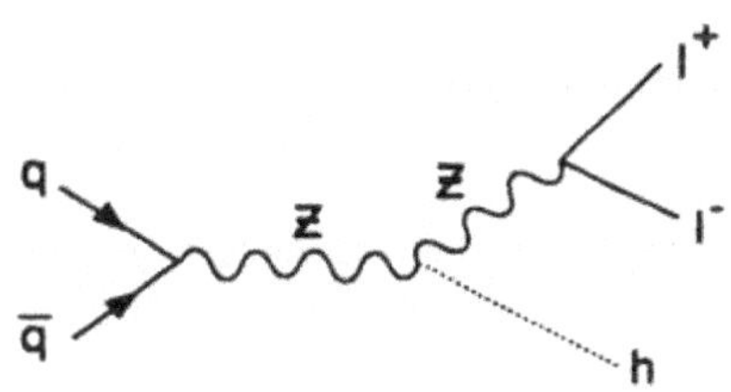

with a cross section $\sigma(Z+h)$. This cross section is about the same size as the process

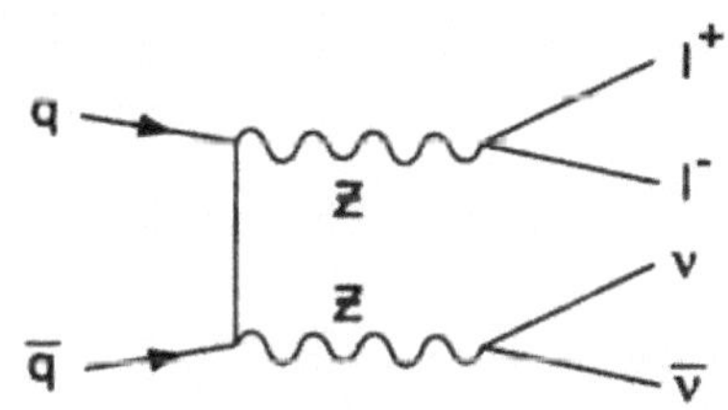

with cross section $\sigma(Z+Z(\to \nu\bar{\nu}))$ for $M_h \sim M_Z$. Since $\sigma(Z+Z)$ will be very accurately normalized at the SSC by other channels, a small increase (few % ?) in its rate will be detectable.

Equivalently one can look in $q\bar{q} \to W(\to \ell\nu) + h$ vs. $q\bar{q} \to W(\to \ell\nu) + Z(\to \nu\bar{\nu})$. Here the number of events is somewhat larger but the extra ν may prevent using this for the basic channel. If a signal were detected it could certainly be confirmed here, since $\sigma(W(\to \ell\nu) + Z(\to \ell^+\ell^-))$ will be well measured. The following table suggests the numbers are satisfactory for the mass region of interest; these are events/SSC year in a typical detector.

	Zh		Wh	
M_h	EVENTS	SIGNAL/$\sqrt{\text{BKG}}$	EVENTS	SIGNAL/$\sqrt{\text{BKG}}$
80	1600	28	11,700	91
110	760	13	5370	41
140	390	6.7	1600	12

Since $\sigma(Z+Z)$ and $\sigma(Z+W)$ are independently normalized, signal/$\sqrt{\text{bkg}}$ is the appropriate statistical measure. If the ZZh or WWh couplings are reduced from their Standard Model ones, or $BR(h \to \text{invisible})$ is less than one, these rates are reduced proportionately.

Can a supersymmetric Higgs boson be detected?

If nature is supersymmetric at the weak scale, there are at least three neutral Higgs bosons $h^\circ, H^\circ, A^\circ$, and a charged pair $H^\pm$. It is essential to be able to detect them and to measure their decay BR. Several groups[16, 17, 18, 19, 20] have analyzed this question in some detail, with the most complete analysis being that of Kunszt and Zwirner. They — appropriately at this stage — assume only non-supersymmetric decays for the scalars, and find that most of the allowed region of parameter space is detectable, but a small region is not, given their assumptions. This region is located inside the area 100 GeV $< M_A <$ 200 GeV and $5 < \tan\beta < 20$, not an unlikely region for models.

When supersymmetric decays are included, the problem region will obviously get a little larger since the non-supersymmetric BR used in the analysis will decrease. On the other hand, some of the supersymmetric modes will be detectable. For example, the interesting analysis of Roberts & Ross[21] (with $M_t = 100$ GeV) gives one set of masses with H° and A° approximately degenerate on the edge of the undetectable region (at about 220 GeV), and their decay $H^\circ, A^\circ \to \widetilde{N}'\widetilde{N}$ (the second lightest and lightest neutralinos) followed by $\widetilde{N}' \to \ell^+\ell^-\widetilde{N}$ ($\ell = e, \mu$) with about 2600 events/SSC standard year.[22] This is a detectable signal, and will allow a significant reduction of the undetectable region on the large M_A side.

Also, the low M_A and the low $\tan\beta$ sides of the region are largely fixed by the upper limits on LEP energy and luminosity. These were chosen as an "optimistic" 190 GeV and 500 fb^{-1}. These are indeed optimistic from the point of view of present discussions about CERN decisions. But LEP could be run up to $\sqrt{s} \simeq 240$ GeV, given the properties of the magnets and the tunnel size. That would allow coverage of much of the presently existing gap, and combined with supersymmetric decay modes may eliminate the gap.

It is not necessary to make a commitment to such a large $\sqrt{s}$ now, but the option should be left open. Perhaps one or more of the scalars will be detected without running LEP up that high, and then the best strategy for proceeding can be worked out.

For example, h° could be detected in the next few years with a mass of 75-85 GeV, very common in models. And once top is detected at FNAL, there is a significant chance that $t \to H^+b$ will occur. It should have a BR in the 2% – 15% range, and $BR(H^+ \to \tau\nu) \approx 1$. If neither is detected, it is clear that LEP must be run up to 240 Gev.

Upper limit on M_h in supersymmetry?

We saw above that in a perturbatively unifiable world, there was an upper limit below about $2M_Z$ on the mass of a Standard Model Higgs boson. What about a supersymmetric world?

In a minimal supersymmetric world, fermions have gauge couplings. Higgs bosons are superpartners of Higgs fermions (Higgsinos), so the scalar ("φ^4") couplings are fixed by gauge couplings. Then the Higgs potential does not depend on new couplings, so the Higgs masses also do not.

In particular, in a minimal supersymmetric theory,

$$M_{h^\circ}^2 \leq \cos^2 2\beta M_Z^2 + K\alpha_2^2 M_t^4 / M_W^2$$

where K is a known constant, the second term arising from loop corrections.[23] Other SUSY Higgs masses satisfy sum rules but do not have upper limits. If $M_t \approx 135$ GeV, $M_{h^\circ} \leq 105$ GeV. This limit is independent of the supersymmetric masses or the scale of supersymmetric breaking.

Discovery of h° would of course be a powerful breakthrough. Conversely, if h° were not found in this range and we knew that nature was not described by a minimal supersymmetric theory it would have a deep effect on the direction of theoretical thinking. Assuming sufficient luminosity (which can be calculated), LEP would have to be run at most at 210 GeV to guarantee detecting or excluding h°; of course M_{h° could be well below its upper limit. These numbers are based on $M_t = 135$ GeV; they may have to be changed if top is not found at about that mass at FNAL next year.

This analysis holds for a minimal supersymmetric theory. What about any supersymmetric theory? Upper limits on M_{h° in non-minimal theories have been studied,[24, 25, 26] and numerical limits given in fairly general ones. Recently an equation has been written for an absolutely general limit,[27] and numerical limits calculated in particular models.[27] Study of an absolutely general numerical limit is underway. It appears that for $M_t \simeq 135$ GeV, the general limit will be about 140 GeV, which could be reached experimentally at LEP.[28]

Getting such a limit requires getting bounds on Higgs self-couplings just as in the Standard Model. These bounds are obtained by requiring perturbative unification, as in the Standard Model bound.

Can we learn about a Higgs boson from precision measurements and radiative corrections?

The increasingly precise measurements of Z properties, M_W, ν reactions, etc. now provide data that tests the Standard Model predictions at the one loop level. The data is sensitive to M_t, and less sensitive to the existence and mass of a Higgs boson, as expected. The effect of a supersymmetric Higgs spectrum is similar to that of Standard Model Higgs boson,[29] both dominated by the light Higgs boson, so effectively any conclusions about Higgs physics apply to both.

The basic question[30] is whether the data implies now or will imply that a light Higgs boson exists. If it did it would be a remarkable result. Most people who have studied the situation agree that a two-variable analysis, in terms of M_t and M_h, gives a fairly good measurement of M_t but does not imply that h exists, though it is encouraging in that direction because the best fit has a small M_h. But once M_t is known accurately, it does appear that the data has achieved a precision that could give significant results on M_h.

Several points are important. First, usually curves of χ^2 vs. M_t for various M_h are shown, or something equivalent. Normally χ^2 is defined so that a shift $\Delta\chi^2$ of 2.7 corresponds to a 90% confidence level, $\Delta\chi^2 = 4$ implies 95% CL, and $\Delta\chi^2 = 6.7$ implies 99% CL. A change in χ^2 of one unit does not imply anything. To draw a conclusion about something as momentous as the existence of a Higgs boson we should require at least that $\Delta\chi^2 > 4$ as M_h increases to the upper limit we assign. Second, the radiative correction from top grows as M_t^2 and becomes so large as M_t increases that the sensitivity to M_h disappears. Figure 1 is a sketch to illustrate the point, based on ref. 31. Note that (1) for $M_t \gtrsim 150$ GeV

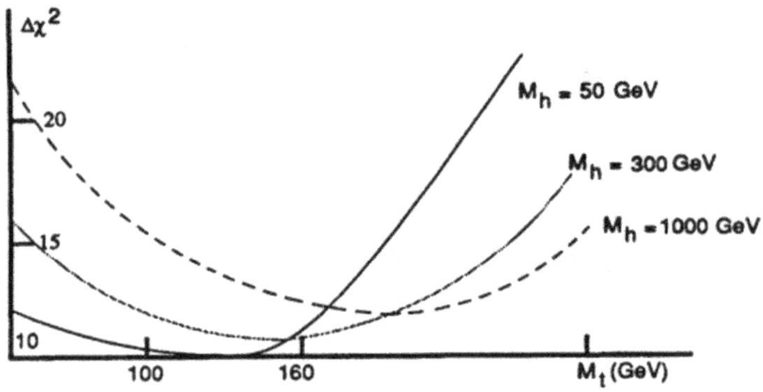

there is very little sensitivity to M_h, while for $M_t \lesssim 130$ GeV the sensitivity to M_h is large. In particular the upper error on the measured M_t is very important, while our ability to learn about M_h is hardly affected by the lower error. (2) If M_t is set at 135 GeV, then[32] the analysis of del Aguila, Martinez, and Quiros implies that $M_h \leq 425$ GeV at 90% CL, $M_h \leq 660$ GeV at 95% CL. These numbers include the effects of the 1991 data.

If $M_t \simeq 165$ GeV measurements must have much higher precision to learn about M_h. Numbers similar to the above could be obtained if the error on $\sin^2\theta_W$ could be reduced to ± 0.0001,[33] while reducing the errors on M_W would not help much.

If it turns out that $M_t \approx 135$ GeV and the precision measurements imply a light Higgs boson, and if the approximate coming together of the $SU(3), SU(2)$, and $U(1)$ force strengths at about 10^{16} GeV is accepted as implying that the parameters of the theory must allow the theory to remain perturbative up to such a scale, then we will have learned that a Higgs boson lighter than about $2M_Z$ exists. That would be remarkable progress.

What kind of Higgs boson is it?

Once h is discovered (directly or indirectly), the next major question will be "what kind of Higgs boson is it?". Standard Model? Supersymmetric? Other? Many tests are possible, almost all very difficult. From indirect evidence based on precision measurements there is no known way to decide.

From direct measurements probably the three most likely ways to tell, and the only three possible at LEP or FNAL as far as I know, are

(i) Observe $t \to H^+b$ at FNAL. Then it cannot be a Standard Model Higgs sector, and is presumably though not certainly a supersymmetric one.

(ii) Observe both h° and A° at LEP, e.g., in $e^+e^- \to Z \to h^\circ A^\circ$. Again, presumably that implies a supersymmetric world.

(iii) Study $e^+e^- \to Z^{(*)} \to h(\to b\bar{b})+$ nothing.[34] In the Standard Model that is due to $e^+e^- \to Z^{(*)} \to Z^{(*)}(\to \nu\bar{\nu}) + h(\to b\bar{b})$. In a supersymmetric world the additional process $e^+e^- \to Z^{(*)} \to \widetilde{N}'(\to \widetilde{N} + h(\to b\bar{b})) + \widetilde{N}$ will occur, with a rate that is model dependent but could be large enough to observe.

We will be lucky if any of these three processes can be detected. One might think that because the cross section $e^+e^- \to Z^{(*)} \to Z^{(*)}(\to f\bar{f}) + h$ is suppressed by a factor $\sin^2(\beta - \alpha)$ in a supersymmetric world, one could simply measure that rate. In practice however, the mere fact that one observes the process at all implies $\sin^2(\beta - \alpha) \approx 1$, and in addition $\sin^2(\beta - \alpha) \approx 1$ in most models. Thus in practice this method will require perhaps a hundred events or more of $Z + h$ to give a result, needing years of data. A number of other even more difficult methods to learn what kind of Higgs boson it is are discussed in ref. 35.

IF HIGGS BOSONS DO NOT EXIST, HOW STRONGLY WILL W_LW_L INTERACT? CAN THE W_LW_L INTERACTION BE STUDIED?

How strongly will W_LW_L interact?

A number of people have argued strenuously[36] that if light Higgs bosons did not exist, the W_LW_L interaction should get very strong, and even resonate. The justification was mainly analogies with $\pi\pi$ scattering, based on arguing that the physics of W_L should be analogous to the physics of pions because both were Goldstone bosons interacting in a non-Abelian gauge theory.

Recently two papers[1, 2] have emphasized that this is not so. As pointed out long ago by Lehmann,[37] the resonant behavior of pions is mainly due to presence of nucleons as well, affecting the pions through loops. Thus the expectation that pion interactions should resemble W interactions is only expected to be valid if major new physics (equivalent to the nucleons) is present in the same energy region where the resonances occur ($M_\rho \approx M_N$).

These papers have studied how W_LW_L interactions might behave in the Standard Model without light Higgs bosons. The paper of Veltman and Veltman (V^2) is a particularly thorough study of how to formulate the Standard Model if there is no light Higgs boson, by taking $M_h \to \infty$, and the implications for experimental W_L interactions. While it has been known[38)] for some time that letting $M_h \to \infty$ numerically in the WW amplitude did not produce much of a W_LW_L interaction until $M_{ww} \gg 1$ TeV, the connection of this to approaches leading to a large W_LW_L interaction was not understood. Now V[2] have shown that depending on how $M_h \to \infty$, either W_LW_L should not resonate at all, or resonate at such a high energy ($M_{ww} \sim 9$ TeV) that it is irrelevant for experimental data. In their words, "... the manifestation of a ρ-like resonance at around 2 TeV becomes questionable if the physics below 1 TeV is described by the effective Lagrangian of the Standard Model in the heavy Higgs limit." That is, building the symmetries of the Standard Model into the W_LW_L interaction does not generate resonant behavior at even 2 TeV, let alone below that scale, unless new physics is present in the region below 1 TeV. That new physics cannot be very hidden if it is to be strong enough to generate resonances.

One implication of these results for thinking about future facilities and detectors is that planners should not assume any resonant or particularly strong interactions of W_L's in the TeV region. While such an effect could occur, it would require additional new physics in the same region. Technicolor theories are an example where this happens. Resonances do occur, but only if lots of technifermions are present, and a number of phenomena might have already shown up such as light technipseudogoldstone bosons or effects on radiative corrections. Many observers feel that such effects and the associated physics are already excluded, though true believers have not softened their position.

A model that leads to a 1 TeV resonance produces about 350 times more W_LW_L events at a TeV than suggested by the analysis of V[2] based on the Standard Model Lagrangian with $M_h \to \infty$; one with a 2 TeV resonance about 30 times more. While such effects are not yet completely excluded, they should not be used for judging how well a new facility can study physics in the TeV region.

Can strong W_LW_L interactions always be detected?

Techniques have been proposed to study W_LW_L interactions in the central region even if they are not very strong. These techniques have not yet been studied with full detector simulations.

Recently Chivukula and Golden suggested[3)] that W_LW_L interactions could get strong in such a way as to have a number of inelastic undetectable channels present, and have the elastic W_LW_L rate suppressed to the level where it cannot be detected in the usual ways. That is an interesting suggestion, and an interesting challenge to detecting W_LW_L whatever form the interactions might take. [Note that the model they use does not fully[6)] exhibit the hoped-for behavior; it gives a central region W_LW_L cross section of about 0.5 pb, which is surely detectable (5000 extra W_LW_L events/SSC year).]

In fact, there is a way[5)] to detect such a situation. If $\sigma_{TOT}(W_LW_L)$ is so large that the elastic channel is suppressed, then by the optical theorem the elastic forward amplitude is very large. These W_L's go down the beampipe ("beampipe W's") rather than into the central region, but that is no problem since W's decay. For a 1 TeV WW interaction the opening angle between $\ell^\pm\nu$ or jj is 17°, and the decay products emerge nicely into a detector with coverage in the region of rapidity $3 \leq \eta \leq 5$. Thus $\sigma_{TOT}(W_LW_L)$ can be measured as a function of M_{WW}.

A consideration of the effects of inelastic channels and unitarity has led to some interesting work here.[4] In a model with a large number of fermions in loops, it was found that the effect of having the contribution of inelastic channels to amplitudes be comparable to that of the tree level led to interesting cancellations and enhancements. In particular, low isospin channels (W^+W^-, ZZ) were enhanced, while high isospin channels (W^+Z, W^+W^+) were suppressed. This could be interpreted to suggest that in strongly interacting theories the W^+W^+ cross section will usually be suppressed.

How good is the effective W approximation?

If there is no light Higgs boson, it would be inconsistent to use perturbation theory to relate WW interactions to experimental data; the effective W method, where W's emitted by quarks are treated as approximately real, is essential.

The corrections to the leading approximate result have been studied recently by Cortese and Petronzio. In outline, they proceed as follows. The process in question is

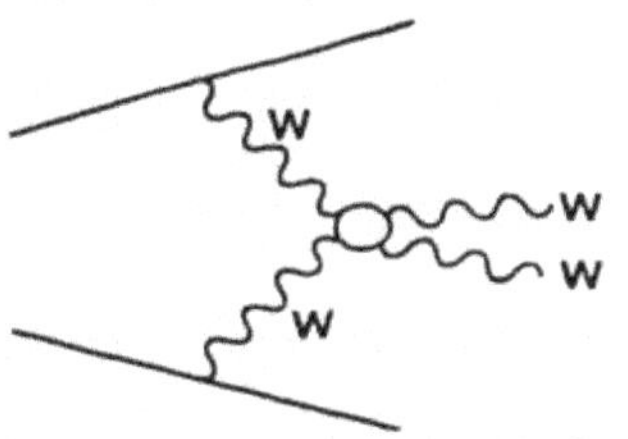

where the W's may attach to a non-perturbative interaction. The full cross section can be written

$$\sigma = \int \frac{d^4K}{K^2 - M_W^2} \frac{d^4K'}{K'^2 - M_W^2} \sigma^{\mu\nu}(K,K')\Gamma_{\mu\nu}(K,K') \tag{1}$$

where $\sigma^{\mu\nu}$ is the $WW \to WW$ cross section, and $\Gamma_{\mu\nu}$ the appropriate polarization tensor for the initial W's. To obtain corrections due to the W's being not quite real, expand

$$\sigma^{\mu\nu}\Gamma_{\mu\nu} = \sigma_0^{\mu\nu}\Gamma_0^{\mu\nu} + \left(K^2 - M_W^2\right)\Gamma_{\mu\nu}\frac{\partial\sigma^{\mu\nu}}{\partial K^2}\bigg|_{\text{on shell}} + \ldots \tag{2}$$

The first term gives the effective W approximation, and there are derivative terms for every variable, including transverse momenta. Using equation (2) in equation (1) and carrying out some integrations results in

$$\sigma = \int dx\sigma_{WW}P + \int \frac{\partial\sigma}{\partial K^2}\bigg)_{\text{on shell}} P'dx + \ldots \tag{3}$$

where σ_{WW} is the on-shell WW cross section, P the usual polarization tensor consisting of known W structure functions, and the second and following terms are corrections involving derivatives of cross sections.

Since derivatives are present, one could go from any theory of WW interactions to predictions for experiment, but one could not go the opposite way and extract $\sigma_{WW}^{\mu\nu}$ from data, contrary to what was originally hoped. Fortunately, the results of Cortese and Petronzio show that in the TeV region where the effective W approximation would be really needed, the approximation is numerically valid to within about 2%. Although the corrections are present in principle, they will not interfere in practice with studying strong WW interactions if we have to do that.

CONCLUDING REMARKS

Probably there are three points that should be emphasized. The combined efforts of many experimenters and theorists over many years may have finally brought us to the stage where within a year the explicit measurement of M_t, plus the precision measurements, imply that a light Higgs boson exists. Depending on the degree of confidence required, and on the final measured value of M_t, it may be possible to conclude that $M_h \lesssim 500$ GeV or so. Adding the apparently valid requirement that the theory should remain perturbative until a scale of order 10^{16} GeV implies that not only does a Higgs boson exist but it is lighter than about $2M_Z$. Of course, if a light Higgs boson exists it is less important to study WW interactions; basically no $W_L W_L$ pairs should be detected in the TeV region.

Second, an upper limit on the mass of the lightest Higgs boson exists in a general supersymmetric perturbatively unifiable theory, and it is not very heavy, perhaps as low as 140 GeV, if $M_t \lesssim 135$ GeV. A better numerical value for this limit will exist soon. Searching experimentally up to that limit can be done at LEP, and should surely be done if a lighter Higgs boson is not discovered. Finding a Higgs boson would of course be of extraordinary importance for the search for deeper understanding of the laws of nature; if one were not found and thus a supersymmetric unifiable theory could not be correct it would have a powerful effect on the direction of future research.

Third, although some details need to be sorted out, basically we know how to get the experimental data needed to learn abut the Higgs sector. With LEP plus the SSC most initial measurements can be made. A complete understanding will probably require NLC as well. Depending on what is found, it may be necessary to run LEP up as high as 240 GeV, or to upgrade the SSC detectors, but eventually we can learn what is needed to tell us which ideas are correct or lead us to new ones.

ACKNOWLEDGEMENTS

I am grateful for discussions with C.-P. Yuan, F. Pauss, M. Quiros, F. del Aguila, C. Llewellyn-Smith, and R. Petronzio.

References

1. H. Veltman and M. Veltman, Acta. Phys. Polon. **B22** (1991) 669.
2. R.S. Willey, Phys. Rev. **D44** (1991) 3646.
3. R.S. Chivukula and M. Golden, Phys. Lett. **B267** (1991) 233.
4. S. Naculich and C.-P. Yuan, JHU-TIPAC-920017, June 1992.
5. G.L. Kane, S. Naculich, and C.-P. Yuan, to be published.
6. S. Naculich and C.-P. Yuan, JHU-TIPAC-920012, June 1992.
7. S. Cortese and R. Petronzio, Phys. Lett. **B276** (1992) 203.
8. See the review chapter on this subject by C.-P. Yuan, in ref. 9.
9. "Perspectives in Higgs Physics", World Scientific, 1992, ed. G.L. Kane.
10. N. Cabibbo, L. Maiani, G. Parisi, and R. Petronzio, Nucl. Phys. **B158** (1979) 295. This effect seems to have first been noticed and commented on by V. Elias, Phys. Rev. **D20** (1979) 262.
11. See the chapters by S. Dawson and Z. Kunszt in ref. 9 for a review.
12. Ling-Fong Li, Ying Liu, and L. Wolfenstein, Phys. Lett. **159B** (1985) 45.
13. Kim Griest and H. Haber, Phys. Rev. **D37** (1988) 719.
14. "A Full Acceptance Detector For SSC Physics at Low and Intermediate Momentum Transfers", J.D. Bjorken, SLAC-PUB-5545, 1991.
15. S.G. Frederikson, N.P. Johnson, G.L. Kane, and J.H. Reid, in preparation.
16. Z. Kunszt and F. Zwirner, ETH-TH-91-7; also the talk of F. Pauss at this conference.
17. J. Gunion and L. Orr, UCD-91-15.
18. V. Barger, M.S. Berger, A. Strange, R.J.N. Phillips, Phys. Rev. **D45** (1992) 4128.
19. J. Gunion, R. Bork, H. Haber, and A. Seiden, SCIPP-91-34.
20. A.H. Baer, M. Bisset, C. Kao, and X. Tata, FSU-HEP-911104, Nov. 1991.
21. R. Roberts and G. Ross, RAL-92-005, Jan. 1992.
22. G.L. Kane and C. Kolda, to be published.
23. For a summary of these see the chapter of H. Haber in ref. 9.
24. H. Haber and M. Sher, Phys. Rev. **D35** (1987) 2206.
25. M. Drees, Int. J. Mod. Phys. **A4** (1989) 3635.
26. P. Binetruy and C. Savoy, Phys. Lett. **B277** (1992) 453.
27. J. Espinosa and M. Quiros, Phys. Lett. **B279** (1992) 92.
28. For studies of what e^+e^- collider can detect one of the neutral SUSY Higgs spectrum, see P. Janot, LAL 92-27, or V. Barger, K. Cheung, R.J.N. Phillips, and A.L. Strange, MAD/PH/704.

29. R. Barbieri, M. Frigeni, F. Giuliani, and H. Haber, Nucl. Phys. **B341** (1990) 309.

30. For a review see the talk of G. Fogli at this conference.

31. F. del Aguila, M. Martinez, and M. Quiros, CERN-TH-6289-92.

32. F. del Aguila and M. Quiros, private communication.

33. A. Blondel, F.M. Renard, and C. Verzegnassi, Phys. Lett. **B269** (1991) 419.

34. H. Haber, G.L. Kane, I. Kani, and M. Quiros, Nucl. Phys. **B283** (1987) 111.

35. See the chapter of G.L. Kane in ref. 9.

36. For reviews, see, for example, M. Chanowitz, Ann Rev. Nuc. Sci. **38** (1988) 323.

37. H. Lehmann, Acta. Phys. Austriaca, Suppl. XI (1973) 139.

38. M.A.C. Duncan, G.L. Kane, and W.W. Repko, Nucl. Phys. **B272** (1986) 517.

SEARCHING FOR RARE EXOTIC Z^0 DECAYS

Michael Dittmar,
Department of Physics, University of California,
Riverside, CA 92521 USA

Abstract

In this paper, experimental signatures for rare exotic Z^0 decays, the latest available results and the future experimental sensitivity are discussed. Branching ratio limits for exotic rare Z^0 decays are currently at a level of a few times 10^{-5} and are close to some model predictions. Several of the studied signatures are essentially background free and might provide the first signs of physics beyond the Standard Model, once several million Z^0 decays are analysed. Despite the so far negative searches, the window for rare Z^0 decays is small but still open. In fact it has just been touched by the LEP experiments.

1 Introduction

Searches for the pair production of new particles with well defined couplings, but unknown masses, like new heavy quarks, leptons and supersymmetric particles, were essentially completed using the 1989 data from the first year of LEP operation with a few 10^4 Z^0 decays [1]. Not only were the results negative, but also not a single exotic event was found in these early searches.

Still all four LEP experiments have continued to search for exotic effects in Z^0 decays, but the results remain negative. Besides the Higgs search[1], searches for flavour changing neutral currents, for compositeness and for single production and decay of exotic particles were performed.

Usually, no definite model predictions exist for such rare decays. Therefore, all signatures with small backgrounds from known Standard Z^0decays should be carefully investigated. However, often previous experiments, like the obtained branching ratio limit for the μ decay to 3 electrons (it has to be smaller than 10^{-12} (!) [2]), exclude essentially any possibility to find some rare Z^0 decays. The above measurement excludes the Z^0 decay into $e\mu$. In contrast to this example, limits for flavour changing neutral currents, which involve the third quark or lepton family are weak. The absence of the τ decay to three charged leptons excludes Z^0 decays into a $e\tau$ or $\mu\tau$ only for branching ratios larger than 10^{-4}. Experimentally a sensitivity of smaller than 10^{-5} exists in direct searches at LEP [3]. Some other rare Z^0 decays, like 4 fermion final states, are allowed within the Standard Model. Their rates might be enhanced due to some exotic particles, with masses of a few 100 GeV. Typical examples are charged Higgse, extra Z's or W's. With the introduction of such particles, some rare decays might be observable and Z^0 branching ratios of $10^{-5\pm1}$ are obtained in some models.

After discussing some general aspects of searches for rare Z^0 decays at LEP, we analyse various search results from ALEPH, DELPHI, L3 and OPAL. Instead of presenting various experimental limits for exotic models, the experimental signatures will be discussed in a model independent way. The current limits obtained from neutral final states, low charged multiplicity events and high charged multiplicity events are estimated.

2 Prospect of finding rare exotic Z^0 decays at LEP

During the first three years of LEP operation (from 1989-1991) about 5×10^5 Z^0 decays have been recorded by each LEP experiment and an order of magnitude more Z^0 events are expected during the next few years.

The sensitivity to a certain branching ratio is obviously inverse to the total number of analysed Z^0 decays, devided by the detection efficiency. If no signal is observed, limits for a certain signature are obtained. For zero background events, a limit at 95% confidence level corresponds to three events. If for example a background of 10 events is estimated, a

[1]For details of Higgs searches we refer to the proceedings of this conference

possible limit is, in the absence of a signal, roughly only a factor of two larger ($2 \times \sqrt{10}$). However, to observe a new particle, a significance of 5 σ is required. In principle, if no background exists, one detected event would be enough. But, since for example a tiny background due to inefficiencies or acceptance gaps in vetoing missing energy and momentum can never be excluded completely, a few (3-5) events are generally needed. In the presence of background, e.g. a signal and background of similar size, a factor of roughly 5-10 more data is required for a discovery (25 signal events above a background of 25± 5).

To enhance possible signal to background ratios, several experimental cuts, finetuned with Monte Carlo techniques, are made. Obviously these cuts have to be chosen carefully, since a first indication of a signal results also in a worse limit for such a process. Due to the tendency to obtain the best limits for a search, one finds many examples where the number of observed events is smaller than the expected number of background events. Reasons might be the limited data statistics and slightly wrong background estimates. Consequently, too hard cuts might remove the first signs of new physics.

The chances to observe new physics are of course best, if consistent exotic models lead to signatures with small backgrounds. Such a scenario has been investigated in a Monte Carlo study for heavy singlet neutrinos (N) [4]. In this paper it has been shown that: a) essentially no constraints exist to produce such particles in Z^0 decays via neutrino mixing and detectable branching ratios; b) the N particle would decay into either a virtual W^* and a charged lepton or into a Z^* and a neutrino followed by standard W^* and Z^* decays. The signature studied, two acoplanar charged particles with large missing energy, was found to be background free. It was concluded that the Z^0 is an ideal place to search for singlet neutrinos with masses in the range from a few GeV to much more than half the Z^0 mass, as shown in figure 1a and b.

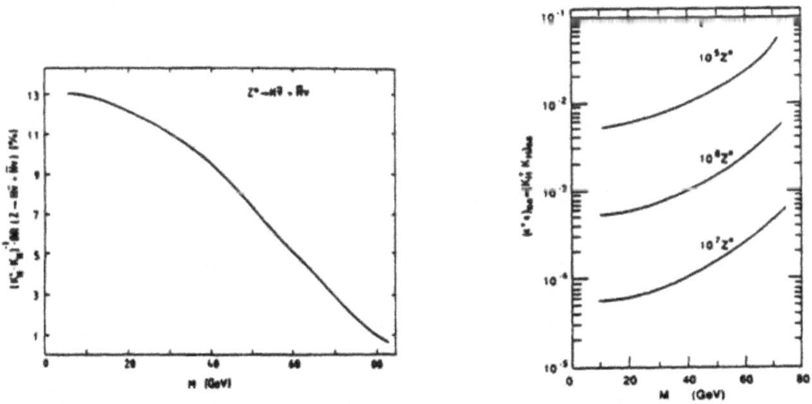

Figure 1: a) The branching ratio for the decay $Z^0 \rightarrow N + \nu$ as a function of its mass, devided by the coupling strength parameter; b) Observable coupling strength for N particles as a function of mass for different number of Z^0 decays.

3 Neutral final states

Events with only neutral particles are the simplest experimental signatures. Searches for events without charged particles and one, two or three energetic isolated photons are discussed below.

No search for heavy long living neutral particles, produced in Z^0 decays, has been performed so far. Such particles could be detected as single energetic neutral clusters in a hadron calorimeter. Perhaps during the future LEP operation searches could obtain limits (signals), with cross sections not much bigger than neutrino interaction cross sections.

3.1 Single Photon final States

Single photon events are expected from inital state bremsstrahlung and a Z^0 decaying to neutrinos. This photon rate depends strongly on the required minimum photon energy and the center of mass energy. Experiments have counted the number of photons with minimal energy requirements, and obtained results for the number of ν families [5], [6].

Single energetic photons could also be produced in decays of exotic neutral particles from Z^0 decays. The energy spectrum of high energy photons from ALEPH is shown in figure 2, similar results have also been presented by the other eperiments [5], [6], [7], [8]. The absence of very high energy single photons has been used to exclude for example the Z^0 decay into excited neutrinos.

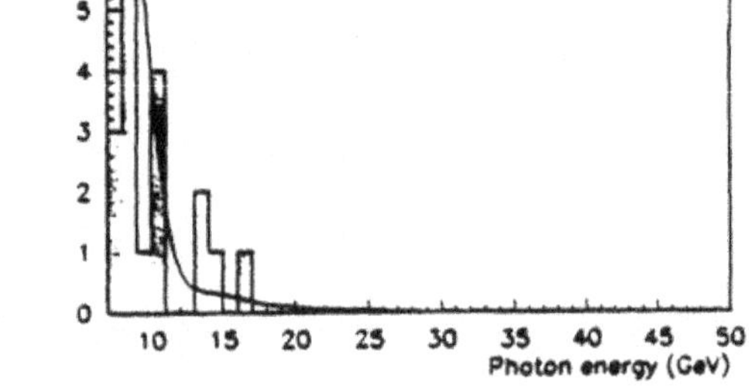

Figure 2: Observed single photon energy spectrum from ALEPH with a luminosity of 19 pb^{-1}.

The event with the highest energy single photon, which has a measured energy of 19.7 GeV, has been reported by the L3 collaboration. The observed number of single photon events with a minimum energy above 10 GeV and 17 GeV are summarized in table 1. No excess for such high energy photons is observed. Since essentially identical cuts are used, the results of the three experiments can be combined. We obtain a limit for the Z^0 Branching Ratio into single high energy photons of roughly 3×10^{-6}, devided by a model dependent efficiency.

Experiment	Hadronic Z^0 decays	N_γ $E_\gamma > 10$ GeV	N_γ $E_\gamma > 17$ GeV
ALEPH	500K	8 (6)	0 (<1)
L3	300K	2 (1.5)	1 (compatible)
DELPHI	300K	0	0
OPAL	130K	1 (1.1)	0 (0.13)

Table 1: Observed number of high energy single photon events for a minimum of 10 GeV and 17 GeV. For ALEPH, L3 and OPAL exptectations from standard sources are given in brackets.

3.2 Events with two neutral clusters

All experiments have analysed their data to measure the rate of two photon events. These events are expected from QED t-channel electron exchange. If the only source of these events is the above QED reaction, these photons are expected to be coplanar and to have the beam energy. The observed events [10] are found to be consistent with this assumption and no acoplanar event with only two neutral clusters has been observed. The absence of such acoplanar events allows to determine branching ratio limits. Previously an efficiency to find heavy particles with such decay characteristics has been estimated to be roughly 40% [11]. Using this efficiency and the combined number of analysed Z^0 decays, the Z^0 branching ratio into neutral massive particles, which decay into photons and invisible particles must be smaller than 5×10^{-6}.

Events of the type $Z^0 \to \gamma\pi^0(\eta)$ might be hidden in the collinear and coplanar 2 photon sample, since, with the existing calorimeters, a 45 GeV π^0 can not be distinguished from a single photon. A signal for the above decay would show up like an enhancement of the two photon cross section at the Z^0 pole.

Experiment	total Luminosity (pb^{-1})	γ acceptance $\|\cos\Theta_{Max}\|$	BR $(Z \to \pi^0\gamma)$ $< n \times 10^{-4}$
L3(prelim.)	14.2	0.97	1.2
ALEPH	8.5	0.95	2.1
OPAL	7.2	0.90	1.4
DELPHI	4.7	0.74	1.5

Table 2: Search results for two photon like final states.

As can be seen from figure 3a-d, the measured cross sections are in very good agreement with expectations and no resonance like production around the Z^0 peak is found. The results are summarized in table 2 [10]. As can be seen from table 2 (and table 3 below) the "best" limits for a very specific process are obtained for restricted polar ranges and relatively hard photon energy cuts. However, a wider range of exotic physics is excluded by more general measurements.

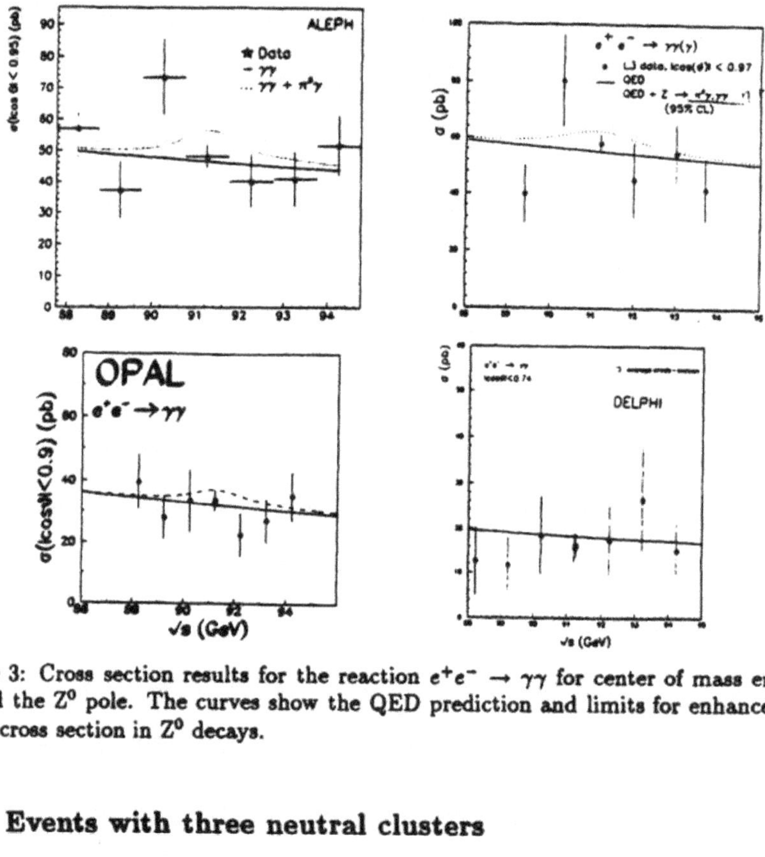

Figure 3: Cross section results for the reaction $e^+e^- \to \gamma\gamma$ for center of mass energies around the Z^0 pole. The curves show the QED prediction and limits for enhancements of the cross section in Z^0 decays.

3.3 Events with three neutral clusters

Some composite models predict Z^0 decays to three photons with BR's as large as 10^{-5} [12]. Small QED background exists for such a process from the above t-channel electron exchange reaction with hard initial state bremsstrahlung. The numbers of three photon events observed by all four experiments are given in table 3 [10]. No excess for center of mass energies close to the Z^0 peak is observed. The combined number of three photon events is 14, in agreement with an expectation of 17. Branching ratio limits are now at roughly 10^{-5}. The ALEPH group has used hard selection cuts, tuned by a specific model, to obtain a Z^0 branching ratio limit of 0.9×10^{-5}. In contrast to this analysis, the L3 group has decided to use very general cuts and obtained with a higher statistics sample a slightly worse limit of 3.3×10^{-5}.

Experiment	used Luminosity (pb^{-1})	$\lvert\cos\Theta_{Max}\rvert$	$E_{min}(\gamma)$ in GeV	number of events observed	expected
L3	14.2	0.97	3	5	8.6
DELPHI	12.5	0.9	5	2	3.3
ALEPH	8.5	0.92	12.2	2	2.5
OPAL	7.2	0.90	4.5	5	3
combined	42.4	-	-	14	17

Table 3: Search results for three photon final states on the Z^0 peak.

4 Exotic low charged multiplicity events

Indirect limits on exotic Z^0 decays are sometimes estimated, by using the agreement between the measured total and partial width and the Standard Model predictions. Using this approach, any additional contribution to the width can now be limited to about 25 MeV, or exotic particles can only be produced with Z^0 branching ratio of less than 10^{-2}. Such indirect conclusions might be incorrect, since most of these exotic particles would limit strongly any predictive power of the Standard Model. Instead of starting a perhaps emotional discussion about such an indirect procedure, we prefer to demonstrate that direct searches reach at least two orders of magnitude stronger limits.

Figure 4a-d show scatter plots, of the visible calorimeter energy versus the measured energy of the tracks plus the energy of isolated clusters without any collinearity cut for events with 2-4 tracks and a minimum visible energy of 5% [13]. Two densely populated and well separated areas are seen in figure 4a. One, where the energy of the tracks and isolated photons and the measured calorimeter energy cluster around the center of mass energy, as expected from e^+e^- final state events. A second well separated region is found with very small energy in the calorimeter and track momenta close to the center of mass energy, as expected from $\mu^+\mu^-$ final state events. The events with zero, one and two identified muons are separately shown in figure 4b-d. As expected, the cluster of e^+e^- events is found only when zero μ's are identified, while most of the $\mu^+\mu^-$ events are found with two identified μ's.

A third very broad region is found with energy and momentum smaller than the center of mass energy. Such an energy deposit is expected from $\tau^+\tau^-$ events. Many proposed exotic Z^0 decays, like the Z^0 decay into a τe, could be found in this region. These events would usually be counted as Z^0 decays into the $\tau^+\tau^-$ channel. Thus, deviations from lepton universality in Z^0 decays, would indicate probably the existence of exotic events in the $\tau^+\tau^-$ event sample. So far, the measured partial leptonic widths are in good agreement with lepton universality. A ratio of the $\tau^+\tau^-$ to $\mu^+\mu^-$ partial Z^0 width of 0.996 ± 0.016 has been determined [14] using the combined 1990 lineshape results from all LEP experiments. Exotic events in the $\tau^+\tau^-$ sample must therefore be produced with a branching ratio of less than about 10^{-3}, roughly a factor of 10 smaller than the above limit from the total Z^0 width. Still, direct searches are much more sensitive and limits of 10^{-5} for some exotic Z^0 decays have already been reached.

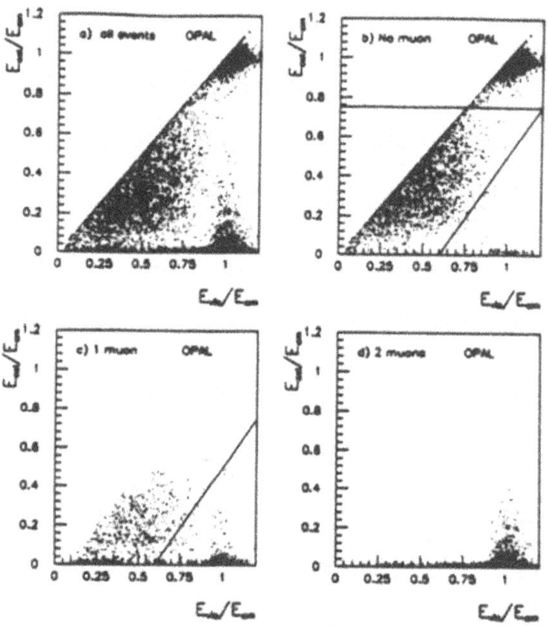

Figure 4: Scatter plots of the observed calorimeter energy (scaled to the center of mass energy) against the scaled visible energy for all two to four track events (a) and separated for events with 0, 1 and 2 identified muons. As expected the lepton pair events cluster in three different regions.

4.1 Acoplanar 2-jet like final states

Figure 5 shows the acoplanarity angle distribution for two jets in the early OPAL low multiplicity data [11].

For this analysis events with less than 5 tracks were clustered into jet cones with an opening angle of less than 20 degrees. Events with isolated electromagnetic clusters above a few 100 MeV were rejected. Due to kinematics of the low τ mass no events without isolated photons can have opening angles smaller than roughly 150 degrees in the plane transverse to the beam. Exotic massive particles, like supersymmetric leptons, with decays into visible and invisible particles would have smaller angles. In the case of only one heavy decaying particle, like the decay $Z^0 \rightarrow N\nu$ with subsequent decays of the N into leptons (see section 2) and small masses of N, the opening angle would get very small, such events are often called monojets.

No acoplanar event has been reported by OPAL, ALEPH and L3 [13], [15], [16]. Four such events are found by the DELPHI group, while a background of two radiative lepton

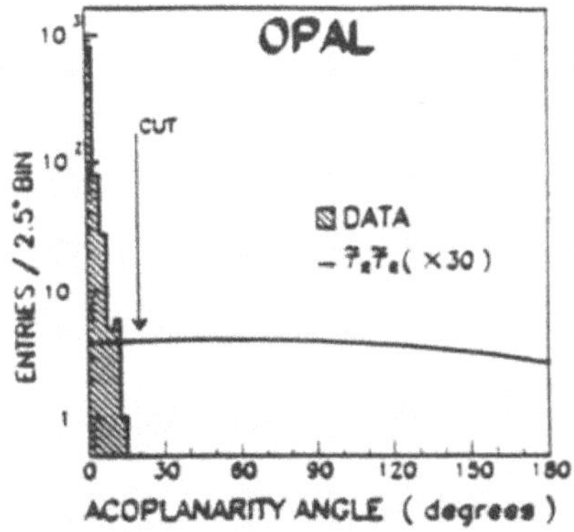

Figure 5: Acoplanarity distribution of low charged multiplicity 2-jet like events (1989 data), the curve shows the expected distribution from the decay products of supersymmetric τ particles.

pair events with undetected photons is expected [17]. The negative results from OPAL, DELPHI, ALEPH and L3, where samples of about 200k, 470k, 680k and 820 Z^0 decays have been analysed, can be used to estimate that the current branching ratio limits per experiment must be smaller than 10^{-5}. One can also estimate that by the end of the 1992 running period a combined BR sensitivity of about 10^{-6} will be reached.

4.2 Searches in the collinear lepton pairs

4.2.1 Search for lepton flavour violation

Many ideas and models are related to the question of neutrino masses, allowing for example the mixing of neutrinos and lepton flavour violation [18]. Several experiments are being performed to observe such neutrino transformations. If such a mixing exists, for example $e\tau$ final states might be produced in Z^0 decays via W^* exchange. The experimental sensitivity for such Z^0 decays has been studied and at best, signals might be seen if the Z^0 branching ratio is larger than 10^{-6} [19]. These events could be detected if an excess of beam energy electrons or muons would be observed back to back with the decay products of τ's. Background stems from $\tau^+\tau^-$ pairs with one τ decaying to a very energetic electron or muon. Obviously, the best sensitivity is obtained with very good energy and momentum resolution at 45 GeV. An example for the electron endpoint spectrum from τ decays together with events from the Z^0 decay to $e\tau$ is shown in figure 6a assuming a total number of 10^7 Z^0 decays. The sensitivity to observe such decays is shown as a function of the total number of Z^0 decays in figure 6b.

About a year after such a measurement was performed by OPAL, all LEP experiments have presented the lepton endpoint spectrum in $\tau^+\tau^-$ events, summarized in table 4 [20]. The L3 group has presented the highest statistics result and obtained the smallest limits,

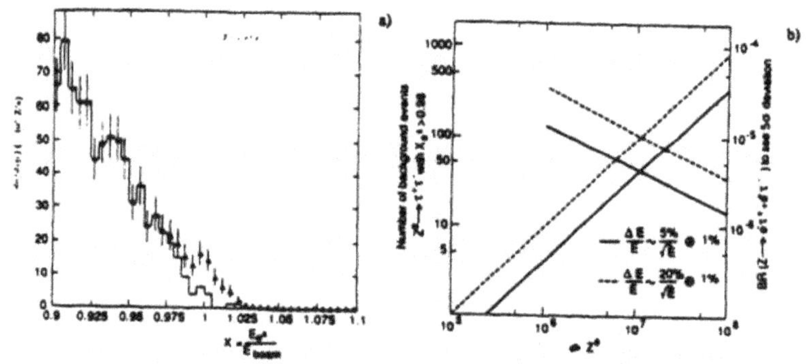

Figure 6: a) Endpoint electron energy spectrum from a MC study of $\tau^+\tau^-$ events in 10^7 Z^0 decays, a signal for $e\tau$ events with a branching ratio of 10^{-5} is overlaid; b) the sensitivity to observe such decays as a function of the number of Z^0 decays.

their endpoint spectra for electrons and muons recoiling against a τ are shown in figure 7a and b [20].

Experiment	Date	Z^0 decays	$e\tau$			$\mu\tau$		
			see	MC	BR limit	see	MC	BR limit
OPAL	Oct 90	200k	1	2	7.2	31	21	35
L3	Aug 91	370k	0	.8	3.4	1	1.6	4.8
ALEPH	Sep. 91	260k	11	6	12.	12	11	10
DELPHI*	Jun 92	430k	5	7.3	7.0	6	6.8	8.0
L3*	Jun 92	600k	2	2.	3.0	1	2	2.9

Table 4: Search results for Lepton Flavour violating Z^0 decays, the * means preliminary results. The branching ratio limit is given in units of 10^{-5}.

No signal or excess has been observed so far. An optimist can use the above results to estimate how many Z^0 decays he or she needs to observe the Z^0 decay into $e\tau$, if a true branching ratio would be 10^{-5}, close to the current limit.

We use the published results obtained by the L3 collaboration for a rough estimate. For about 7×10^6 Z^0 decays, background and signal would have roughly the same size. One would expect to see a total of 28 events, while only 14 background events are expected, a signal of about five standard deviations. The other experiments would see at best a signal between two and four standard deviations. But, since their resolutions might still improve and because of a lucky fluctuation, such a signal might be observed by any LEP experiment.

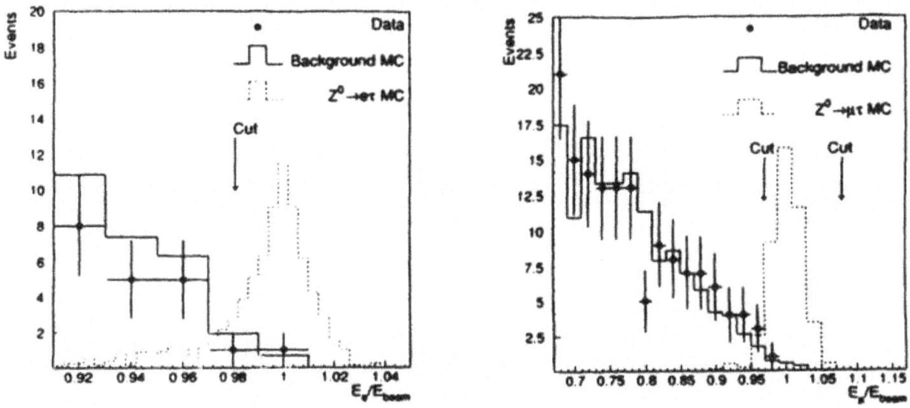

Figure 7: a) Endpoint electron and muon energy spectrum in $\tau^+\tau^-$ pair candidates from the L3 analysis with roughly 600k Z^0 decays, a possible signal is overlaid.

4.2.2 Search for pair produced light neutral particles

No theoretical model, with Z^0 decays into a pair of light unstable neutral particles N and branching ratios much smaller than the neutrino pair production, is known to us and (perhaps consequently) no such search has been published so far. Under the assumption, that the N particle would decay fast enough into a virtual Z^* and a neutrino, a very clean signature, back to back like narrow jets, each with **even number of charged particles** and large momentum imbalance would be seen in a detector. Knowing the high quality tracking systems of all four experiments, only very small backgrounds can be expected. Regardless of a very unlikely discovery, we consider such an analysis as very interesting.

4.3 Three body final states

4.3.1 Search for the Z^0 decay into $\ell^+\ell^-X$

The ALEPH collaboration has searched for a light invisible scalar particle X produced in Z^0 decays into Z^*X [21]. They analysed in detail the events with a small acoplanarity angle of the $\mu^+\mu^-$ and e^+e^- pairs. Once events with isolated photons were rejected, no such event was found in a sample of roughly 260k Z^0 decays. Such scalar particles could be excluded down to Z^0 branching ratios of 10^{-4}. A high statistics update of such a measurement might be promising. The other experiments have also investigated such a signature when searching for a very low mass Higgs. They have stopped their search however, once limits of roughly 1/10 of the Standard Model Higgs cross section were obtained [22].

4.3.2 Search for Z^0 photon couplings

Radiative events with energetic photons have been used previously by all LEP experiments to search for excited leptons and quarks. OPAL has presented a measurement of the photon yield with energies as low as 1 GeV, using roughly 2×10^5 Z^0 events [13]. An excess of photons, compared to the QED bremsstrahlung photons [23], [24] could demonstrate anomalous Z^0 - photon couplings, predicted within some composite Z^0 models, as discussed for example in [12].

In this mesurement, the yield of isolated photons in lepton pair events is compared to the KoralZ Monte Carlo program, which includes first order QED final state bremsstrahlung [24]. Furthermore, using e^+e^- , $\mu^+\mu^-$ and $\tau^+\tau^-$ events it is shown, that for large angles of the photon with respect to the closest charged lepton, identical photon yields are measured for the different types of lepton pair events, as expected from QED. Figure 8a shows the yield of photons in $\mu^+\mu^-(\gamma)$ events for separation angles smaller than 20 degrees as a function of the photon energy, the data are in good agreement with the Monte Carlo prediction. From this agreement one concludes, that photon bremsstrahlung is well described by the KoralZ program. Figure 8b and 8c show the angular distribution of the photon, with respect to the closest lepton, for scaled photon energies between $x_\gamma = 0.02 - 0.1$ and $x_\gamma = 0.1 - 0.3$. Figure 8d shows the isolated photon yield as a function of the photon energy. Very good agreement with QED predictions is observed in all cases and no exotic source is needed.

Numbers for observed and predicted photon yields are given in table 5. After subtracting the predicted number of bremsstrahlung photons, 95% confidence level limits on the number of excess photons are obtained. Correcting this maximum allowed excess, for efficiency and the number of analysed Z^0 decays, limits on exotic three body Z^0 decays to a photon and a pair of leptons are calculated. Recently, the DELPHI collaboration has reached, without calculating limits, similar conclusions from their $\mu^+\mu^-(\gamma)$ events [25].

Data on peak			
$Z \to \ell^+\ell^-\gamma$	$e^+e^-(\gamma)$	$\mu^+\mu^-(\gamma)$	$\tau^+\tau^-(\gamma)$
$N^{coll}_{\ell^+\ell^-}$	3698	3124	2661
N_γ (Data)	123	108	103
N_γ (MC)	117 ± 6	99 ± 5	90 ± 3
$N_\gamma/N^{coll}_{\ell^+\ell^-}$ (Data) ($\times 10^{-2}$)	3.3± 0.3	3.5 ± 0.3	3.9 ± 0.4
$N_\gamma/N^{coll}_{\ell^+\ell^-}$ (MC) ($\times 10^{-2}$)	3.2± 0.2	3.2 ± 0.2	3.4 ± 0.1
Limits for $Z \to \ell^+\ell^-\gamma$			
limit N^{extra}_γ (95% c.l.)	28	30	33
$BR(Z \to \ell^+\ell^-\gamma)$ limit	5.2×10^{-4}	5.6×10^{-4}	7.3×10^{-4}

Table 5: Numbers for collinear lepton pair events, for events with isolated photons with x_γ above 0.02 in the data and the Monte Carlo. The calculated 95% confidence level limits for the possible photon excess and the calculated Z^0 branching ratio limits for anomalous three body Z^0 decays are also given.

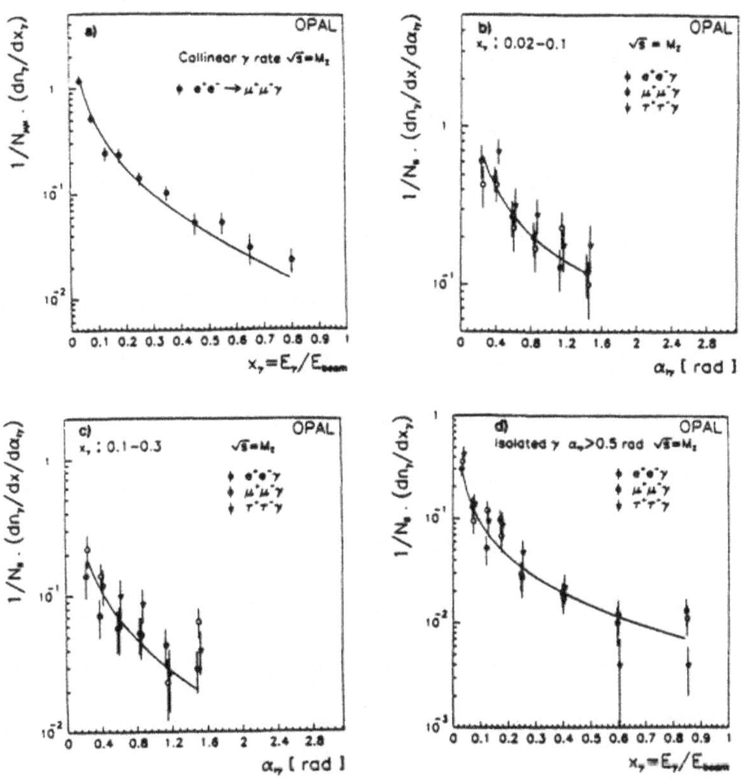

Figure 8: The measured photon yield for the three types of lepton pairs as a function of energy and separation angle. The curve is the KORALZ Monte Carlo prediction.

4.3.3 Measurement of acollinear $\mu^+\mu^-$ pair events

The well defined e^+e^- center of mass energy allows to identify events with hard initial state photon bremsstrahlung and undetected photons. The energy of such photons can be calculated with good accuracy, assuming three body kinematics and that the photons are produced along the beam line. From the photon energy, the reduced center of mass energy $\sqrt{s'}$ of the reaction can be determined. The rate of such events compared to Monte Carlo predictions provides indirect cross section measurements at the reduced center of mass energy. Seven such events could be clearly identified by OPAL [13]. The measured $\sqrt{s'}$ of these events is shown in table 6, together with expectations and cross sections for the far uncovered center of mass energies between TRISTAN and LEP.

The measured energy distribution of candidate events and the cross section result are shown in figure 9a and b.

$\sqrt{s'}(GeV)$	40–60	60–70	70–78	78–84	$<\sqrt{s'}>= 75$
$N_{events}^{measured}$	2	0	2	3	5
$N_{events}^{expected}$	0.9	1.6	2.7	3.8	8
$\sigma_{Born}^{expected}$ (pb)	36	23	23	36	24
$\sigma^{measured}$ (pb)	$80\pm^{76}_{36}$	< 43 (95% c.l.)	$17\pm^{16}_{8}$	$28\pm^{21}_{11}$	$15\pm^{8}_{6}$

Table 6: Observed and expected number of acollinear $\mu^+\mu^-$ events for the different effective dilepton centre-of-mass energies $\sqrt{s'}$ and the corresponding cross sections.

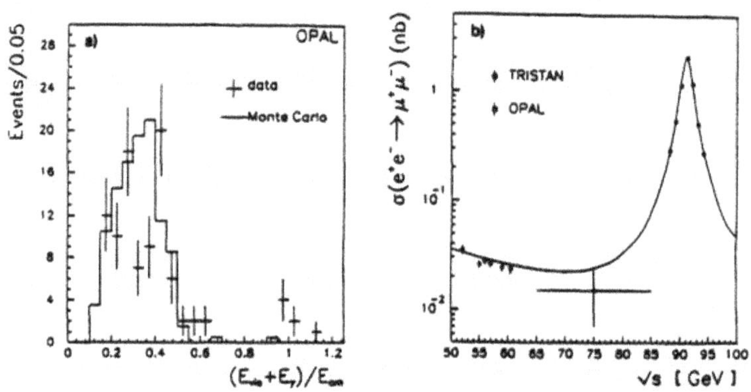

Figure 9: a) The total scaled event energy (including the calculated photon energy for acollinear $\mu^+\mu^-(\gamma)$ events; b) the cross section measurements for e^+e^- annihilation into $\mu^+\mu^-$ pairs are shown from TRISTAN to LEP energies including this OPAL cross section measurement at an average center of mass energy of 75 GeV. The curve is the Standard Model prediction.

4.4 Four body final states

ALEPH [26] and L3 [27] have analysed different four body final state signatures and presented results, which differ by about three standard deviation from expectations.

4.4.1 $\ell^+\ell$–V events

The ALEPH group has investigated in detail low multiplicity events of the type lepton lepton plus two charged tracks. A surprising excess of such events was observed in the $\tau^+\tau$–X channel, the numbers are shown in table 7. Soon after the ALEPH result was published, the other LEP experiments have analysed their data for such an effect. Their results were negative and also the new ALEPH data, with a factor of about 3 more statistics, do not show an excess [28].

Experiment	data sample	eeV see	eeV expect	μμV see	μμV expect	ττ V see	ττ V expect
ALEPH	89-90	10	7.2	10	6.6	15	3.2
ALEPH	89-91	18	18.2	18	15.2	18	7.4
OPAL	90-91	21	27.7	18	15.3	11	9.5
L3	91	8	6.3	5	4.	4	4.
DELPHI	90-91	11	9.9	9	9.3	8	3.9
combined		58	61	50	43.7	41	25

Table 7: Observed and expected number of lepton lepton plus two track events.

4.4.2 Lepton pair events with high mass $\gamma\gamma$ pairs

Using their 90 and 91 data sample, the L3 collaboration has found two events of the type $\mu^+\mu^-\gamma\gamma$ with a $\gamma\gamma$ mass of 58.8 and 59 GeV. Such events are expected from second order QED Bremsstrahlung, with however a probability of only 0.15% for the number of analysed events. During the 92 running period of LEP two more events, one $\mu^+\mu^-\gamma\gamma$ and one $e^+e^-\gamma\gamma$ with high invariant $\gamma\gamma$ masses of 62 GeV and 60 GeV were seen. These four events deviate with a significance of about 3 standard deviations from the QED expectations of 1 event. Figure 10 shows the observed mass spectrum compared to a Monte Carlo calculation.

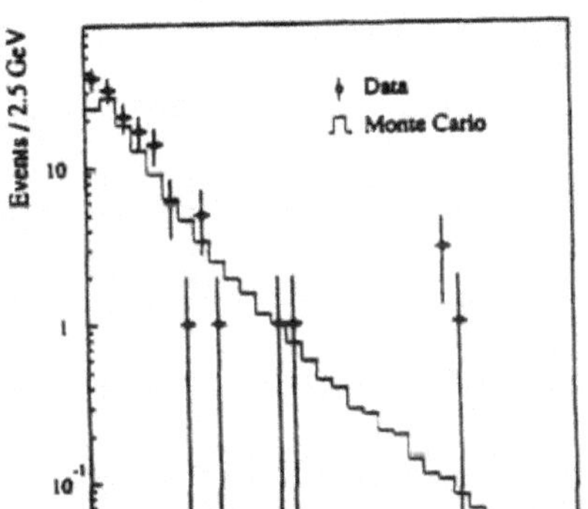

Figure 10: $\gamma\gamma$ invariant mass spectrum from the L3 analysis of double radiative lepton pair events, the histogram shows the expectation.

In reference [13], OPAL has previously published an indirect statement about double radiative lepton pair events, saying that six lepton pair events in their 1990 data sample have two hard photons. This indicates that no particular feature of these events was found. The DELPHI collaboration has also looked for such events, but no event with a mass above 40 GeV is seen [29].

We conclude, that both observed three standard deviation effects, found with 4-body final states, look unfortunately like strong statistical fluctuation.

5 Exotic high charged multiplicity events

Searches for exotic particles with hadronic signatures are much more difficult and not very competitive. Some reasons are listed below:

- Particles with dominant hadronic couplings should be produced abundantely at high energy hadron colliders, where masses of up to 200 GeV can be reached now.
- Because of high backgrounds from QCD multi jet events, searches in hadronic final states without leptons have very high background and bad resolution. Such a search has been performed for example by the ALEPH collaboration [15]. Their obtained jet-jet mass spectrum is shown in figure 11. The histogram shows the background expectations (dashed line) and the effect of a 41 GeV charged Higgs, H^+ particle which decays to two jets. No such signal is observed at high masses, however at a mass of about 36 GeV, a small peak is visible. But, since a H^+ is expected to have other decay modes like the decay to $\tau\nu$, it would have shown up in the search for acoplanar low charged multiplicity events. Therefore, because of other searches a H^+ with a mass of 36 GeV can be excluded and the peak in the jet-jet system is assumed to be a fluctuation.

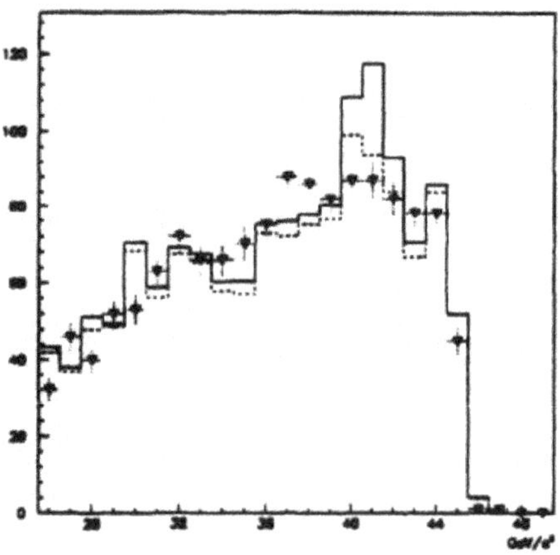

Figure 11: Mass distribution of two jets in multi jet hadronic events obtained by ALEPH.

- In the case of heavy quarks, or quark like objects with weak decays, a much cleaner signature exists for leptonic decays of the virtual W^* or Z^*. For example, leptoquarks with semileptonic decays and isolated leptons are easy to find in many reactions. Therefore, such limits from LEP are not very competive with similar searches at the TEVATRON or HERA colliders. Nevertheless, very detailed searches for hadronic events with isolated leptons have been performed by all LEP experiments. The background levels were found to be very low, between 0 and 2 events, but usually not even a single event has been found [30].

5.1 $Z^0 \rightarrow b\bar{s}$

Another type of search for flavour changing neutral currents has been proposed for the Z^0 decay into a b + s quark [31]. Hadronic flavour-changing Z^0 decays would occur within the Standard Model with branching ratios of about 10^{-9}. New physics could increase this rate considerably. Here we investigate the signature shown in figure 12a [31].

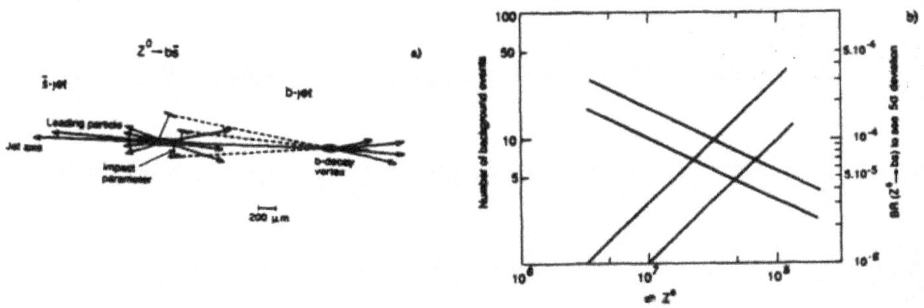

Figure 12: The signature of Z^0 decays into $b\bar{s}$ (a) and the sensitivity as a function of the number of recorded Z events (b).

A b-hadron at Z energies has an average decay length of about 2 mm. Its decay products would not come from the interaction point, leading to large impact parameters. The new microvertex detectors with small beam pipes obtain impact parameter resolutions of about 20 μm (about 200 μm for secondary vertices from b decays) as indicated in figure 12a. The particles from the non b-jet should come from the interaction point, which reduces the background from $b\bar{b}$ events by about a factor of 10. In addition, the requirement of a stable high x (=E/E_{beam}) particle ($x > 0.6(0.7)$) reduces the background from $b\bar{b}$ events by a factor of about 10^{3-4}. Figure 12b shows the sensitivity to such rare decays as a function of the total number of Z^0 events.

6 Summary and outlook

Searches for rare and exotic events have been performed by the four LEP experiments and are discussed in terms of experimental signatures. No clear effects have been found in samples of up to 10^6 Z^0 decays. In most cases very similar experimental cuts were used by the experiments, allowing to obtain combined branching ratio limits and sensitivities of a few $\times 10^{-5}$ and smaller.

Many of the investigated signatures, with low charged multiplicity have very small backgrounds. Thus, from the next order of magnitude increased LEP luminosity, Z^0 branching ratio limits or signals in regions, allowed by several models, can be reached.

Having in mind, that one has to wait many years before the future super colliders will start hunting for new particles, the best chance to observe first signs of exotic physics might be detailed Z^0 studies using the future high statistic Z^0 data.

7 Acknowledgements

It is a pleasure to thank the organizers of this conference and my colleagues in the LEP experiments for providing me with the partially preliminary information. Especially I want to thank D. Schlatter, M. Green (ALEPH), F. Richard (DELPHI), J. Qian (L3) and my colleagues within OPAL. I am also grateful to D. Haztifotiadou for her comments on an early version of this paper.

References

[1] Particle Data Group, J.J. Hernandez et al., Phys. Lett. **B239** (1990).

[2] SINDRUM Collaboration, W. Bertl et al., Nucl. Phys. **B260** (1985) 1.

[3] M. Dittmar CERN-PPE/90-106 and Proceedings of the Second International Conf. on Medium and High Energy Physics in Taiwan, May 8-18 (1990) 89.

[4] M. Dittmar, A. Santamaria, M. C. Gonzalez-Garcia and J.W.F. Valle, Nucl. Phys. **B332** (1990)1-19.

[5] OPAL Collaboration, M.Z. Akrawy et al., Z. Phys. **C50** (1991) 373-384 and private communication G. Wilson.

[6] L3 Collaboration, B. Adeva et al. Phys. Lett. **B252** (1990) 525-532, Phys. Lett. **B275** (1992) 209-221 and A. Weber, internal L3 note (preliminary), July 15; contribution to the 26th international Conference on High Energy Physics, Dallas (August 92).

[7] ALEPH Collaboration, D. Decamp et al., Phys. Lett. **B250** (1990) 172-182 and ALEPH internal note ALEPH 92-128; contribution to the 26th international Conference on High Energy Physics, Dallas (August 92).

[8] DELPHI Collaboration, internal note July 9; DELPHI 92-85 PHYS 196; contribution to the 26th international Conference on High Energy Physics, Dallas (August 92).

[9] Invited talk by T. Azemoon at the 26th international Conference on High Energy Physics, Dallas (August 92).

[10] The latest numbers on two photon like final states are given in [9].

[11] OPAL Collaboration, M.Z. Akrawy et al., Phys. Lett. **B240** (1990) 261.

[12] See for example: F. Boudjema et al., Z Physics at LEP1, eds. G. Altarelli, R. Kleiss and C. Verzegnassi (1989) Vol. 2 page 202-206.

[13] OPAL Collaboration, P. D. Acton et al., Phys. Lett. **B273** (1991) 338-354.

[14] see for example: Michael Dittmar, Proceedings of the 26th Rencontres de Moriond, Electroweak Interactions and Unified Theories Les Arc, 10-17 March (1991) 45.

[15] ALEPH Collaboration, D. Decamp et al., CERN-PPE 1991/149.

[16] A. Rubbia, internal L3 note (preliminary), July 29; contribution to the 26th international Conference on High Energy Physics, Dallas (August 92).

[17] DELPHI Collaboration, internal note June 15; DELPHI 92-95 PHYS 206; contribution to the 26th international Conference on High Energy Physics, Dallas (August 92).

[18] See for example J. F. W. Valle to appear in *Progress in Particle and and Nuclear Physics*, ed. A. Faessler, FTUV/90-36.

[19] M. Dittmar and J.W.F. Valle FTUV/90-19 and CERN/91-02.

[20] OPAL Collaboration, M.Z. Akrawy et al., Phys. Lett. **B254** (1991) 293-302;
L3 Collaboration, B. Adeva et al., Phys. Lett **B271** (1991) 453-460 and contribution to the 26th international Conference on High Energy Physics, Dallas (August 92) see [9];
ALEPH Collaboration see [15]
DELPHI Collaboration, internal note June 24; DELPHI 92-69 PHYS 180; contribution to the 26th international Conference on High Energy Physics, Dallas (August 92).

[21] ALEPH Collaboration, D. Decamp et al., Phys. Lett. **B262** (1991) 139-147 and [15].

[22] DELPHI Collaboration, P. Abreu et al. CERN-PPE/90-193;
OPAL Collaboration, P.D. Acton et al, Physics Lett. **B268** (1991) 122-136;
L3 Collaboration, B. Adeva et al., Phys. Lett. **B252** (1990) 518-524.

[23] J. Fleischer and F. Jegerlehner, Zeit. Phys. **C26** (1985) 629-638.

[24] S. Jadach, B.F.L. Ward, Z. Was, R.G.S. Stuart, and W. Hollik, KORALZ, to be published in Comp. Phys. Comm., CERN-TH-5994-91.

[25] DELPHI Collaboration, internal note June 25; DELPHI 92-76 PHYS 187; contribution to the 26th international Conference on High Energy Physics, Dallas (August 92).

[26] ALEPH Collaboration, D. Decamp et al., Phys. Lett. **B263** (1991) 112-122.

[27] L3 Collaboration, O. Adriani et al., CERN-PPE/92-152

[28] OPAL Collaboration, P. Acton et al., CERN-PPE/92-57;
DELPHI Collaboration, internal note June 30; DELPHI 92-84 PHYS 195; contribution to the 26th international Conference on High Energy Physics, Dallas (August 92);
the latest results from ALEPH and L3 are summarized in [9].

[29] DELPHI Collaboration, internal note June 25; DELPHI 92-82 PHYS 193; contribution to the 26th international Conference on High Energy Physics, Dallas (August 92).

[30] ALEPH Collaboration, see [15];
L3 Collaboration, B. Adeva et al., Phys. Lett. **B261** (1991) 169-176;
OPAL Collaboration, G. Alexander et al., Phys. Lett. **B263** (1991) 123-134;
DELPHI Collaboration, internal note July 7; DELPHI 92-102 PHYS 210; contribution to the 26th international Conference on High Energy Physics, Dallas (August 92).

[31] D. Cocolicchio and M. Dittmar; CERN-TH-5753/90 and CERN/91-02.

PREDICTIONS FOR THE MASSES OF THE HIGGS BOSON AND THE TOP QUARK

PER OSLAND*
Department of Physics, University of Bergen, Allégt. 55
N-5007 Bergen, Norway

and

TAI TSUN WU§
Gordon McKay Laboratory, Harvard University
Cambridge, Massachusetts 02138, U.S.A.
and
Theoretical Physics Division, CERN
CH-1211 Geneva 23, Switzerland

ABSTRACT

By theoretical arguments involving the absence of certain divergences in the Standard Model, the masses of the Higgs boson and the top quark have been found to be related to those of the W and the Z by $m_H^2 = 9m_Z^2 - 6m_W^2$, and $m_t^2 = \frac{5}{2}m_Z^2 - m_W^2$. From the known masses of the W and the Z, it is found that $m_H =$ 190 GeV and m_t = 120 GeV. This value of m_t is within the range favored by existing phenomenology. The Higgs boson of this mass has a width of 1 GeV, with the major branching ratios of 77%, 22% and 1% into W^+W^-, ZZ and $b\bar{b}$, and is quite easy to detect at SSC and LHC.

Experimentally, the Standard Model of Glashow, Weinberg and Salam[1] has been verified to high accuracy, especially at LEP.[2] In this model, the photon, the gluon, and the three neutrinos are massless, while there are twelve massive fundamental particles: the gauge particles W and Z, the Higgs boson H,[3] the three leptons e, μ, τ and the six quarks u, d, c, s, t and b. Of these twelve, only the Higgs boson H and the top quark t have not been observed experimentally, leaving their masses m_H and m_t as two major unknown parameters in the Standard Model. Thus, most of particle physics can be said to be understandable in terms of the seventeen dimensionless parameters of the Standard Model.

It is perhaps reasonable to believe that these seventeen dimensionless parameters are not independent of each other. We would like to present here one specific line of investigation that leads to the theoretical determination of the masses of the Higgs boson and the top quark.

* Work supported in part by the Norwegian Research Council for Science and the Humanities.

§ Work supported in part by the U.S. Department of Energy under Grant DE-FG02-84ER40158.

In order to motivate this line of investigation, let us begin by recalling some of the wisdom about quantum field theory that has been accumulated during the past half a century.

(1) In the span of a few years beginning in 1946, renormalization for quantum electrodynamics was carried out by Tomonaga, Schwinger, and Feynman[4] to one loop. Only three years later, Dyson[5] extended this procedure to *all* orders. A remarkable feature of the work of Dyson is the fact that no regularization or cutoff is ever introduced. This is also true of the other papers of this era.[6]

(2) In 1971, supersymmetry[7] was discovered. It has the property that quadratic divergences are eliminated by arranging fermions and bosons into pairs. Theories with supersymmetry have attracted a great deal of attention because of the well-justified belief that divergences, especially quadratic ones, are undesirable.

(3) Also in 1971, 't Hooft[8] and then 't Hooft and Veltman[9] succeeded in showing that Yang-Mills non-Abelian gauge theories[10] can be renormalized to one loop. In the work of 't Hooft and Veltman, the dimensional regularization of Bollini and Giambiagi[11] plays a central role.

(4) In 1974, Georgi and Glashow[12] pointed out that the Standard Model has the nice feature of the cancellation of anomalies.[13] If the anomalies did not cancel, then renormalization would be spoiled.

A number of observations can be made on the basis of these developments; let us list the following which are directly relevant to our specific line of investigation.

(i) From Item (1) on renormalization, all divergences in masses and coupling constants are considered to be irrelevant to physics. On the other hand, according to Item (2), not all infinities are equal: some are more menacing than others.

(ii) The cancellation of anomalies in the Standard Model[12] is a case where renormalization requires certain equalities to be satisfied between the finite parameters of the theory.

(iii) It is especially striking that the philosophy about regularization has changed so dramatically between the two eras roughly a quarter of a century apart. In the first era beginning in the forties, regularization was considered to be undesirable; in the second era since the seventies, regularization was considered to be essential.

In connection with Item (iii), the following points are noteworthy. On the one hand, it is not possible to use dimensional regularization for the purpose of renormalizing the Standard Model beyond one loop or at most two loops, mainly because of the well-known difficulties with γ_5. On the other hand, if the introduction of regularization, any regularization, is to be justified, it must accomplish something that has not been accomplished without regularization. What can this be?

With the question phrased this way, the following answer suggests itself. In the work on the renormalization of quantum electrodynamics in the forties and the fifties, it was difficult to confirm the massless nature of the photon by a direct computation. Indeed, if one computes the one-loop correction to the photon mass by applying the Feynman rules in the most straightforward way, the answer is infinity rather than zero. [Remember that the photon mass does not have a counter term.]

On the other hand, if dimensional regularization is used, this one-loop correction to the photon mass is indeed zero.

The present line of investigation[14] is motivated by the following considerations. By Item (2), we consider the cancellation of certain divergences in the supersymmetric theories to be highly desirable. Yet, the doubling of the number of fundamental particles in supersymmetric theories seems to be too high a price to pay. We would therefore like to be able to remove some of the divergences in the Standard Model without introducing extra particles. This can be accomplished by calculating the coefficients of the divergences to be removed, and set them to zero. Such a procedure was first suggested by Stückelberg,[15] even before the development of renormalization theory. Of course he was handicapped by not having the Standard Model to work with! Setting these coefficients of divergent terms to zero leads to relations between the finite parameters of the Standard Model, rather similar to the cancellation of the anomaly as discussed under Item (ii). By Item (i), since it is not clear at present which divergences are more menacing, we are allowed to choose the divergences to be removed. Our choices are directed toward the determination of the masses of the Higgs boson and the top quark. It is hoped that further relations to be found in the future will give a deeper understanding of the other parameters of the Standard Model.

Since the coefficients of divergent quantities are to be calculated, it is for the present purpose unavoidable to introduce regularization. In the Standard Model, unlike quantum electrodynamics, there are quadratic divergences. Let δ^{-1} be a suitable momentum cut-off, then the Higgs tadpole yields a term of the order δ^{-2}. Since the Higgs boson couples universally to the masses, the coefficient of this δ^{-2} term is proportional to the sum of the squares of all masses. Fortunately, as is well-known from supersymmetric theories,[7] the contributions of the bosons and the fermions have opposite signs. In this connection, the quadratic mass formula obtained by Ferrara, Girardello and Palumbo[16] by introducing soft breaking of supersymmetry is especially noteworthy. In the case of the non-supersymmetric Standard Model, the Higgs tadpole does not contain any inverse power of δ in the one-loop amplitudes if the following condition is satisfied:

$$\begin{aligned} &m_e^2 + m_\mu^2 + m_\tau^2 + 3(m_u^2 + m_d^2 + m_c^2 + m_s^2 + m_t^2 + m_b^2) \\ &\quad = \tfrac{3}{2} m_W^2 + \tfrac{3}{4} m_Z^2 + \tfrac{3}{4} m_H^2 . \end{aligned} \tag{1}$$

This relation was first given by Veltman.[17] It is interesting that in the present consideration this relation (1) is obtained entirely within the Standard Model, while the argument of Veltman requires using knowledge from supersymmetry.[18] It is of course unclear what supersymmetry has to do with the non-supersymmetric Standard Model. Note also that this relation *cannot* be obtained with dimensional regularization, but apparently can be obtained with dimensional reduction.[18]

Let us discuss this Eq. (1) in some more detail. First, when it is satisfied, there is no quadratic divergence to one loop, not only for the Higgs tadpoles, but for all self-masses, including that of the H. Secondly, it should be emphasized that

Eq. (1) is *not* an exact relation, since there are radiative corrections due to two-loop diagrams. Let us again take the Higgs tadpole as an example. For one-loop diagrams, a typical quadratically divergent integral is of the form

$$\int \frac{d^4k}{k^2}.$$

Thus, the one-loop tadpoles contribute a term of the form $C_1\delta^{-2}$. Similarly for the two-loop diagrams, a typical quadratically divergent integral is of the form

$$\int \frac{d^4k\, d^4l}{k^2 l^2 (k-l)^2},$$

with the corresponding contribution $C_2\delta^{-2}$. To two-loop approximation, the condition for the absence of divergences is then

$$C_1 + C_2 = 0. \tag{2}$$

Since C_1 is given by Eq. (1), this is more explicitly of the form

$$\begin{aligned} m_e^2 + m_\mu^2 + m_\tau^2 + 3(m_u^2 + m_d^2 + m_c^2 + m_s^2 + m_t^2 + m_b^2) \\ - \tfrac{3}{2}m_W^2 - \tfrac{3}{4}m_Z^2 - \tfrac{3}{4}m_H^2 + g^2F = 0, \end{aligned} \tag{3}$$

where F is from the two-loop tadpole diagrams, and g is a coupling constant.

If dimensional regularization were used, then the two typical quadratically divergent integrals for the one-loop and two-loop Higgs tadpole diagrams would be

$$\int \frac{d^dk}{k^2} \quad \text{and} \quad \int \frac{d^dk\, d^dl}{k^2 l^2 (k-l)^2}.$$

Since the leading pole of the first integral is at $d = 2$ while that for the second one is at $d = 3$, the one-loop and two-loop Higgs tadpole contributions, respectively, are of the forms

$$\frac{C_1'}{d-2} \quad \text{and} \quad \frac{C_2'}{d-3}.$$

The absence of divergences then seems to lead to the two conditions $C_1' = 0$ and $C_2' = 0$. These are much stronger than the condition (2) and thus cannot be correct. This is an additional reason, apart from the γ_5 problem, why dimensional regularization has to be avoided.

In Eq. (3), the radiative correction g^2F is proportional to the squares of the coupling constants, and is expected to be of relative order 10^{-3}, or perhaps somewhat larger. With the known masses, we have $3m_b^2/(\frac{3}{2}m_W^2) \simeq 8 \times 10^{-3}$ and $3m_c^2/(\frac{3}{2}m_W^2) \simeq 10^{-3}$. Thus, the $3m_b^2$ *may* be larger than the radiative corrections g^2F, while the $3m_c^2$ is at most of the order of the radiative corrections g^2F. [Remember that the shift of the Higgs field by its vacuum expectation value leads to a rearrangement of the perturbation series. Hence, there is no clear distinction between the small mass terms and the radiative corrections.] The conclusion is therefore reached that the condition (3), without radiative corrections, gives simply

$$m_t^2 + m_b^2 = \tfrac{1}{2}m_W^2 + \tfrac{1}{4}m_Z^2 + \tfrac{1}{4}m_H^2. \tag{4}$$

With this mass relation (1) or (4), or more precisely the corresponding relation (3) with radiative corrections, there are only logarithmic divergences, i.e., terms proportional to $\ln\delta$. This desire to remove all divergences except logarithmic ones is in accordance with the basic theme of supersymmetry.[7] In the Standard Model it is possible to go further. Let us for definiteness take the one-loop electron self-mass diagrams. For small δ, there are not only terms proportional to δ^{-2} but also terms of the order of $\ln\delta$. By Eq. (1), terms of order δ^{-2} can be removed. However, these terms need not be made exactly zero; instead it can be easily arranged that there is a small remainder such that it cancels the $\ln\delta$ terms. The point here is that the various coupling constants must also be allowed to depend on δ; it is only necessary for the deviations in the coupling constants to approach zero as δ approaches zero.

This finiteness of the electron self-mass encourages us to proceed further. Since the self-masses are from two-point functions, the natural next step is to study three-point functions. In the Standard Model, there are two distinct types of three-point functions, those due to gauge couplings and those due to Yukawa couplings. Similar to the case of quantum electrodynamics, in the case of the gauge couplings, it is not possible to remove the terms proportional to $\ln\delta$ by any condition like Eq. (1). Attention is therefore turned to the Yukawa couplings of the form fermion-fermion-Higgs.

Since the purpose here is to give a theoretical determination of the masses of the Higgs boson and the top quark, and Eq. (4) gives one relation between these two quantities, the remaining task is to find a second relation. If the fermion in the fermion-fermion-Higgs coupling is a quark, then the condition derived from the absence of divergences contains both the electroweak and the strong coupling constants, in addition to the Cabibbo-Kobayashi-Maskawa matrix.[19] Thus, such a condition cannot give a relation between the masses only. The only possibility is to study the lepton-lepton-Higgs vertex. Since the neutrino-neutrino-Higgs vertex is zero in the Standard Model, we are left with the three cases of the eeH, the $\mu\mu H$ and the $\tau\tau H$ vertices, which are very similar. For definiteness the eeH is to be studied here. This is to be contrasted with earlier works.[20,21]

Since only gauge invariant quantities are physically meaningful, all three external lines are put on mass shell. For the coupling constant renormalization of this eeH vertex, there are three contributions, from the proper vertices, the electron wave-function renormalization, and the Higgs wave-function renormalization. As shown in Fig. 1, there are thirteen one-loop proper vertex diagrams. Since these diagrams are at most logarithmically divergent without regularization, it is sufficient to pick out these logarithmically divergent pieces, and this can easily be accomplished.

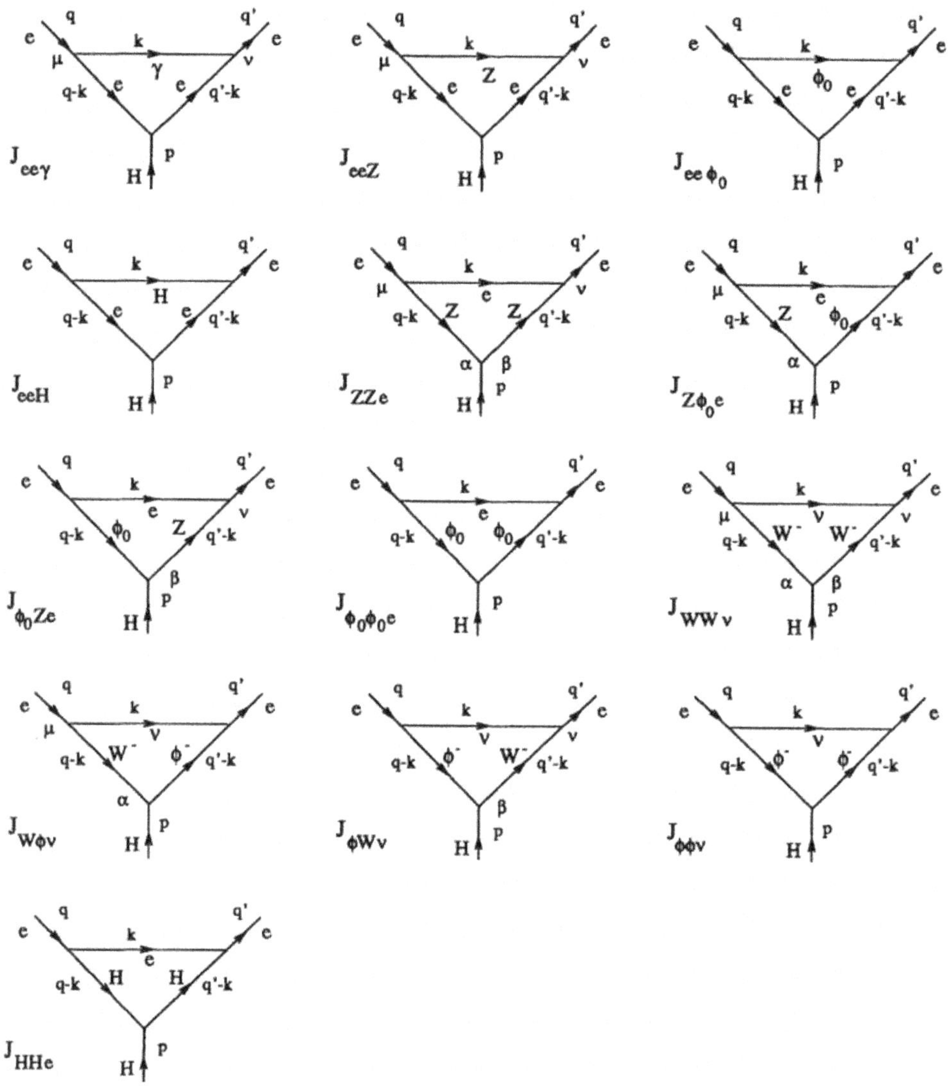

Figure 1. The thirteen one-loop *eeH* proper vertex diagrams.

As an example, consider the second diagram, where the internal loop consists of two electron lines and one Z line. The integral for this diagram is

$$J_{eeZ} = \frac{1}{(2\pi)^4} \int d^4k \frac{i}{k^2 - m_Z^2} \left[-g^{\mu\nu} + \frac{(1-\xi)k^\mu k^\nu}{k^2 - \xi m_Z^2} \right] \left(\frac{-ig}{4\cos\theta_W} \right)^2 \left(-\frac{igm_e}{2m_W} \right)$$
$$\times \gamma_\nu (1 - 4\sin^2\theta_W - \gamma_5) \frac{i}{\not{q}' - \not{k} - m_e} \frac{i}{\not{q} - \not{k} - m_e} \gamma_\mu (1 - 4\sin^2\theta_W - \gamma_5), \tag{5}$$

in the R_ξ gauge.[22] A straightforward calculation gives the result that the divergent part of this J_{eeZ} is given by

$$-\frac{g^3 m_e}{2m_W} \frac{(1 - 4\sin^2\theta_W)^2 - 1}{(4\cos\theta_W)^2} i\Lambda(3 + \xi), \tag{6}$$

where Λ is the logarithmically divergent integral

$$i\Lambda = \frac{1}{(2\pi)^4} \int \frac{d^4k}{(k^2 - M^2 + i\epsilon)^2}. \tag{7}$$

In Eq. (7), any positive mass M may be used.

This consideration can be applied not only to the other twelve diagrams of Fig. 1, but also to the diagrams with the loop on the external lines. The logarithmically divergent part of the sum of contributions is found to be

$$\frac{g^3 m_e}{8m_W^3} i\Lambda \left[3(m_t^2 + m_b^2) - \tfrac{15}{2} m_Z^2 + 3m_W^2 \right], \tag{8}$$

where terms of the order of radiative corrections are again neglected, as discussed above. Note that, as expected, this sum does not depend on ξ. Therefore the eeH coupling constant renormalization can be rendered finite by imposing the second mass relation

$$m_t^2 + m_b^2 = \tfrac{5}{2} m_Z^2 - m_W^2. \tag{9}$$

Unlike Eq. (1), in Eq. (9) the only mass that is not experimentally known is m_t. Thus,

$$m_t = \left[\tfrac{5}{2} m_Z^2 - m_W^2 - m_b^2 \right]^{\frac{1}{2}}. \tag{10}$$

The substitution of Eq. (10) into Eq. (1) then gives the elegant relation

$$m_H^2 = 9m_Z^2 - 6m_W^2. \tag{11}$$

With the experimentally known masses,[23] it follows from Eq. (10) and Eq. (11) that m_t = 120 GeV and m_H = 190 GeV. This theoretically determined mass of the top quark is in the range favored by existing phenomenology.[24,2] For this Higgs particle with a mass of 190 GeV, its properties are Γ_H = 1 GeV, BR($H \to W^+W^-$) = 77%, BR($H \to ZZ$) = 22%, and BR($H \to b\bar{b}$) = 1%. It is easy to detect at SSC and LHC even without enhanced luminosity. With a luminosity of 10^{34} cm^{-2}s^{-1}, the discovery of a Higgs boson of 190 GeV through the decay $H \to ZZ \to$ four leptons is expected to take only a couple of days. Thus, with 3% of this luminosity, it would take about two months.

Acknowledgments

One of us (TTW) thanks Professors Martin Perl and Gregor Herten for their hospitality at this Workshop. We are indebted to Dr. Conrad Newton for checking some of our calculations. For helpful discussions, it is a pleasure to thank Professors Carlo Becchi, John Ellis, Raoul Gatto, Howard Georgi, Maurice Jacob, Kenneth Johnson, and especially Professors Harry Lehmann, Carlo Rubbia and Chen Ning Yang. Part of the work was carried out when one of us (TTW) was visiting the Theory Division of CERN.

References

1. S. L. Glashow, *Nucl. Phys.* **22** (1961) 579;
   S. Weinberg, *Phys. Rev. Lett.* **19** (1967) 1264;
   A. Salam, in *Elementary Particle Theory: Relativistic Groups and Analyticity (Nobel Symposium No. 8)*, ed. N. Svartholm (Almqvist and Wiksell, Stockholm, 1968), p. 367;
   S. L. Glashow, J. Iliopoulos, and L. Maiani, *Phys. Rev.* **D2** (1970) 1285.
2. The LEP Collaborations – ALEPH, DELPHI, L3 and OPAL, *Phys. Lett.* **B276** (1992) 247.
3. F. Englert and R. Brout, *Phys. Rev. Lett.* **13** (1964) 321;
   P. W. Higgs, *Phys. Lett.* **12** (1964) 132; *Phys. Rev. Lett.* **13** (1964) 508;
   G. S. Guralnik, C. R. Hagen and T. W. B. Kibble, *Phys. Rev. Lett.* **13** (1964) 585;
   P. W. Higgs, *Phys. Rev.* **145** (1966) 1156;
   T. W. B. Kibble, *Phys. Rev.* **155** (1967) 1554.
4. S.-i. Tomonaga, *Progr. Theor. Phys.* **1** (1946) 27;
   J. Schwinger, *Phys. Rev.* **73** (1948) 416; **74** (1948) 1439; **75** (1949) 651;
   R. P. Feynman, *Phys. Rev.* **74** (1948) 1430; **76** (1949) 749, 769.
5. F. J. Dyson, *Phys. Rev.* **75** (1949) 1736.
6. J. C. Ward, *Proc. Phys. Soc. (London)* **A64** (1951) 54;
   A. Salam, *Phys. Rev.* **82** (1951) 217;
   T. T. Wu, *Phys. Rev.* **125** (1962) 1436;
   R. L. Mills and C. N. Yang, *Suppl. Progr. Theor. Phys.* **37** and **38** (1966) 507.
7. Yu. A. Gol'fand and E. P. Likhtman, *JETP Lett.* **13** (1971) 323;
   D. V. Volkov and V. P. Akulov, *JETP Lett.* **16** (1972) 438;
   J. Wess and B. Zumino, *Nucl. Phys.* **B70** (1974) 39.
8. G. 't Hooft, *Nucl. Phys.* **B33** (1971) 173.
9. G. 't Hooft and M. Veltman, *Nucl. Phys.* **B50** (1972) 318.
10. C. N. Yang and R. L. Mills, *Phys. Rev.* **96** (1954) 191.
11. C. G. Bollini and J. J. Giambiagi, *Phys. Lett.* **B40** (1972) 566;
    G. 't Hooft and M. Veltman, *Nucl. Phys.* **B44** (1972) 189;
    J. F. Ashmore, *Nuovo Cim. Lett.* **4** (1972) 289;
    G. M. Cicuta and E. Montaldi, *Nuovo Cim. Lett.* **4** (1972) 329.

12. H. Georgi and S. L. Glashow, ***Phys. Rev. Lett.*** **32** (1974) 438.
13. S. L. Adler, ***Phys. Rev.*** **177** (1969) 2426;
S. L. Adler, W. A. Bardeen, ***Phys. Rev.*** **182** (1969) 1517;
J. S. Bell and R. Jackiw, ***Nuovo Cim.*** **51** (1969) 47.
14. P. Osland and T. T. Wu, Bergen preprint 1992-02, Harvard preprint DAS-92/1; CERN preprints TH.6385, TH.6386, TH.6387 (February 1992), ***Zeitschrift f. Physik C***, in press;
P. Osland and T. T. Wu, CERN preprint TH.6567, ***Phys. Lett. B***, in press.
15. E. C. G. Stückelberg, ***Nature*** **144** (1939) 118.
16. S. Ferrara, L. Girardello and F. Palumbo, ***Phys. Rev.*** **D20** (1979) 403.
17. M. Veltman, ***Acta Physica Polonica*** **B12** (1981) 437.
18. W. Siegel, ***Phys. Lett.*** **84B** (1979) 193;
D. M. Capper, D. R. T. Jones and P. van Nieuwenhuizen, ***Nucl. Phys.*** **B167** (1980) 479.
19. N. Cabibbo, ***Phys. Rev. Lett.*** **10** (1963) 531;
M. Kobayashi and T. Maskawa, ***Prog. Theor. Phys.*** **49** (1973) 652.
20. T. H. Bergstrøm and K. Olaussen, ***Arkiv for Det Fysiske Seminar i Trondheim***, No 4 - 1980;
R. Decker and J. Pestieau, ***Lett. Nuovo Cim.*** **29** (1980) 560;
D. Atkinson, A. Khare and W. J. Schoenmaker, Groningen preprint 1980.
21. R. Decker and J. Pestieau, ***Modern Phys. Lett.*** **A4** (1989) 2733;
M. Capdequi Peyranère, J. C. Montero and G. Moultaka, ***Phys. Lett.*** **B260** (1991) 138;
I-H. Lee and S. D. Drell, in ***Bég Memorial Volume***, eds. A. Ali and P. Hoodbhoy (World Scientific, Singapore, 1991).
22. K. Fujikawa, B. W. Lee and A. I. Sanda, ***Phys. Rev.*** **D6** (1972) 2923.
23. Review of Particle Properties, ***Phys. Rev.*** **D45** (1992) S1.
24. J. Ellis and G. L. Fogli, ***Phys. Lett.*** **B231** (1989) 189; **B232** (1989) 139;
P. Langacker, ***Phys. Rev. Lett.*** **63** (1989) 1920;
G. Altarelli, CERN preprint CERN-TH-6206-91;
J. Ellis, in ***Proc. 1991 LP-HEP Conference***, Geneva 1991 (World Scientific, 1992), vol. 2, p. 27.

THE LAST NEUTRINO EXPERIMENTS?

F. VANNUCCI
LPNHE, Univ. Paris 6 and 7

Abstract

Neutrino physics at accelerators has evolved from the study of neutrino interactions to the search for exotic phenomena occurring beyond the Standard Model. Experimental results on new neutrino states and on new properties (specially the problem of mass) are reviewed. The future of the field is sketched.

1 Introduction

Thirty years of very successful experiments at accelerators have taught us possibly everything about neutrino properties. The main achievements have been in particular: the discovery of the second neutrino, the neutral currents, complete studies of nucleon structure and precise tests of the electroweak theory.

One of the most important questions which remained, namely the number of neutrino species, has been recently answered at LEP[1]:

$$N_n = 3.00 \pm 0.05$$

The statement is clear: there exist 3 and only 3 neutrinos. But this result must be analysed carefully: it only refers to light and stable, left-handed neutrinos having standard couplings to the Z^0.

So Nature seems to offer us a straightforward and elegant scenario. The neutrinos are elementary particles of zero mass, zero electric charge and spin 1/2. There are 3 such neutrinos, each of them associated to its charged lepton partner. No clear evidence exists beyond these simple facts.

Is it the end of neutrino physics? It could well be, but neutrino physicists remain... and they want to continue their exploration of neutrino properties. They are justified by theorists who find the scenario too naive, in particular strict masslessness seems unnatural to most of them. In fact neutrino masses arise in most extensions of the Standard Model, and they offer one of the easiest ways to evade a robust theory which numbers only successes so far.

Thus one can say that the era of the "Standard Model neutrinos" is over, and the new endeavour is to look for exotic properties. The search has started in two main directions: are there new neutrinos beyond the three already known?, are there new properties of the existing neutrinos.

This presentation is limited to studies at accelerators. To be complete one should not forget all the efforts going on in other environments: double beta decays, reactors, astophysical neutrinos.

2 New neutrinos?

Various searches have looked for neutrinos beyond the three known ones. But with the LEP result on the number of flavours, such searches are limited to either large masses or small couplings to the Z^0.

2.1 Heavy neutrinos at LEP

The limit of 3 neutrino species comes from considerations of the Z^0 lineshape parameters (see Fig. 1). The Standard Model predicts a total width of:

$$\Gamma_{tot} = 2484 \pm 11 \text{ MeV}$$

and a partial width for each neutrino generation of:

$$\Gamma_{\nu\nu} = 166 \pm 1 \text{ MeV}$$

The theoretical uncertainties allow for a variation of m_t, m_H and α_s within their respective allowed ranges.

Experimentally LEP measures the total width of the Z^0, but also the partial leptonic and hadronic widths. By comparison, one extracts the invisible width dominated by neutrinos. This is the basis for the 3 generations. Any new heavy lepton L not decaying in the detector would give an extra contribution to this invisible width. Thus, non standard channels are restricted by the maximum excess invisible width allowed within measurement errors. The limit coming from the LEP experiments combined together[2] gives:

$$\Gamma_{LL} \leq 16 \text{ MeV} \qquad (95\% \text{ CL})$$

This is less than 10% of a normal neutrino width.

New neutrinos are excluded up to almost the whole available phase-space. Dirac and Majorana neutrinos having a different mass behaviour, the exact mass limits are:

$$M_L \leq 45.2 \text{ GeV} \qquad \text{for a Dirac particle}$$
$$M_L \leq 40.5 \text{ GeV} \qquad \text{for a Majorana particle.}$$

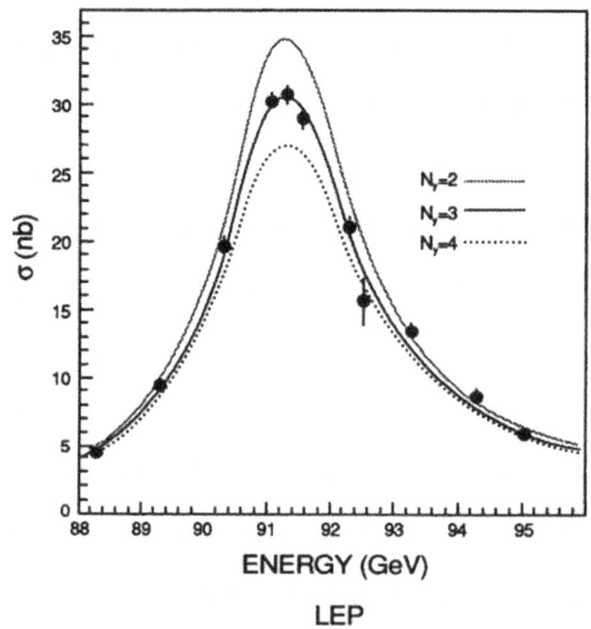

Figure 1: The Z^0 lineshape at LEP

2.2 Direct searches

The limits just discussed apply to heavy neutral leptons with full coupling to the Z^0. To test smaller couplings, direct searches are more sensitive. For example, excited neutrinos (which arise in composite models) have been looked for in the production mode[3]:

$$Z^0 \to \nu^* \nu^*$$

with the subsequent decay: $\nu^* \to \nu \gamma$.

The signal results in two high-energy photons seen in the detector. A search for such a signature, beyond the expected background, gives no event. This can be interpreted as a limit on a possible coupling of excited neutrinos to the Z^0. The result is:

$$g_{Z\nu^*\nu^*} \leq 0.02 * g_{Z\nu\nu} \qquad (95\% \text{ CL})$$

up to the kinematical mass limit. Searches for the decay $Z^0 \rightarrow \nu^*\nu$ allow to extend the mass range, at the price of less constrained couplings .

2.3 Decays of heavy neutrinos

The previous discussion related to purely radiative decays of excited neutrinos. More generally, massive neutrinos are expected to decay, with decay modes which depend on the mass.

Below a mass of 1 MeV, decays are only radiative with one or two photons emitted., but predictions are too small to allow for experimental tests. Above 1 MeV the favored decay mode becomes:

$$\nu_H \rightarrow \nu_e\, e^+e^-$$

The lifetime is easily computed:

$$\tau = 2.7\ 10^{-11} /(m_\nu)^5 / |U_{eH}|^2 \qquad (s)$$

It depends on the mass of the decaying neutrino m_ν expressed in GeV, and on the mixing of this neutrino with the electron U_{eH}. With increasing masses, other decay modes open up:

$$\nu_H \rightarrow \nu_e\, e^+\, \mu^-, \quad \nu_\mu\, \mu^+\, \mu^-, \quad e^-\, \pi^+$$

For much larger masses (at LEP), the generic decay mode:

$$\nu_H \rightarrow l^-\, W^{+*}$$

will give 1 lepton and 1 jet for each heavy neutrino.

Experiments have searched for decaying neutrinos at the CERN PS, the SPS, and at e^+e^- colliders[4]. Limits exist for the maximum coupling U_{lH}^2 as a function of the neutrino mass m_ν. They are summarized in Fig. 2. In the whole range between 20 MeV and 45 GeV, there are no massive neutrinos coupled to known leptons with a coupling greater than $\sim 10^{-6}$. Some masses show much lower limits.

2.4 The latest hot news

The history of neutrino physics has often be enlivened by puzzles and open questions. The latest enigma comes from the CCFR experiment at Fermilab which claims the observation of double vertex events in its ~20 m long steel and scintillator detector: a first interaction is followed by a second one, occurring in time, some distance downstream. This cannot be explained away by punch-through or uncorrelated neutrino interactions. Furthermore, the signal appears only in the case of double neutral current topology (no muon observed). There is no excess if one of the vertices is of the charged current type.

For a distance between the two events greater than 10 interaction lengths, the excess is claimed to be[5]:

39 double events for 13 expected.

One possible explanation requires the existence of a heavy neutrino produced in the first neutral current interaction mixed with the ν_μ at a level of 10^{-3}, followed by its decay or reinteraction some distance away. The mass could be around 2 GeV. This is a weak spot in Fig. 2, where the limit is rather soft...

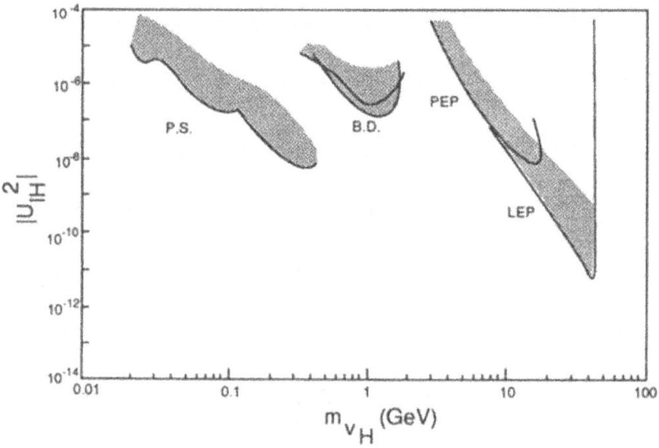

Figure 2: Limits on U_{lH}^2 as a function of mass for massive neutrinos

3 New neutrino properties

The quest for massive neutrinos has been going on for many years. Together with the mass, other fundamental parameters can be questioned: The Dirac or Majorana nature of the neutrinos, their mixing, their magnetic moment, charge radius, electric charge... These theoretical questions are paralleled by experimental puzzles. Some are on the vane but new ones are popping up. They concern solar and atmospheric fluxes, the 17 keV and now double vertices.

3.1 Magnetic moments

Magnetic moments would give anomalous electromagnetic interactions. This would contribute to ne scattering, on top of the electroweak cross-section. In particular, the signal of isolated electrons would be concentrated at low energies. No signal has been found at LAMPF[6], and the limits for ν_e and ν_μ are:

$$\mu(\nu_e) \leq 1.1\ 10^{-9}\,\mu_B$$

$$\mu(\nu_\mu) \leq 7.4\ 10^{-10}\,\mu_B$$

where μ_B refers to one Bohr magneton.

The same search has been made at high energy, in the CERN beam dump experiment using the BEBC bubble chamber as a detector [7]. It is possible to calculate the flux of ν_τ expected via D_s decays (a few % of n_m). No excess single electron being seen, the limit is put at:

$$\mu(\nu_\tau) \leq 5.4\ 10^{-7}\,\mu_B$$

We are still very far from the interesting region, since a 1 eV neutrino is expected to have a magnetic moment of $10^{-19}\,\mu_B$! Anyhow, it is amazing that one can put a limit on properties of a particle which has never been seen.

3.2 Neutrino masses

The best limits from direct measurements obtained so far have been presented at Neutrino 92. They are:

$$m(\nu_e) \leq 7.2 \text{ eV}$$

$$m(\nu_\mu) \leq 270 \text{ keV}$$

$$m(\nu_\tau) \leq 31 \text{ MeV}$$

The first limit comes from a new Mainz tritium β decay measurement, the last comes from a reanalysis of ARGUS data with the new value of the τ mass from Beijing.

Are neutrinos massless after all? Theorists answer that a zero mass is unprobable, and experimentalists will never be able to prove an identically zero mass. So the quest goes on...

3.3 Neutrino oscillations

At accelerators the most sensitive way of attacking the mass problem is via oscillations. The probability of such a phenomenon between two neutrinos is given by:

$$P = \sin^2 2\,\theta \sin^2(\pi R/L)$$

with θ the mixing angle between the neutrinos R the distance over which the experiment is sensitive and L the oscillation length:

$$L = 2.5\, E\,(\text{GeV}) / \delta m^2\,(\text{eV}^2) \quad (\text{km})$$

where $dm^2 = m_1{}^2 - m_2{}^2$, m_1 and m_2 being the two masses entering the problem.

The results are presented as exclusion regions in the plane ($\sin^2 2\,\theta, \delta m^2$). The present limits for the three physical channels are shown in Fig. 3. The only new result of the past two years comes from Brookhaven[8]. The contours reach already the very low level of 1 eV for maximum mixing, and future experiments plan to explore further new regions of the plane:

- at CERN two experiments have been approved to push the limit on the mixing for the $\nu_\mu - \nu_\tau$ oscillation. One (Chorus) uses 800 kg of emulsions, the other one (Nomad) relies on kinematics to detect the ν_τ.
- at Fermilab, projects exist both for the ν_μ–ν_τ search with emulsions, and for a long baseline experiment able to push the limit in δm^2 by up to 2 orders of magnitude.
- at Kaon, the meson factory recently approved for Triumf, the much increased flux makes possible an oscillation search over as much as 100 km, pushing correspondingly the δm^2 limit.
- LHC and SSC will open a new field: high energy ν_τ beams will be available at a level as high as 10% of the total neutrino flux.

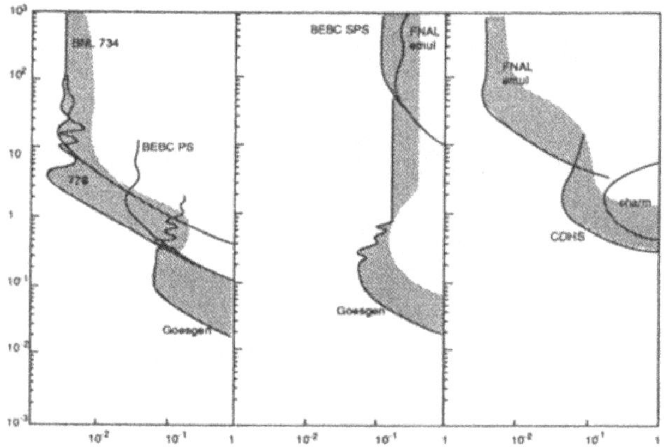

Figure 3: Excluded regions for the oscillations ν_μ–ν_e, ν_μ–ν_τ, ν_e-ν_τ obtained at accelerators and reactors.

3.4 A cosmological hint?

Oscillation searches were not driven by predictions but led by the availability of the necessary neutrino beam and detector. They have been, up to now, stabs in the dark. However for the next round of experiments, some light may come from cosmology[9].

The argument goes as follows:

- the Big Bang scenario predicts the existence of fossil neutrinos with a density of 100/cm^3 for each flavor. This means that neutrinos are about 10^9 times more abundant than baryons in the Universe.

- there exists some dark matter, seen only through gravitation, and representing 10 to 100 times the baryonic mass.

- if this dark matter is made of neutrinos, it is straightforward to calculate their mass. With the hierarchical prejudice (see-saw mechanism), this gives a mass for the ν_τ of between 10 and 100 eV.

Such a mass is extremely tiny for a direct search (6 orders of magnitude smaller than the present limit). On the other hand it is quite large for oscillations: it does not require large distances to see an effect. The difficulty is to push the limit in mixing, which reaches already the level[10] of 5 10^{-3}.

3.5 An experimental challenge

Two experiments are in preparation at CERN[11] to search for the oscillation ν_μ–ν_τ down to mixings of a few 10^{-4}. The experimental difficulty lies in detecting a few ν_τ interactions characterized by the production of a τ lepton which decays almost immediately (typically after 1 mm).

- Chorus uses the very precise technique of emulsions to see the τ track. To minimize scanning, it needs a sophisticated tracking device which selects candidate events in a spectrometer and reconstructs back into the emulsion stacks potential τ fragments.

- Nomad uses a technique entirely based on kinematics. An event with a τ is signed by a missing p^T (due to the escaping neutrinos in the τ decay) which align with the τ fragments in the transverse plane. This requires a detector able to reconstruct as accurately as possible all the secondary particles charged and neutral. A simulated event is shown in Fig. 4a, to be compared with a typical event as seen in the CDHS detector of earlier times (Fig. 4b). The apparatus is composed of a very light target (5 m long radiation length) put inside the old UA1 magnet, supplemented with TRD, electromagnetic calorimeter and muon analyser. The precise information gathered on the recorded events should also make it possible to study many rare processes.

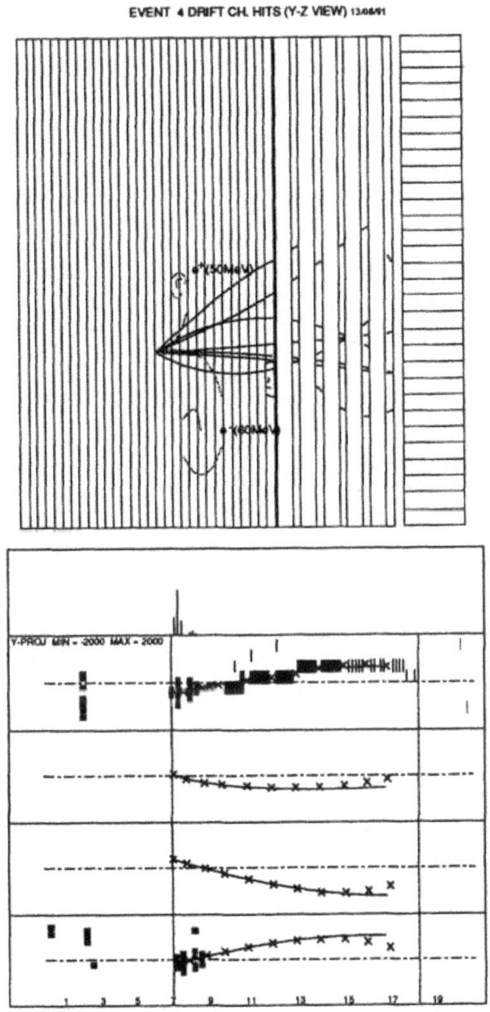

Figure 4: An event as seen in the Nomad detector (4a) and in the CDHS detector (4b).

4 Conclusion

The death of neutrino physics at accelerators, vanquished by the Standard Model, has been wildly exagerated. Neutrino experiments are reacting by trying to prove the theory wrong. Indeed, the neutrino sector remains one of the most probable terrain to look beyond the Standard Model. The stakes are high: if a signal of massive neutrinos is found, neutrino experiments will rise again for many years of active work. If the planned searches only push by a new order of magnitude the present limits, we may reach the level of hopelessness and the brillant chapter of neutrino physics at accelerators may close for a while.

References

1. The LEP collaborations, *Phys. Lett.* **B276** (1992) 247.
2. M. Bardadin-Otwinowska, Preprint CERN-PPE/90-6;
   M. Davier, EPS-LP Conference, Geneva, August 1991, 151.
3. D. Decamp et al., *Phys. Lett.* **B250** (1990) 172.
4. G. Bernardi et al., *Phys. Lett.* **B166** (1986) 479;
   J. Dorenbosch et al., *Phys. Lett.* **B166** (1986) 473;
   G. Bernardi et al., *Phys. Lett.* **B203** (1988) 332;
   P. Abreu et al., *Phys. Lett.* **B274** (1992) 230.
5. S. Mishra, Talk at Neutrino 92, Granada, June 1992.
6. D.A. Krakauer et al., *Phys. Lett.* **B252** (1990) 177.
7. A.M. Cooper-Sarkar et al., *Phys. Lett.* **B280** (1992) 153.
8. L. Borodowsky et al., *Phys. Rev. Lett.* **68** (1992) 274.
9. H. Harari, *Phys. Lett.* **B216** (1989) 413;
   J. Ellis et al., Preprint CERN-TH 6569/92.
10. N. Ushida et al., *Phys. Rev. Lett.* **57** (1986) 2897.
11. N. Armenise et al., CERN-SPSC/90-12 (1990);
    P. Astier et al., CERN-SPSC/91-21 (1991).

SEARCHES FOR MASSIVE NEUTRINOS IN NUCLEAR BETA DECAY*

JOHN A. JAROS
Stanford Linear Accelerator Center
Stanford University, Stanford, CA 94309

ABSTRACT

The status of searches for massive neutrinos in nuclear beta decay is reviewed. The claim by an ITEP group that the electron antineutrino mass > 17 eV has been disputed by all the subsequent experiments. Current measurements of the tritium beta spectrum limit $m_{\bar{\nu}_e} < 10$ eV. The status of the 17 keV neutrino is reviewed. The strong null results from INS Tokyo and Argonne, and deficiencies in the experiments which reported positive effects, make it unreasonable to ascribe the spectral distortions seen by Simpson, Hime, and others to a 17 keV neutrino. Several new ideas on how to search for massive neutrinos in nuclear beta decay are discussed.

I. Introduction

The questions of whether neutrinos have mass, and whether the lepton families mix as do the quarks, are fundamental, with broad implications in particle physics and cosmology. Two very provocative experiments of the 1980's highlighted these questions. The first was a report from Lubimov's group at ITEP in Moscow in 1980 [1] that the electron antineutrino had a rest mass between 15 and 40 eV. The second was John Simpson's claim in 1985 [2] that a kink in the tritium beta spectrum was evidence that the electron antineutrino has a 17 keV component.

This paper reviews the recent experimental results on both these topics. Although the Lubimov result has been seriously questioned for some time, it is only within the past year that several of the experiments it provoked have reported their final results. These results provide the best direct limits on the electron antineutrino mass to date.

On the 17 keV neutrino front, this year has seen reanalysis of earlier results, thoughtful critiques of existing experiments, and decisive new results. I have chosen to include results that appeared after the time of the Trieste Workshop in this paper for the sake of completeness.

* Work supported by Department of Energy contract DE-AC03-76SF00515.

The organization of this paper is as follows. The first section reviews measurements of the tritium β spectrum near its endpoint, and the resulting mass limits on the electron antineutrino. The second section reviews experiments on 17 keV neutrinos, and focuses on new results and critical interpretations. This field of research has produced contradictory results and controversy; consequently it has stimulated new ideas on how to search for massive neutrinos in nuclear beta decay. These are briefly discussed in the third section of the paper.

II. Measurements of the Tritium Spectrum Near the Endpoint

A. *Signature of Massive ν_e*

The beta decay of tritium, $^3H \rightarrow {}^3He^+ e^- \bar{\nu}_e$, is distinguished by its very low Q value, 18.57 keV, which makes it maximally sensitive to the kinematic effects of finite neutrino mass. The transition is superallowed, a mix of Fermi and Gamow-Teller transitions. The electron energy spectrum, dN/dE, is determined by phase space and Coulomb corrections, with its normalization coming from the nuclear matrix element and the Fermi coupling strength

$$\frac{dN}{dE}\left(m_\nu, E_0^i\right) = K|M|^2\, F(E)$$

$$\times pE(E_0^i - E)\left[(E_o^i - E)^2 - m_{\bar{\nu}}^2\right]^{1/2} . \quad (1)$$

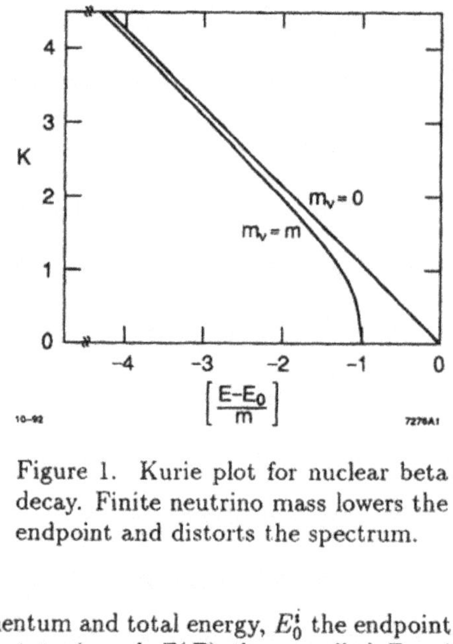

Figure 1. Kurie plot for nuclear beta decay. Finite neutrino mass lowers the endpoint and distorts the spectrum.

Here p and E are the electron momentum and total energy, E_0^i the endpoint energy corresponding to a particular final state i, and $F(E)$ the so-called Fermi function, which accounts for the Coulomb interaction between the electron and the Helium nucleus in the final state. Additional Coulomb corrections, which account for interactions with the atomic electrons, are inconsequential near the endpoint.

The spectral shape near the endpoint is sensitive to neutrino mass. This is shown on the Kurie plot of Fig. 1, where $K = \left\{dN/dE/(F(E)pE)\right\}^{1/2}$ is plotted against the electron energy. The endpoint shifts to $E_0 - m_\nu$, and the rate near the endpoint is depressed. The distortions will be washed out unless the energy resolution is comparable to m_ν. Only decays within about 3 m_ν of the endpoint are of use in searching for an effect, and these are a minute fraction of all decays, $\sim 10^{-10}\ (m_\nu/\mathrm{eV})^3$. Thus, beside good energy resolution, these experiments require great sensitivity and very low backgrounds.

B. *Multiple Final States*

An additional effect complicates the electron spectrum near the endpoint. The atomic electrons which orbit the tritium nucleus are no longer in a unique eigenstate after the tritium decays. In monatomic tritium, atomic physics calculations [3] predict that about 70% of the decays result in a ground state helium ion, 25% in an excited state 40 eV higher in energy (*shake-up*), and the remainder in continuum, fully ionized states above 60 eV (*shake-off*). The pattern is more complicated in molecular tritium, although it is still thought to be reliably calculated. For tritiated, complex organic molecules, the calculations are less robust. In all these cases, the ion can be left in any one of several energy states; hence, a whole range of endpoint energies must be considered, and the spectrum becomes the weighted sum over the final states

$$\frac{dN}{dE_{th}} = \sum_i w_i \frac{dN}{dE} \left(m_\nu,\ E_0^i\right) . \tag{2}$$

Figure 2 shows the resulting Kurie plot.

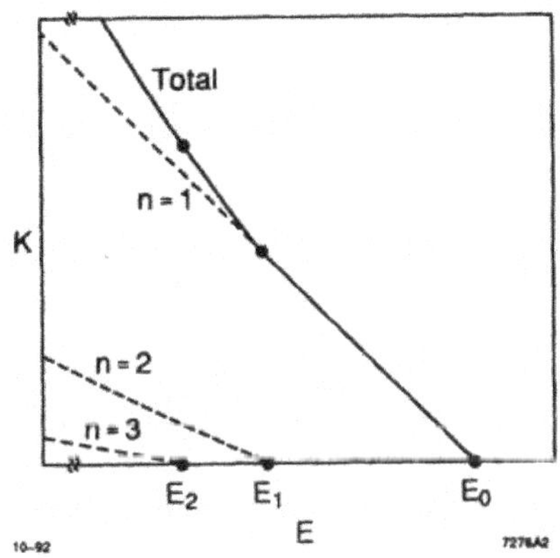

Figure 2. The spectrum of final states complicates the Kurie plot, giving rise to slope discontinuities.

C. *Fits to the Data*

It is common practice to account for small variations in detection efficiency with energy by multiplying the theoretical spectrum by a *shape factor*, e.g., $1 - \alpha(E - E_0)$. Detector resolution and energy loss effects are typically measured with internal conversion sources, and incorporated in a response function. This is convoluted with the shape-corrected spectrum to predict the measured spectrum. The parameters m_ν^2, E_0, α, and the normalization are adjusted to give a best-fit to the data.

D. *ITEP Results*

The ITEP group studied the endpoint of the tritium spectrum using a novel torroidal magnetic spectrometer with 45 eV FWHM resolution. The complex organic molecule valine ($C_5H_{11}NO_2$) was tritiated to serve as a source. Roughly 10^5 counts were accumulated in the last 100 eV of the spectrum. The group concluded [1] that the electron antineutrino mass was finite, $14 \leq m_\nu \leq 46$ eV (99% C.L.). This result was criticized because resolution effects and the complex final state effects were not

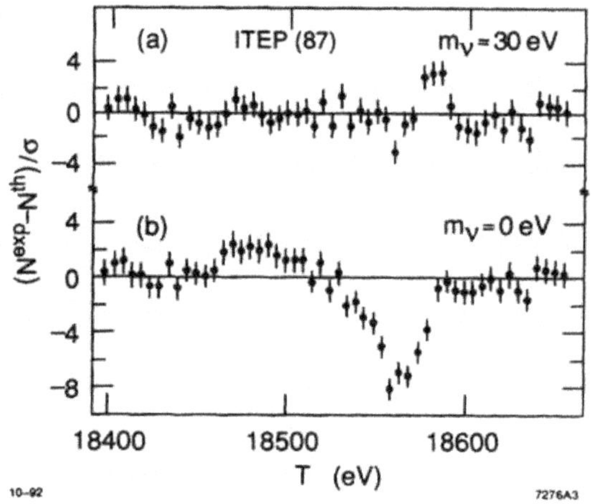

Figure 3. Tritium beta spectrum measured by the ITEP group[4] compared to fits with m_ν = 30 eV and m_ν = 0 eV. Finite neutrino mass was indicated.

correctly included. The final report from the group [4] addressed these criticisms with a refined apparatus, resolution function measurements, and detailed final state calculations. Their result, shown in Fig. 3, stills evidences finite neutrino mass. They conclude $m_\nu = 30.3^{+2}_{-8}$ eV. The vagueries of the atomic physics make it difficult to assign a systematic error. Instead the authors quote the range of m_ν for which their measured endpoint energy is consistent with the He–T mass difference. They find $17 < m_\nu < 40$ eV. This is the lone experiment reporting a finite electron antineutrino mass in the 20 eV range.

E. Recent Results

The strong statistical, if questionable systematic, weight of the ITEP result stimulated several groups to repeat the experiment.

Groups at Zurich and Tokyo were the first to report results, and have continued to refine their work in the last year or so. Both have sources similar to the solid valine source of the Moscow experiment. Zurich uses T_2^+ ions embedded into a thin carbon layer (which they treat as CH_3T) evaporated onto an aluminum backing; Tokyo uses tritiated anachidic acid ($C_{20}H_{40}0_2$). Calculations of the spectrum of final states in these complex molecules have been made. Figure 4 demonstrates the sensitivity of the deduced neutrino mass to the assumed corrections. The Zurich group uses a magnetic spectrometer of the torroidal design used at ITEP: the Tokyo

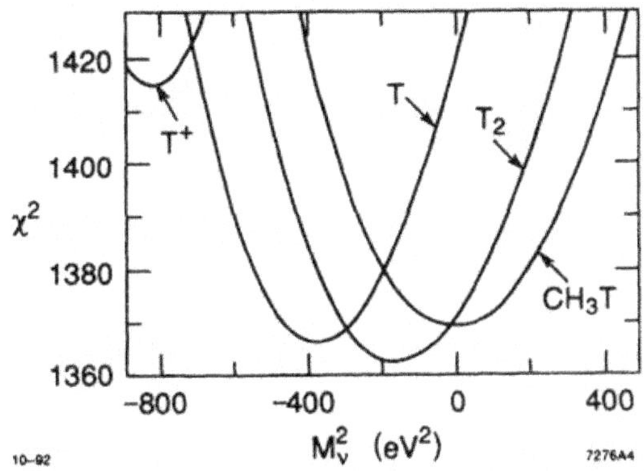

Figure 4. χ^2 versus m_ν^2 for fits to the tritium beta spectrum, with various assumptions about the spectrum of final states. The figure is from Ref. [5].

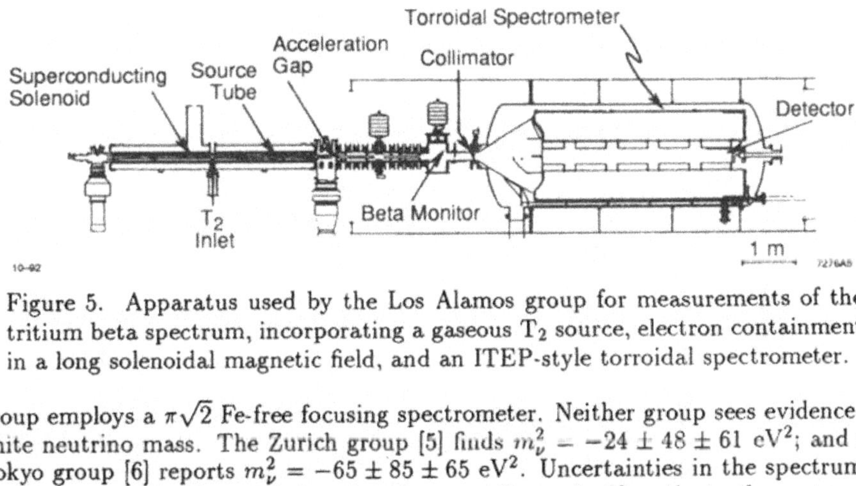

Figure 5. Apparatus used by the Los Alamos group for measurements of the tritium beta spectrum, incorporating a gaseous T_2 source, electron containment in a long solenoidal magnetic field, and an ITEP-style torroidal spectrometer.

group employs a $\pi\sqrt{2}$ Fe-free focusing spectrometer. Neither group sees evidence for finite neutrino mass. The Zurich group [5] finds $m_\nu^2 = -24 \pm 48 \pm 61$ eV^2; and the Tokyo group [6] reports $m_\nu^2 = -65 \pm 85 \pm 65$ eV^2. Uncertainties in the spectrum of final states and in the resolution function contribute significantly to the systematic error.

In order to reduce the uncertainties associated with final state effects, a Los Alamos group undertook the development of a gaseous tritium source. Their apparatus, which incorporates an ITEP-style torroidal spectrometer is shown in Fig. 5. Two additional advantages accrue from this design: (1) good electron acceptance, provided by capturing and transporting decay electrons in a strong solenoidal field; and (2) a well-understood source where backscattering is essentially non-existent, and energy loss calculable and measurable. Handling gaseous tritium is

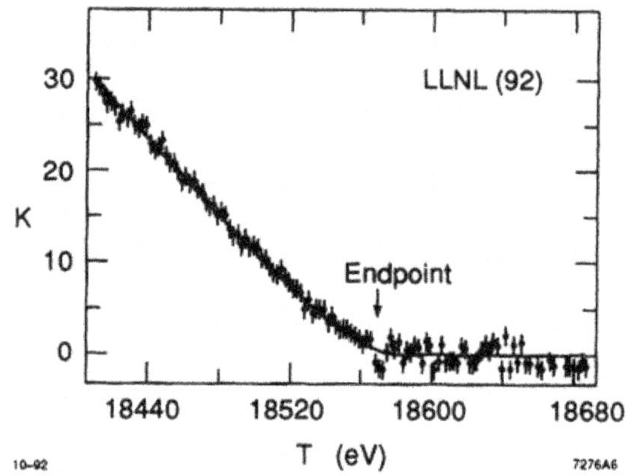

Figure 6. Kurie plot for T_2 decays near the endpoint, measured with the Livermore torroidal spectrometer [8].

tricky, requiring safety precautions, differential pumping, and very high vacuum in the spectrometer in order to avoid spurious volume and surface decays overwhelming the signal. A group at Livermore has constructed a similar device, and improved on the spectrometer's resolution. The Los Alamos group finds [7] $m_\nu^2 = -147 \pm 68 \pm 41$ eV^2 from a fit with roughly 10 k counts in the last 100 eV of the spectrum and 10 eV resolution. The Livermore group [8] has a preliminary result $m_\nu^2 = -72 \pm 41 \pm 30$ eV^2, and is continuing to improve their experiment. Their Kurie plot is shown in Fig. 6. Near the endpoint the data falls slightly above the fitted curve; the same is true of the Los Alamos data, and in fact is true of most of the current experiments. In consequence, the best fit value of m_ν^2 is slightly negative, which may indicate uncorrected systematics.

Final state effects are also manageable in frozen tritium targets, which has prompted the work of a second Livermore group [9] and a group at Mainz [10]. The latter experiment has recently reported results. It utilizes a so-called solenoid-retarding spectrometer, which does integral measurements of the spectrum with 2π acceptance and 6 eV resolution. Their result is shown in Fig. 7. In contrast to other experiments, they fit only the last 137 eV of the spectrum, which minimizes their systematic error. Their target has roughly 20 monolayers of tritium on an aluminum backing; hence backscattering corrections must be carefully understood, and energy loss in the target corrected. They find $m_\nu^2 = -39 \pm 34 \pm 15$ eV^2.

F. Conclusions

Figure 8 summarizes the recent results. All of the experiments are in conflict with the ITEP claim that the electron antineutrino mass is > 17 eV, although

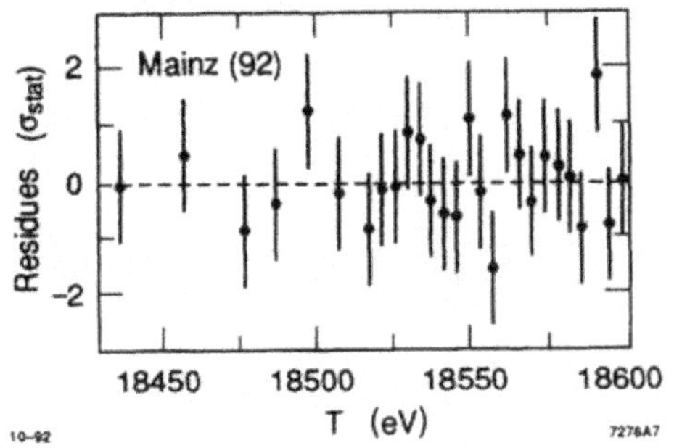

Figure 7. The endpoint region of the tritium spectrum, measured by the Mainz group [10].

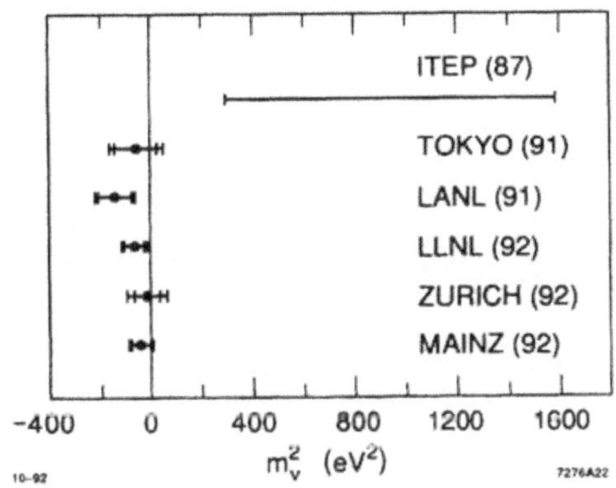

Figure 8. Summary of recent results on the electron antineutrino mass.

no decisive explanation has been offered for the Russian result. The best limit at present is from the Mainz group, $m_\nu < 7.2$ eV (95% C.L.). The recent experiments all find the endpoint energy near 18570 eV, in good agreement among themselves, and consistent with the measured mass difference of T and He^3 from cyclotron resonance measurements [11]. The Russian experiment reported an endpoint energy 18580.8 ± 4 eV. As noted above, all experiments find the best fit value of m_ν^2 to be slightly negative, though not significantly so. This coincidence has prompted some uneasiness, however, and begs further experimental refinement. The Mainz and Livermore experiments look capable of pushing the sensitivity to a few eV. Pushing the limits by another decade looks improbable with this technique, requiring a tenfold improvement in resolution and a thousandfold increase in source strength.

III. Status of Searches for the 17 keV Neutrino

A. Kinks in Beta Spectra

The possibility that the neutrino emitted in β decay might be the admixture of several mass eigenstates was explored in influential papers by Shrock and by McKellar [12] in 1980. For simplicity, imagine the electron antineutrino to be the superposition of two mass eigenstates,

$$|\bar{\nu}_e\rangle = \cos\theta\,|\bar{\nu}_1\rangle + \sin\theta|\,\bar{\nu}_2\rangle\ . \tag{3}$$

For suitable $m_{\nu 2}(Q > m_{\nu 2} > m_{\nu 1})$, and assuming the $\bar{\nu}_1$ is essentially massless, the resulting electron spectrum has two components,

$$\frac{dN}{dE} = \cos^2\theta\,\frac{dN}{dE}\,(m_\nu = 0) + \sin^2\theta\,\frac{dN}{dE}\,(m_\nu = m_{\nu 2})\ . \tag{4}$$

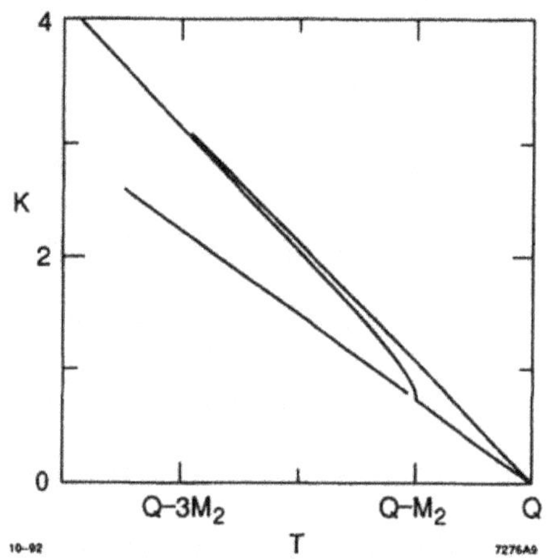

Figure 9. Distortion in the Kurie plot arising from the admixture of a massive neutrino in beta decay.

The Kurie plot is modified to that shown in Fig. 9. It develops a distinctive kink at the threshold for massive neutrino emission, $Q - m_{\nu 2}$. Shrock suggested, among other tests, to scan common beta spectra for such kinks. From the kink position and magnitude, the neutrino mass and mixing are derived.

The experimental signature depends on exactly how the data is fit. If one assumes a linear Kurie plot, and fits to the spectrum above the kink, i.e., to the region where the spectrum is genuinely one component, the data will show a departure from

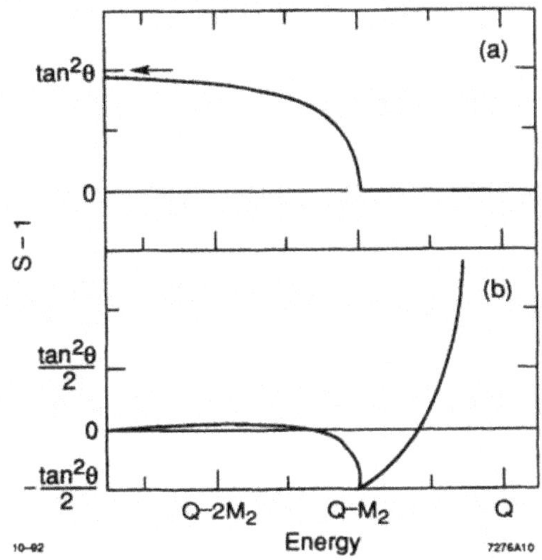

Figure 10. Shape factor signatures for massive admixed neutrinos depend on the fitting strategy: (a) shape factor when the data are fit above the heavy neutrino threshold; (b) shape factor when the data are fit throughout the kink region. The fit assumes a single massless neutrino in both cases.

the fit at T $= Q - m_{\nu 2}$. In terms of the shape factor, S, which is the ratio of the observed spectrum to the single-component expectation,

$$S = 1 + \tan^2\theta\left[1 - \frac{m_{\nu 2}^2}{(Q-T)^2}\right]^{1/2} . \tag{5}$$

This is plotted in Fig. 10(a). In the case of the 17 keV searches described below, the spectral region beyond the kink may be poorly known. It is a region of low rate, potentially high background, and limited extent, making its extrapolation dubious. An alternative signature arises from a different fit strategy, in which the entire spectrum (kink and all) is fit to the one-component theory, and deviations to this fit are sought. Fig. 10(b) shows how the threshold appears in such a fit. The exact shape is *experiment-dependent*, in that it depends on the extent of the region fit and the statistical weight of the points. But generally speaking, the kink in the spectrum extends over a few keV, and the maximum deviation is $\lesssim \tan^2\theta/2$. To see the effect of a 1% admixture of a 17 keV neutrino in a beta spectrum, it is required (1) that the resolution be good (FWHM < 4 keV); (2) that there be sufficient statistics to see a .5% effect significantly ($N \sim 10^6$/keV); and (3) that instrumental effects not distort the spectrum at a level comparable to that of the effect. It is extremely difficult to measure a spectral shape to the 0.1% level required to see a 0.5% effect with certainty, as the confused history of this subject indirectly attests.

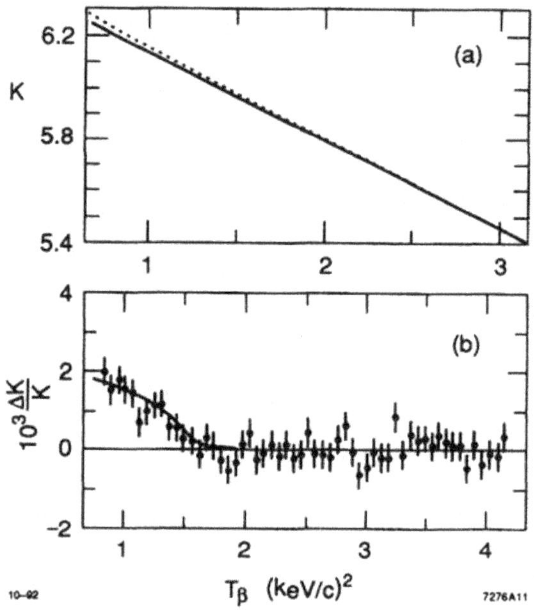

Figure 11. (a) Kurie plot for tritium as measured by Simpson [2]; (b) shape factor for T^+ implanted in Germanium, Simpson and Hime [15].

B. *Simpson's Result in Tritium*

Simpson [2] undertook precision measurements of the tritium beta spectrum using a novel technology. He implanted tritium ions in a silicon detector, which served as a total calorimeter for the emitted β's. The spectrum he obtained is shown in Fig. 11(a), and was well described by theoretical expectation above 1.5 keV without resorting to arbitrary shape factors. Simpson boldly interpreted the deviation in the spectrum at 1.5 keV as evidence for the emission of a 17.1 keV neutrino with mixing probability 3%. He subsequently revised this estimate to $1.1 \pm 0.3\%$ [13], after properly including screening and exchange corrections and the effects of the silicon environment.

C. *Experiments Con and Pro*

Simpson's original paper provoked a spate of experiments which sought, unsuccessfully, kinks in the spectra of ^{35}S and ^{63}Ni. Early results came from magnetic spectrometers [14] (Princeton, ITEP, Cal Tech, Chalk River), silicon beta calorimeters [15] (Bombay, Tokyo), and silicon photon calorimeters [16] (CERN, Zagreb), measuring the internal bremstrahlung spectrum associated with electron capture. (This spectrum also shows characteristic distortions, depending on neutrino mass and mixing). The results are summarized in Table I. Most of this work was not strong enough statistically or robust enough systematically to refute the revised claim for 1% (not 3%) mixing. The Chalk River data, however, strongly refuted even the revised claim. See Fig. 12.

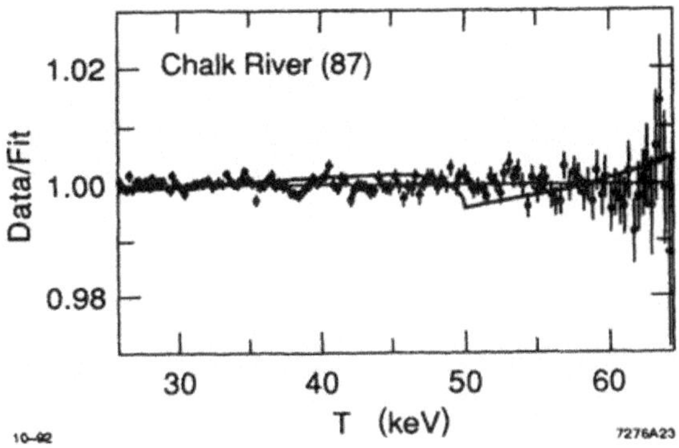

Figure 12. Shape factor in ^{63}Ni (Q = 67 keV) from the Chalk River experiment [16]. The fit includes a linear shape factor to correct for experimental distortions. The curve shows the shape expected for a 1% admixture of a 17 keV neutrino.

Simpson repeated the tritium measurements with Hime using a hyper-pure germanium detector, which was annealed after implantation to cure possible radiation damage. The crystal was calibrated in situ with photons, and the analysis incorporated the revisions mentioned above. The results of this work are shown in Fig. 11(b). There is a statistically significant spectral distortion near 1.5 keV. Simpson and Hime [13] interpreted this as evidence for a 16.9 ± 0.1 keV neutrino, admixed to the level of $\sin^2\theta = 1.1 \pm 0.14\%$.

Following their work in tritium, Hime and Simpson undertook an experiment in ^{35}S to confirm the effect. This experiment [17] used a silicon detector, but relied on an external source. Hime and Jelley [18] repeated the experiment at Oxford, and improved the definition of acceptance and the measurement of the electron response function. Both experiments saw statistically significant spectral distortions consistent with a 1% admixture of 17 keV neutrino. The Oxford results for ^{35}S are shown in Fig. 13. An LBL group [19] using a high purity germanium detector that had been doped with ^{14}C, claimed a similar result, finding $\sin^2\theta = 1.2 \pm 0.3\%$. An indication of 17 keV neutrinos was also reported by a Zagreb group [20] which quoted $\sin^2\theta = 1.6 \pm 0.8\%$. The coincidence of these results, made by several groups in studies of several nuclei, resurrected the possibility that a 17 keV neutrino did exist, mixed at the 1% level in nuclear beta decay.

Table I. Results of 17 keV Neutrino Experiments

Experiment	Isotope	$\sin^2\theta \times 100$	m_ν (keV)	Ref. No.
SOLID STATE β CALORIMETERS				
Guelph (85)	^3H	2–3	17.1	[2]
Reanalyzed		1.10 ± 0.30	17.1 ± 0.1	[13]
INS-Tokyo (85)	^{35}S	< 0.15 (90% C.L.)	17	[15]
Bombay (85)	^{35}S	< 0.60 (90% C.L.)	[17]	[15]
Guelph (89)	^3H	1.11 ± 0.14	16.9 ± 0.1	[13]
Guelph (89)	^{35}S	0.73 ± 0.14	16.9 ± 0.4	[17]
Oxford (91)	^{35}S	0.78 ± 0.09	17.0 ± 0.35	[18]
LBL (91)	^{14}C	$-1.4 \pm 0.45 \pm 0.15$	17 ± 2	[19]
Update		1.2 ± 0.3	1.71 ± 0.6	[19]
Argonne	^{35}S	$-.04 \pm .08 \pm .08$	17	[23]
SOLID STATE γ CALORIMETERS				
CERN (86)	^{125}I	< 2.0 (90% C.L.)	17	[16]
Zagreb (88)	^{55}Fe	< 1.6 (95% C.L.)	15–45	[16]
Zagreb (91)	^{71}Ge	1.6 ± 0.8	17.1 ± 1.3	[20]
GAS PROPORTIONAL CHAMBER CALORIMETER				
Oklahoma (92)	^3H	< 0.4 (99% C.L.)	17	[27]
MAGNETIC SPECTROMETERS				
Princeton (85)	^{35}S	< 0.40 (99% C.L.)	17	[14]
ITEP (85)	^{35}S	< 0.17 (90% C.L.)	17	[14]
Cal Tech (85)	^{35}S	< 0.25 (90% C.L.)	17	[14]
Chalk River (87)	^{63}Ni	< 0.28 (90% C.L.)	17	[14]
Cal Tech (92)	^{35}S	< 0.27 (90% C.L.)	17	[21]
INS-Tokyo (92)	^{63}Ni	< 0.10 (95% C.L.)	17	[22]

In response, several magnetic spectrometer groups returned to the problem. The Princeton and ITEP groups reanalyzed their earlier data, succeeded in reducing the magnitude of the shape corrections needed to describe the measured spectra, and left their original conclusions essentially unchanged. The Cal Tech group has repeated their original experiment several times, correcting field instabilities, achieving better control of systematics, and getting more probable fits as a result. From their studies of ^{35}S they conclude [21] that $\sin^2\theta < 0.27\%$ (90% C.L.). None of these experiments has seen any indication of a 17 keV neutrino.

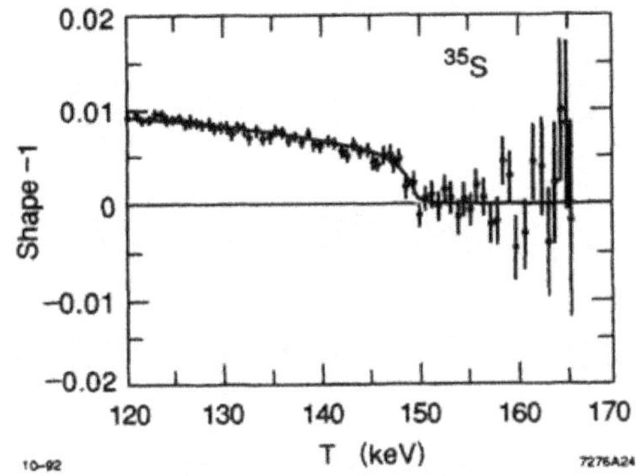

Figure 13. Shape factor for ^{35}S measured by Hime and Jelley[20].

D. Two New Cons

Two new, and very incisive experiments have recently reported null results in searching for the 17 keV neutrino. The first was a magnetic spectrometer experiment performed at the Institute for Nuclear Studies at the University of Tokyo [22], which studied ^{63}Ni decays with the $\pi\sqrt{2}$ iron-free spectrometer used in the tritium endpoint studies mentioned above. The principal modification to the apparatus was the addition of a 30-channel proportional wire chamber to the focal plane, which permitted a much larger statistic to be accumulated, roughly 2.4×10^9 counts in a relatively narrow 20 keV band around the expected threshold. Data was taken in a series of short runs in which the spectrometer setting was chosen randomly. The response function was measured using a ^{109}Cd source incorporated in a Nickel substrate to mimic energy loss and backscattering in the primary source. Each of the thirty channels is treated individually in the analysis, with a separate response function and individual shape factor. A global fit to the data is performed in which channel-to-channel normalization and shape are allowed to vary, but the mixing parameter is held common, as is the heavy neutrino mass. Their result is shown in Fig. 14(a). They find $\sin^2\theta = 0.00018 \pm 0.00033 \pm 0.00033$. The very high statistic, and the good resolution, make the experiment unique in that it is *kink-sensitive*. The data shown in Fig. 14(b), in which a 1% admixture of 17 keV neutrino has been assumed in the fit, mirror the expected signature, demonstrating the superb sensitivity of this experiment. No reasonable instrumental distortion could hide the sharp threshold expected. The probability of the first fit is good, indicating that channel-to-channel systematic differences are negligible to nearly the 10^{-4} level.

The second experiment [23] was performed at Argonne National Labs, and utilizes a small superconducting solenoid to capture a large fraction of the decays from a weak on-axis ^{35}S source and focus them on a silicon detector 40 cm distant.

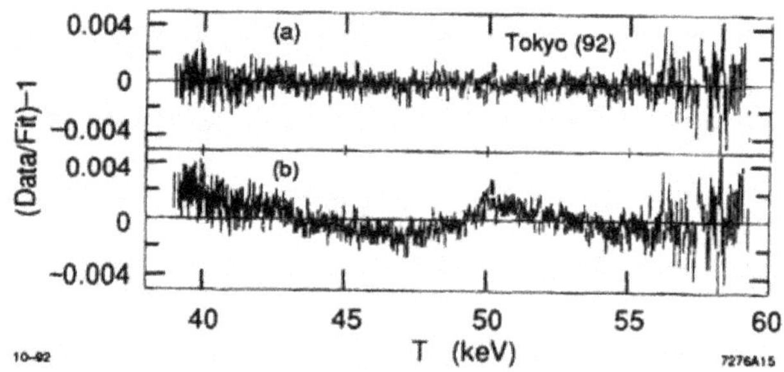

Figure 14. Shape factors for ^{63}Ni measured by the INS–Tokyo group assuming (a) no 17 keV neutrino, and (b) 1% mixture of 17 keV neutrino.

The electron energy is measured calorimetrically in the silicon. ^{139}Ce is used to calibrate the detector and measure the electron response function. The data are fit without an arbitrary shape correction and show no evidence of a kink: $\sin^2\theta = -0.0004 \pm 0.0008 \pm 0.0008$. See Fig. 15(a). Figure 15(b) demonstrates the sensitivity of the experiment to a small spectral distortion. The authors contaminated their ^{35}S source with roughly 1% of ^{14}C, which has Q = 156 keV, roughly 11 keV below that of ^{35}S. The nominal fit is a pure, one-component Fermi spectrum; the data follows instead the distorted shape expected when the admixture is taken into account. This demonstrates sensitivity to 1% effects.

E. *Assessment*

Have any experiments shown conclusively the presence or absence of a 17 keV neutrino?

Consider first the *shape-sensitive* experiments. With the exception of the INS Tokyo experiment, all the 17 keV experiments are sensitive to potential spectral distortions that might mimic or mask an effect. This is true because their statistical sensitivity derives from the detection of a change of slope in the Kurie plot in the vicinity of the 17 keV threshold, and not from detection of the *kink* itself. To prove that a shape distortion is not instrumental, it is necessary to measure the electron response function (shape and normalization) as a function of energy in an independent experiment. That is, one must measure the instrumental distortions. Strictly speaking, no experiment has done so (although the Argonne experiment has come close), so *all* experiments must address the possibility that shape factors are present in their spectra, perhaps *making*, perhaps eliminating a 17 keV neutrino. As Simpson [24] has argued, and Bonvicini [25] has demonstrated in a careful statistical analysis of these experiments, the presence of unknown shape factors reduces sensitivity to spectral anomalies and makes estimation of systematic errors questionable. Thus the

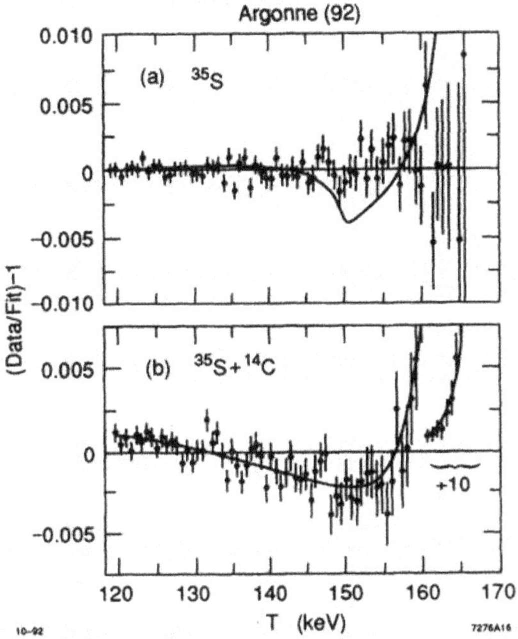

Figure 15. (a) Shape factor for ^{35}S measured by the Argonne experiment. The curve shows the distortion expected from the 17 keV neutrino. (b) Shape factor for a ^{35}S source with a small mixture of ^{14}C. The data falls nicely on the curve, demonstrating sensitivity to small spectral distortions.

Chalk River and Cal Tech results argue strongly against the existence of a 17 keV neutrino, but do not convincingly exclude it. On the other hand, the conclusions of Hime and Jelley may be significantly changed if their experiment is reanalyzed with a (presumably) more accurate response function. That is Piilonen and Abashian's claim [26] in their reanalysis, in which they carefully Monte Carlo the Hime/Jelley experiment, accounting for energy loss, scattering, backscattering, fluorescence, etc. They conclude a spectral distortion persists in the data, but one that bears no resemblance to a massive neutrino threshold.

The recent Argonne experiment is also shape sensitive, but it has several advantages over the Hime/Jelley experiment in regard to its measurement of the electron response function. First, it has very large acceptance for electrons and–thanks to magnetic confinement—no need for material apertures. This permits the use of extremely thin sources, and great enough detector/source separation that the photon acceptance is negligibly small. Thus, it is unlikely that photo-induced processes will be misinterpreted as part of the electron spectrum, and edge scattering and energy loss in the source are non-issues. Second, it has profitably used a ^{139}Ce source to constrain its model of the electron response function, the advantage being that several lines are used, which bracket the kink region. When this response function is convoluted with the Fermi spectrum, it accurately describes the data without recourse to arbitrary shape factors or massive neutrinos. This strongly disagrees with earlier silicon calorimeter work. It would be an airtight case if the normalization, as well as the shape of the electron response function, had been measured at several energies.

The INS experiment, with its high statistics, is very compelling. It essentially is the addition of thirty independent experiments, each with the statistical power of the *shape-sensitive* experiments. What is remarkable is that no subtle systematic distortion seems to be present at the 10^{-4} level, so that the sum of all the spectra is distortion-free, and the global fit to the no-17-keV hypothesis has a respectable goodness of fit. A minor weakness is that there has been no demonstration that the experiment can see a small spectral distortion, as was done in the Argonne and in the Cal Tech experiments. It is a very strong result, nevertheless.

Given these two very strong null results, and appreciating that the experiments reporting positive effects have not demonstrated the absence of instrumental shape corrections, it is not reasonable to ascribe the spectral distortions seen by Simpson, Hime, and Jelley, and others to a 17 keV neutrino. These distortions are not yet explained, however. In particular, there has been little experimental study of the distortions seen in the tritium spectrum. A recent study using a proportional chamber at Oklahoma [27] reports no effect, but must contend with uncertain background subtractions. Studies underway at Carleton [28], also using a proportional chamber calorimeter, and Livermore [29], using the beautiful torroidal spectrometer constructed for the endpoint measurements, will address this issue, and are eagerly awaited.

The experiments which have ruled out the existence of 17 keV neutrinos have used their data to search for kinks across the beta spectrum, and impose limits on the mixing strength of massive neutrinos as a function of neutrino mass. These limits are summarized in Fig. 16.

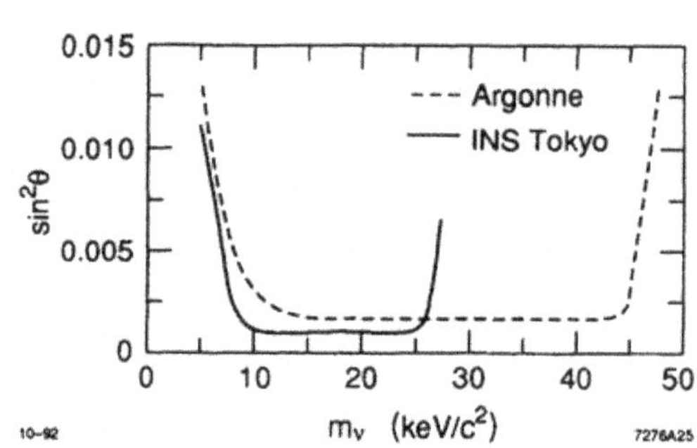

Figure 16. 95% confidence limit on the mixing strength for heavy neutrinos in beta decay versus neutrino mass.

IV. New Beta Decay Experiments

A. *Motivation*

The 17 keV neutrino controversy has sparked renewed interest in studies of nuclear beta decay, with several groups proposing new techniques. While the proximate motive for this work has disappeared, these advances in instrumentation promise to reduce instrumental distortions in spectral measurements. This could have significant impact in experiments which seek physical shape effects, like those expected in ^{14}C decay, and could allow more sensitive kink searches in future experiments. New techniques will also be used to study the tritium spectrum at low energies, where Simpson's anomalies have not been confirmed. Here we consider a few alternative technologies and a new approach to the problem.

B. *Gas Calorimeters*

The proportional chamber calorimeter at Carleton has already been mentioned. The device is a large, 10 atm, single-wire proportional counter using standard gases with a small admixture of tritium. High pressure and gold-coating on the chamber walls minimize backgrounds due to decays of adsorbed tritium. The resolution is relatively poor ($\Delta E/E = 15\%$), but adequate. The detector serves as a total absorption calorimeter, analogous to the implanted solid state detectors, but unlike these detectors, it can be calibrated throughout its volume, and can measure backgrounds with the tritium removed. Gain stability and uniformity and backgrounds are the major concerns.

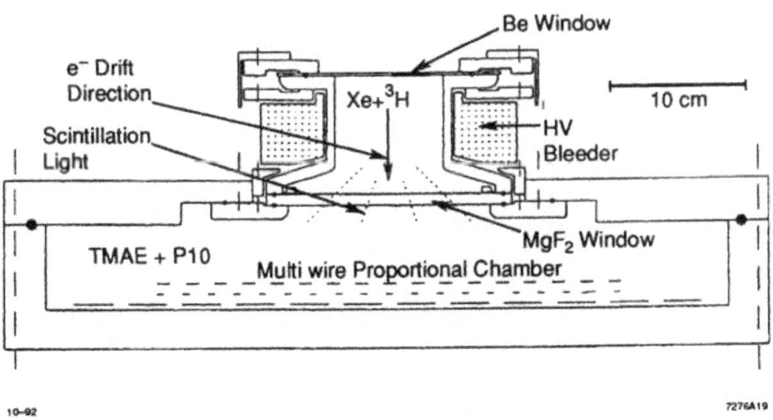

Figure 17. Gas scintillation proportional chamber for study of T_2 decays at Delft [32].

Gas scintillation proportional detectors are being developed by groups at Berkeley [30] and Delft [31]. The Delft detector is shown in Fig. 17. It consists of a decay region filled with tritiated xenon gas, in which a 5 kV/cm field is established. In this high field, ionization drifting toward the collection grid excites scintillations in the gas. The resulting photons pass through a MgF_2 window, and are detected in a proportional chamber filled with TMAE and P10. The time development, pulse-height, and position are recorded to measure the location and energy of the decay. The fact that many photons are released per primary electron gives the device good energy resolution. The Berkeley device measures ^{14}C decays by mixing radioactive CO_2 with xenon, operates at high pressures, and uses waveshifter fibers for readout. Both are total absorption calorimeters with *implanted sources*. The technologies will take time to mature, but are promising.

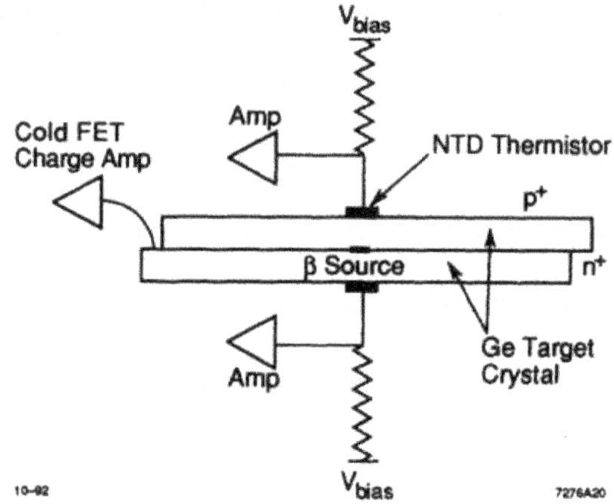

Figure 18. Proposed Cryogenic solid state detector for simultaneous phonon and ionization measurements from Ref. 32.

C. *Cryogenic Detectors*

Cryogenic solid-state detectors can measure energy lost into ionization and phonons simultaneously. At 20° mK, very low fields can fully deplete the detectors, allowing charge collection without the generation of much thermal noise. Since the detector's heat capacity is proportional to the cube of the temperature, small (thermal) energy depositions lead to measurable temperature changes. A schematic detector is shown in Fig. 18. The temperature signal, which develops on a millisecond time scale, is detected by a thermister. A Berkeley group has achieved 1.8 keV resolution at 60 keV [32] in a test device. The ionization signal is also read out. As the diagram shows, a sandwich geometry is proposed that should completely contain the energy loss from the source. Unlike the ionization signal, the phonon signal should not suffer degradation due to detector dead-layers. Background determination and calibration are straightforward in this detector, unlike implanted solid state detectors. An Oxford group [33] is also investigating the technique.

D. *Missing Mass Measurements*

The approaches described above have a common strategy: measure the beta spectrum with very high precision in order to search for small anomalies. A different idea has been independently proposed by experimenters from SLAC [34], Stony Brook [35], and Texas [36]: reconstruct the mass missing in beta decay by measuring simultaneously the momenta of the decay electron and the recoiling nucleus. If the initial nucleus is nearly at rest, measurements of modest precision allow a reconstruction of neutrino mass with enough resolution to distinguish a 17 keV neutrino from one that is massless. Tritium is the preferred nucleus because its low Q value exaggerates the kinematic differences between massive and massless neutrinos.

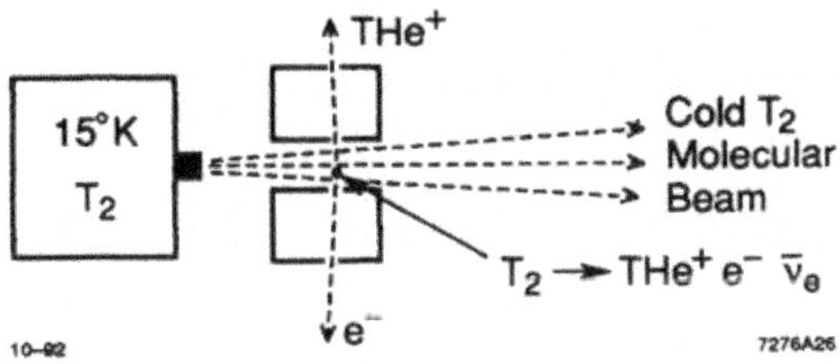

Figure 19. Schematic diagram of experiment to measure recoil mass in molecular tritium decay.

A schematic of the proposed SLAC experiment is shown in Fig. 19. An effusion beam is created from low-pressure, 15°K T_2 gas with a multichannel array. Skimming and aggressive differential pumping remove most of the unwanted gas. Decays are sampled from a small region, which is kept free of electric and magnetic fields. Collimators select back-to-back decays. The source is sufficiently tenuous that the chance of Van der Waals scattering on exiting is small. For electrons in the energy region of interest ($T < 1.5$ keV), the back-to-back topology accepts a large fraction of 17 keV decays and rejects most 0-mass decays. Figure 20 shows how clearly the electron and ion momenta are correlated with this back-to-back cut, and what a huge effect the presence of a 17 keV neutrino has on the kinematics. Unlike the spectral measurements, this experiment seeks to detect a gross effect.

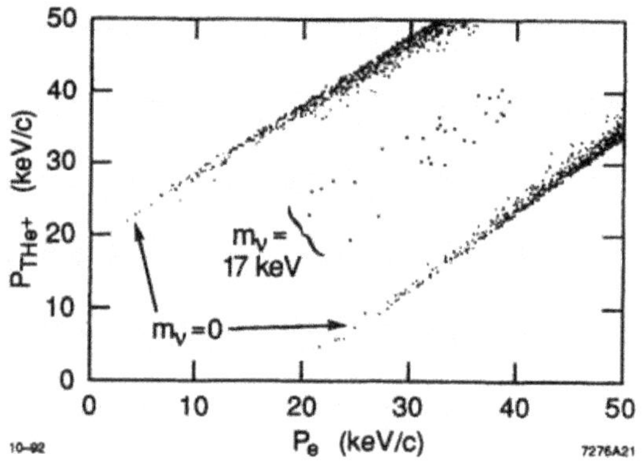

Figure 20. Correlations between ion and electron momenta for near back-to-back tritium decays ($\cos\theta < -.99$) for massless neutrinos (sharp bands, top and bottom) and 17 keV neutrinos (central band). Clean kinematic separation is possible.

Simulated results from this experiment are shown in Fig. 21. The Monte Carlo assumes an effective source of 10^{12} T_2 molecules, 30 msr acceptances in each spectrometer arm, and 2% momentum resolution. The electron rigidity is measured in a magnetic field; the ion velocity determined by time-of-flight. Roughly 10 decays/day appear in the 17 keV peak if 1% mixing obtains.

The experiment is not without technical challenges, most deriving from the difficulty in handling a radioactive gas and dealing with backgrounds originating from tritium adsorbed on vacuum vessel and detector surfaces. If spectral distortions persist in the low-energy tritium β spectrum, this experiment could discriminate new particle production from subtle atomic affects, and it could search for massive neutrinos with good sensitivity.

V. Conclusions

Precision measurements of the tritium beta spectrum near the endpoint have failed to confirm the claim by a Russian group that the electron antineutrino mass is ~ 20 eV. The best of these experiments presently limit the neutrino mass to be less than 7 eV, and could lower this bound to a few eV, nature willing.

The spectral distortion reported by Simpson, Hime, Jelley, and others as evidence for a 17 keV neutrino, is not seen in a new round of experiments with improved sensitivity and improved systematics. Although the original effects are not yet explained, it is no longer reasonable to ascribe them to an exotic, heavy neutrino.

Experimental techniques are being developed which may bring new precision and sensitivity to β decay studies.

The study of nuclear beta decay, which is nearly 100 years old, still generates excitement and hot debate, and still addresses questions of fundamental importance to particle physics. These experiments have provided our best direct knowledge of the electron antineutrino mass, and have searched for neutrino mixing with great sensitivity.

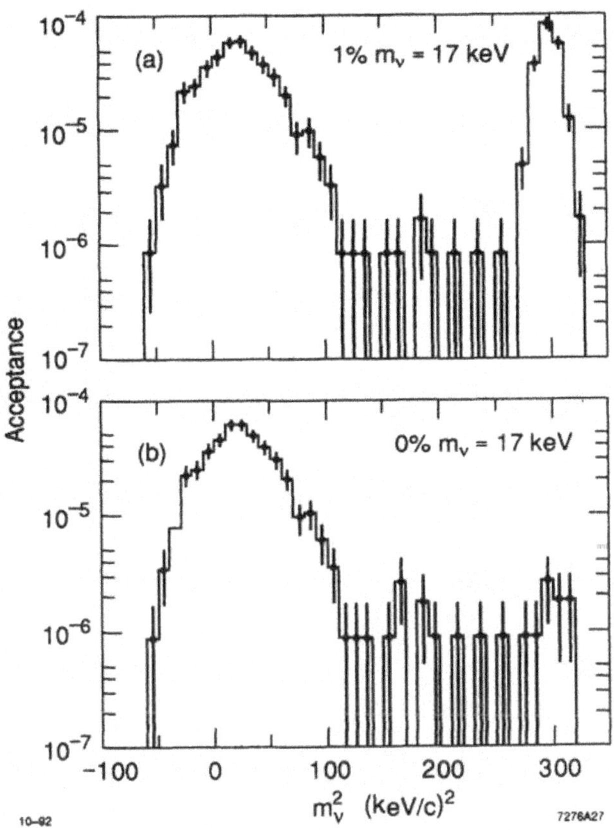

Figure 21. Simulated recoil mass spectra for tritium missing-mass experiment. The acceptance is plotted versus m_ν^2 for (a) 1% admixed 17 keV neutrino and (b) 0% 17 keV neutrino.

Acknowledgements

It is a pleasure to thank Martin Perl and Gregor Herten for organizing, and Luciano Bertocci and the International Center for Theoretical Physics for hosting, a very stimulating conference. I would also like to thank R.G.H. Robertson, D. Decman, W. Stoeffl, M. Mugge, O. Fackler, S. Freedman, and J. Shirai for introducing me to their work; and M. Swartz for many discussions, mostly concerning our tritium missing-mass experiment.

References

1. V. A. Lubimov et al., Phys. Lett. **94B**, 266 (1980).
2. J. J. Simpson, Phys. Rev. Lett. **54,** 1891 (1985).
3. See for example, O. Fackler et al., Phys. Rev: Lett. **55**, 1388 (1985); I. G. Kaplan et al., Phys. Lett. **B161**, 389 (1985).
4. S. Boris et al., Phys. Rev. Lett. **58**, 2019 (1987).
5. Zurich Group, reported at the 15th Int. Conf. on Neutrinos and Neutrino Astrophysics, Granada, Spain, June 1992. Earlier results were reported in Fritschi et al., Phys. Lett. **B173**, 485 (1986).
6. H. Kawakami et al., Phys. Lett. **256B**, 105 (1991).
7. R.G.H. Robertson et al., Phys. Rev. Lett. **67**, 957 (1991).
8. W. Stoeffl, American Physical Society Meeting, Washington, D.C., April 1992.
9. M. Mugge and O. Fackler have constructed a high resolution electrostatic spectrometer for studying a frozen T_2 target at Lawrence Livermore National Laboratory.
10. Mainz Group, H. Backe et al., reported at the 3rd Int. Sym. on Weak and Electromagnetic Interactions in Nuclei, Dubna, June 1992.
11. R S. Van Dyck, Jr., NFFS 6–AMCO 9, Bern, Kastel Kues, 1992.
12. R. E. Shrock, Phys Lett. **96B**, 159 (1980); B. McKellar, Phys. Lett. **97B**, 93 (1980).
13. A. Hime and J. J. Simpson, Phys. Rev. **D39**, 1837 (1989).
14. Magnetic spectrometer experiments:
    T. Altzitzoglou et al., (Princeton), Phys. Rev. Lett. **55**, 799 (1985);
    A. Apalikov et al., (ITEP) , JETP Lett. **42**, 289 (1985);
    J. Markey and F. Boehm (Cal Tech), Phys. Rev. C **32**, 2215 (1985);
    D. W. Hetherington et al., (Chalk River), Phys. Rev. C **36**, 1504 (1987).
15. Silicon Calorimeter Experiments:
    V. M. Datar et al., (Bombay), Nature **318**, 547 (1985).
    T. Ohi et al., (Tokyo), Phys. Lett. **160B**, 322 (1985).
16. IBEC Experiments:
    M.J.G. Borge et al., (CERN), Phys. Scr. **34**, 591 (1986).
    I. Zlimen et al., (Zagreb), Phys. Scr. **38**, 539 (1988).
17. J. J. Simpson and A. Hime, Phys. Rev. **D39**, 1829 (1989).

18. A. Hime and N. A. Jelley, Phys. Lett. **257**, 441 (1991).
19. B. Sur et al., Phys. Rev. Lett. **66**, 2444 (1991); updated by E. Norman, Workshop on the 17 keV Neutrino Question, Berkeley, December 1991.
20. I. Zlimen et al., Phys. Rev. Lett. **67** 560 (1991).
21. T. J. Radcliffe et al., in Proc. 12th Moriond Workshop, Les Arc, France, O. Fackler and J. Thran Thranh Van, eds., (Editions Frontieres, Paris, 1992).
22. H. Kawakami et al., KEK Preprint 92–14; INS Rep–921; submitted to Phys. Lett. B.
23. J. L. Mortara et al., submitted to Phys. Rev. Lett., September 1992.
24. For example, see Ref. [17].
25. G. Bonvicini, CERN–PPE/92–54.
26. L. Piilonen and A. Abashian, VT–IHEP–92/06.
27. M. Bahran and G. Kalbfleisch, Phys. Lett. **B291**, 336, 1992.
28. C. K. Hargrove, in Workshop on the 17 keV Neutrino Question, Berkeley, December 1991.
29. W. Stoeffl, *ibid.*
30. S. Weiss, *ibid.*
31. W.J.C. Okx et al., in Proc. 1992 Vienna Wire Chamber Conf., Nucl. Instrum. Methods.
32. B. Ellman, in Workshop on the 17 keV Neutrino Question, Berkeley, December 1991.
33. N. E. Booth et al., OUNP–91–29.
34. M. Swartz, in Workshop on the 17 keV Neutrino Question, Berkeley, December 1991.
35. G. Finocchiaro and R. Shrock, ITP–SB–91–62.
36. S. Cook et al., Phys. Rev. **D46**, R6 (1992).

SEARCHES FOR MASSIVE STABLE PARTICLES FROM MONOPOLES TO QUARKS

Giorgio Giacomelli
Dipartimento di Fisica dell'Università di Bologna and INFN, Sezione di Bologna
40126, Bologna, Italy

Abstract. In this paper will be first described some accelerator searches for magnetic monopoles, free quarks and free gluons. Then will be discussed underground searches for rare particles, such as cosmic magnetic monopoles, nuclearites, WIMPs and particles from Cygnus-X3. A brief discussion on bulk matter searches follows.

1. Introduction

Physicists, in their quest for simplicity, tried always to find new particles, which were assumed to be more fundamental and more elementary. In the 60's and 70's the "classical searches" concerned the "well known undiscovered particles", such as the Dirac monopole, free quarks, intermediate vector bosons, heavy leptons, a.s.o. [1]. The Standard Model of the electroweak and strong interactions brought a new level of simplicity and of understanding at a smaller space scale. Several particles predicted by the Standard Model have yet to be discovered. Moreover many physicists consider the present Standard Model of particle physics as a low energy limit of a more fundamental theory, which should reveal itself at higher energies; it is possible that there should be two or more new energy thresholds.

The first energy threshold, possibly associated to supersymmetric particles or to substructures of quarks and leptons, could be around few TeV and could be revealed with the new generation of colliders.

The second threshold could be associated with the Grand Unification of the electroweak and strong interactions and of quarks with leptons. This unification is expected to appear at extremely high energies, $> 10^{14}$ GeV. It is unthinkable to reach these energies with accelerators; instead these energies were readily available in the first instants of our Universe, for $t < 10^{-35}$ s. One may now look in cosmic rays for "fossil" particles, left over from the Big Bang [2]. GUT magnetic monopoles, the lightest supersymmetric particles, nuclearites, a.s.o. , are examples of possible fossil particles.

In this paper will be described and discussed some recent searches for magnetic monopoles, free quarks and gluons performed at accelerators and searches for monopoles, WIMPs, nuclearites and particles from Cygnus X-3 performed in underground experiments. A brief discussion on bulk matter searches will also be made. All searches yielded only upper limits. Some limits quoted have been updated.

2. Searches at accelerators

2.1. Magnetic monopoles. Magnetic monopoles were hypothesized by P.A.M. Dirac in 1931 to explain the quantization of the electric charge. Since then, many experimental searches for these "classical" monopoles have been performed at nearly every new high energy accelerator, employing a variety of direct and indirect methods [3]. The searches are based on the hypothesis that the poles should have a large magnetic charge, $g = n\ g_{Dirac} = n\hbar c/2e$, with $n = 1, 2, ...$ In the c.g.s. symmetric system of units $g_D = 68.5\ e$. Fast poles should ionize heavily and their detection would therefore be easy. In particular in plastic nuclear track detectors, such as CR39, a monopole would leave a "latent" track in the form of ruptured polymeric bonds along the path of the pole. This track would be easily detected after appropriate chemical etching and scanning, even in the presence of large background [3-5]. The experimental method is thus simple: one surrounds an intersection region with stacks of CR39 sheets and after a relatively long exposure, the sheets are etched and scanned.

Fig. 1 shows a compilation of present upper limits obtained from experiments at high energy accelerators up to a mass of 0.9 TeV [4-6]. Notice that in indirect experiments one has to hypothesize that magnetic monopoles may be trapped in a ferromagnetic material, that they can be extracted and accelerated by a strong magnetic field.

The significance of the limits depends on the physical process by which the monopoles are presumed to be produced. For Dirac monopoles in e^+e^- collisions the most obvious mechanism is monopole pair production via the electromagnetic interaction [7]. For a single photon process the lowest order cross section σ_D for a pointlike monopole of mass m scales with the cross section for unlike-sign dimuons ($e^+e^- \rightarrow \mu^+\mu^-$) times a phase space factor:

$$\sigma_D(m; s) = \left(\frac{g_D}{e}\right)^2 \sigma_{\mu\mu}(> 2m; s)\sqrt{1 - \frac{4m^2}{s}} \tag{1}$$

In this case the limits may be expressed as limits on the quantity $R_D = \sigma(m)/\ \sigma_D(m)$, which would be expected to be of order unity for pointlike Dirac monopoles with magnetic charge g_D at energies above threshold. The present limits on R_D are shown in Fig. 2. If the naive lowest order limit is valid one may conclude that pointlike Dirac monopoles with mass below 44.9 GeV are ruled out. However it has been speculated that monopoles may have non pointlike structure, resulting in a reduction of the cross section by orders of magnitudes.

2.2. Free quarks. Present theories assume that quarks and gluons are confined inside hadrons. It is nevertheless worthwhile to keep an open mind and continue the search for free quarks, and also for free gluons. All searches for free quarks at accelerators are based on the lower ionizing power of fractionally charged quarks. Among the many searches performed I shall single out one search performed in high energy neutrino and antineutrino interactions [8], one in an RF separated secondary beam at the CERN-SPS [9], one search in nucleus-nucleus collisions [10] and the possibilities at LEP [11].

The first type of search was performed using the CERN-SPS wide-band (anti)neutrino beam in the Big European Bubble Chamber (BEBC) filled with deuterium. In a bubble chamber filled with a light liquid, one may easily search with high detection efficiency for any low ionizing secondary track produced over the whole solid angle in any primary interaction vertex and close to the vertex. This last point may be important if free quarks have a large interaction cross section with matter. Fig. 3 shows the upper limits obtained in such type of search by various experiments.

In the second type of experiments, one searches for exotic particles in a high intensity secondary beam produced at a high energy accelerator by interaction of the extracted primary proton beam with a target of low Z material (usually beryllium). The search is conducted using scintillation counters, threshold and differential Cherenkov counters, with pulse height and time-of-flight analyses. In one experiment was added a physical separation of different particles by means of superconducting radiofrequency separators. Fig. 4 shows a bidimensional plot of charge q versus $(m/q)^2$ for the particles in a beam produced at 0° in p-Be collisions at 210 GeV at the CERN-SPS [9].

A quark-gluon plasma may be formed in heavy nuclei collisions. It thus may be relevant to search for nuclear fragments with fractional charge produced in high energy nuclear collisions. Fig. 5 shows the charged fragments produced in silicon-copper collisions at 200 GeV/nucleon measured with the CR39 technique [10].

2.3. Recent searches at LEP. Many searches have been made recently at LEP by the four experiments ALEPH, DELPHI, L3 and OPAL. The searches concern the missing items of the third family (the t quark, the ν_τ lepton) and the Higgs particle, possible members of a fourth family, and items beyond the standard model, like supersymmetric particles, charged Higgs, Technicolor and others. Other searches concern free quarks and free gluons. All searches were made in e^+e^- collisions at c.m. energies from 88 to 93 GeV, around the Z resonance. For pair produced particles, all limits obtained are at the kinematic limit of 45 GeV.

The searches relevant to this paper concern free quarks, free gluons, the lightests supersymmetric particle, and stable singly charged or neutral particles.

No search on free quark has been published, but all detectors have this capability, as illustrated in Fig. 6 [11]. One may assume that the limit is at the kinematic limit for pair produced particles, with a cross section level $\sim 10^{-37} cm^2$.

Other speakers at this Workshop have reported on the searches for supersymmetric particles at LEP and the limits quoted [12].

Let me recall briefly a search for pair produced stable charged heavy (SCH) particles in Z^0 decays. The search concerned two-prong events, where both charged tracks have long flight times. Possible candidates for these events would be a fourth generation lepton lighter than its neutrino, mirror fermions, supersymmetric scalar leptons with mass smaller than the photino and the stable heavy hadrons hypothesized by Frampton and Glashow. The search performed by OPAL concerns the direct production of a pair of SCH leptons [13]. The trigger for such a search was based on the track trigger of the central detector and on the time of flight counters, requiring long flight times. Since no signal was found, the experiment excludes new charged heavy leptons in the mass range 18.5-42.8 GeV, see Fig. 7.

Many extensions of the Standard Model predict the existence of neutral heavy leptons. The existence of a stable new neutral heavy lepton would reflect in an increase of the invisible width of the Z^0. Thus one may immediatly place a limit from the measured Γ_{inv}. A more sensitive direct search yielded a nul result [14].

2.4 Free gluons. The existence of free gluons was suggested in a phenomenological model [15] as a possible explanation for the observation of cosmic ray muons from the direction of Cygnus-X3 ("cygnets"). In this model $SU(3)_c$ colour symmetry would be broken so as to produce free gluons and not free quarks. The free gluon would have a mass smaller than 100 MeV and its production would be regulated by a potential well of the type

$$V(r) = 3\sigma r e^{-\mu r} \tag{2}$$

with $V_{max} = 3\sigma/\mu e$ at a confining radius $R_m = 1/\mu$. σ is the string tension, about 1 GeV/fm. The free gluon would be stable if colour symmetry is violated locally.

OPAL has searched for free gluons in multihadronic events [16]. The free gluon would in general be isolated from the produced quark antiquark pair which fragments in two jets of hadrons. Assuming that a free gluon has an interaction rate typical of hadrons, say 25 mb, they would cross the central detector and interact in the electromagnetic and hadron calorimeters. In this hypothesis a free gluon would simulate a neutral hadron isolated from the charged hadrons. Two candidate events were selected by these selection criteria. A Monte Carlo analysis shows that this number of events is expected from ordinary hadrons. A simulation of free gluon production allows to estimate the barrier height $V_m = 47$ GeV, see Fig. 8

3. Searches in underground experiments

The searches for rare exotic particles in underground experiments require large detecting areas, good spatial resolution and redundancy of information. The MACRO experiment, located in hall B of the Gran Sasso Laboratory, will have three types of detectors [17]: three horizontal layers of liquid scintillators, 14 layers of limited streamer tubes and a track-etch detector, with 3 layers of CR39. The sides of the detector are closed by scintillators and streamer tubes. When completed MACRO will have a large area ($S \simeq 1000\ m^2$, $S\Omega > 10000\ m^2 sr$), good space resolution (achieved with the streamer tubes, $\delta_x \sim \delta_y \sim 1.2\ cm$), good timing and pulse height informations from the scintillators and eventually very good spatial resolution with the track etch detector. MACRO is presently running with the lower six supermodules.

3.1. Search for magnetic monopoles. Grand Unified Theories (GUT) of strong and electroweak interactions predict to existence of magnetic monopoles [3] with large mass ($\sim 10^{16} GeV$), with unit Dirac charge and also with 3 units [18]. These theories leave completely open the question of monopole abundance. The existence of the galactic magnetic field leads to an astrophysical constraint, the so called Parker bound, with an upper limit on the monopole flux at the level of $\sim 10^{-15} cm^{-2} s^{-1} sr^{-1}$ (the limit becomes $\sim 3 \times 10^{-16}$ for $n = 3$ poles).

A recent experiment used 2000 m^2 of CR39 nuclear track detector [19]. As already stated, a passing monopole would leave a latent track in the form of broken polymeric chemical bonds. This trail is more sensitive to chemical etching than the bulk of the material. By etching the plastic, visible etch cones are made on each surface of the sheet with a half angle $\Phi = \arcsin(v_B/v_T)$, where v_B is the bulk etch rate of the material and v_T is the etch rate along the track. If the sensivity $S = v_T/v_B$ is high enough, a heavy etching may make the sheet very thin and the two etch cones on both surfaces join making a hole. This allows fast scanning. If a candidate is found, it may be cross checked in a second sheet of CR39, etched normally. No signal was found.

The present status of the upper limits on cosmic magnetic monopoles is given in Fig. 9 [19,20].

3.2. Searches for nuclearites (strange quark matter). Nuclearites would be chuncks of nuclear matter made of approximately an equal number of u, d and s quarks. This form of matter could be more tighthy bound than ordinary nuclear matter, because of the Pauli exclusion principle: the energy per quark of a 3-flavour system (s,d,u) should be lower than for a 2-flavour system (u,d) because it is easier to fill low energy quantum states. The Pauli effect should overcome the effect due to the higher mass of the s-quark if $m_s - m_d < 250\ MeV$. Strange quark matter would be stable for any Z. Nuclearites would have been formed at the cosmic time $t \sim 10^{-6}\ s$ (temperature $T \simeq 200\ MeV$) when quarks formed protons and neutrons.

The energy losses of nuclearites in matter are due to blackbody radiation from the moving nuclearites; in size the energy loss would be similar to that of monopoles (the mechanisms of energy loss are quite different). Assuming that the nuclearites have galactic velocities ($\sim 250\ Km/s$), an underground detector may be reached from all sides by nuclearites with $m_N > 0.1\ g$, and from above by those with $3 \times 10^{-11} - 0.1\ g$. The present best limits are at the level of $6.4 \times 10^{-16} cm^{-2} s^{-1} sr^{-1}$ for the downward nuclearites [19,21,22]. In terms of flux versus mass of nuclearites the limits are shown in Fig. 10.

3.3. Search for underground muons from Cygnus-X3. In 1985 there were two reports of the possible detection with underground detectors of high energy muons coming from the direction of Cygnus-X3 and having the proper phase. (A possible explanation was the emission of neutral particles ("cygnets") with traveled in a straigh line and interacted in the earth atmosphere). The present situation is illustrated in Fig. 11 [23]: larger detectors located at underground laboratories of different rock cover and measuring at different times did not confirm the findings. Cygnus-X3 is a highly fluctuating source. It may thus still be possible that there were fluctuations and that one has to watch the source for longer periods of time.

3.4. Neutrinos from WIMPs annihilating in the Sun or in the Earth. The presence of dark matter (DM) in our galaxy is suggested by the rotation curves of stars in the galaxy halo (another interpretation could be a breakdown of Newton's law). The nature of the dark matter remains a mistery. In what follows we shall consider *cold heavy dark matter* in the form of Weakly Interacting Massive Particles (WIMPs), typically with masses of tens of GeV. These could be detected directly with detectors of the germanium type (high resolution, low background) through coherent scattering of the WIMPs with the nuclei of the detector. Other methods are indirect. We are here interested in WIMPs which could be trapped inside the Sun or the Earth, would annihilate and would give a flux of high energy neutrinos.

The Kamioka, Frejus, IMB, Nusex and MACRO detectors have looked for neutrinos coming from the Sun. No signal was detected. Together with LEP results and with the results of direct searches these experiments have essentially excluded the Dirac heavy neutrino as an important component of the dark matter. Majorana neutrinos are also excluded, al least for $m < 80$ GeV.

As far as high energy neutrinos from the Earth, the observed upper going muons may come from atmospheric neutrinos from the other side of the Earth or from WIMPs, which should have been captured by the Earth, thermalized and concentrated close to the Earth center. Eventually the density would be large enough so that annihilations of WIMPs and antiWIMPs would happen.

One interesting WIMP candidate is the supersymmetric neutral particle with the smallest mass which would be stable because of R-conservation. In the Minimal Supersymmetric Standard Model (MSSM) there are four neutralino mass eigenstates which can be expressed as linear combinations of superpartners of the gauge bosons and of the neutral Higgses:

$$\chi = c_1\tilde{\gamma} + c_2\tilde{z} + c_3\tilde{H}_1^\circ + c_4\,\tilde{H}_2^\circ \tag{3}$$

The stable neutralino is the lowest mass linear superposition of photino $\tilde{\gamma}$, zino $\tilde{z}$ and higgsinos $\tilde{H}$. For certain parameter combinations the neutralino may coincide with the photino ($\tilde{\gamma}$) or the higgsino ($\tilde{H}^\circ$).

There are three independent parameters in the MSSM; they may be taken to be the W_3 mass parameter M_2, the Higgs mixing parameter μ and the parameter $tg\,\beta = v_2/v_1$, where v_1, v_2 are the vacuum expectation values for the two Higgs.

Neutralino masses below 20-40 GeV are excluded by the LEP experiments. The interpretation of the searches for neutralinos from the Earth necessitates many hypotheses: the neutralino dark matter density near the Earth is taken as $\rho_z = 0.3\ GeV/cm^3$ and their velocity $v_z \sim 300\ km/s$. Capture of neutralinos by the Earth would be due to coherent interactions with nuclei and the capture rate is particularly high for neutralinos with masses close to those of the atomic nuclei in the Earth (30-80 GeV). Neutralinos and anti-neutralinos should then concentrate toward the center and annihilate into neutrinos. They would lead to a steady flux of high energy neutrinos coming from the center of the Earth in a relatively narrow cone of about 15°. The calculation of these processes requires a complicated Monte Carlo which includes evaluation of the contribution from atmospheric neutrinos.

Limits were obtained by Kamiokande and MACRO, see Figure 12 [24]. Until now no signal has been detected which can be attributed to neutralino. The results may be used to place constraints in the models parameters. It may be that the limits are already at such levels that in part of the parameter space the neutralino hypothesis cannot provide the total amount of invisible matter of the galactic halo.

4. Quark searches in stable matter

Several experiments performed searches for fractionally charged particles in stable matter.

In Millikan type experiments two measurements are performed: the first one is used to determine the radius and thus the mass of the droplets (of μm dimensions); the second measurement determines the charge by observing the limiting velocity in an electric field which makes the droplet move upward. Only few μg of material have been analyzed. In order to improve the limits one has to perform a sort of mass production Millikan type experiment, for instance with new aerosol generators [25].

In a different series of experiments a ball of diamagnetic or ferromagnetic material (of mm dimension) is levitated in a proper magnetic field configuration. The electric charge on the ball is measured via the behaviour of the ball in an alternated horizontal electric field. Several *mg* of material have been analyzed by different authors. Apart from one experiment all others gave negative results. Thus one may establish an upper limit of less than 1 quark in 10^{22} nucleons [26]. (Enrichment methods may yield better limits).

5. Conclusions

At present there are no indications of deviations from the predictions of the Standard Model of particle physics. But the motivations to search for the missing items of the Standard Model and for new physics beyond the Standard Model remain completely valid.

Acknowledgements. I would like to acknowledge many colleagues for their cooperation and for discussions. I thank Mrs. B. Simoni for typing and editing this manuscript.

References

[1] G. Giacomelli (Searches for missing particles) Invited paper at the Int. Meeting on Frontiers of Physics, Singapore (1978); IFUB 78/16 and Proceedings of the Meeting.

[2] G. Giacomelli (Post Collider Physics) in "Techniques and Concepts of High Energy Physics", St. Croix (1988) and DFUB 88/22.

[3] G. Giacomelli (Magnetic Monopoles) *Rivista Nuovo Cimento* 7 (1984) 1.
N. Craigie et al. (Theory and detection of magnetic monopoles in gauge theory) World Scientific (1986).

[4] M. Bertani et al. (Search for magnetic monopoles at the Tevatron Collider) DFUB 90/3, *Europhys. Lett.* 12 (1990) 613. See this paper for other references on monopole searches at accelerators.

[5] G. Giacomelli (Neutrino astrophysics in deep underground experiments) Invited paper at the 4th Int. Symp. on *"Neutrino Telescopes"*, Venezia (1992).

[6] J. M. Barkov et el., (Search for Dirac Monopoles in pN collisions at 400 GeV/c) CERN/EP 83-194 (1983).

[7] K. Kinoshita et al., (Search for highly ionizing particles in e^+e^- annihilations at $\sqrt{s} = 91.1$ GeV) *Phys. Lett.*, to be published (1992). See this paper for further references for monopole searches at e^+e^- colliders.

[8] D. Allasia et al (Search for fractionally charged particles in (anti)neutrino-deuterium interactions) *Phys. Rev.* D37 (1988) 219.

[9] A. Bussière et al. (Particle production and search for long lived particles in 200-240 GeV/c p-N collisions) *Nucl. Phys.* B174 (1980) 1.

[10] S. P. Ahlen at al., (Improvements in the CR39 polymer for the MACRO experiment) XV *Int. Conf. Part. Tracks in Solids*, Marburg, Germany (1990).

[11] G. Giacomelli (Results from the OPAL experiment at LEP) DFUB 90/07 (1990).
K. Ahmet et al., (The OPAL detector al LEP) *Nucl. Instr. Meth.* A305 (1991) 275.

[12] A. Lopez-Fernandez et al. (Searches for SUSY particles al LEP/SLC and LEP 200), *This Workshop.*

[13] M.Z. Akrawy et al. (Searches for pair produced stable singly charged heavy particles in Z^0 decays) *Phys. Lett.* B 252 (1990) 290.

[14] M.Z. Akrawy et al.(Limits on neutral heavy lepton production from Z^0 decay) CERN-EP/90-72, *Phys. Lett.* B247 (1990) 448.

[15] R. L. Rinfret and P.S. Watson, *Phys. Lett.* B20 (1988) 177.

[16] P. D. Acton et al., (Search for free gluons in hadronic Z^0 decays) *Phys. Lett.* B278 (1992) 485.

[17] M. Calicchio et al. (The MACRO detector at the Gran Sasso Laboratory) *Nucl. Instr. Meth.* A264 (1988) 18.

[18] G. Giacomelli et al., (Detection of multiply charged magnetic monopoles from superstring theories) MACRO Int. Memo 3/87 (1987).

[19] S. Orito et al. (Search for supermassive relics with a 2000 m^2 array of plastic track detectors) *Phys. Rev. Lett.* 66(1991) 1951.

[20] M. Calicchio et al. (Search for magnetic monopoles with the MACRO detector) Contributed paper at the *1990 Int. Cosmic Ray Conf.*, Adelaide (1990).

[21] S. Nakamura et al. (A new limit on the flux of strange matter) *Phys. Lett.* B263 (1991) 529.

[22] S. P. Ahlen et al. (Cosmic ray search for strange quark matter with the MACRO detector) *Nucl. Phys. B* (Proc. Suppl.) 24B (1991) 191.

[23] S. P. Ahlen et al. (A search for sidereal anisotropies with underground muons and an all-sky survey for muon point sources) LNGS-91/08 (1991).

[24] A. Bottino et al. (Direct versus indirect searches for neutralino dark matter) DFTT 35/91 (1991)

[25] M. P. Morigi (Proposta di un esperimento alla Millikan in continua per la ricerca di cariche frazionarie) Tesi di Dottorato, University of Bologna (1991).

[26] Review of particle properties, *Phys. Lett.* B239 (1990) 1.

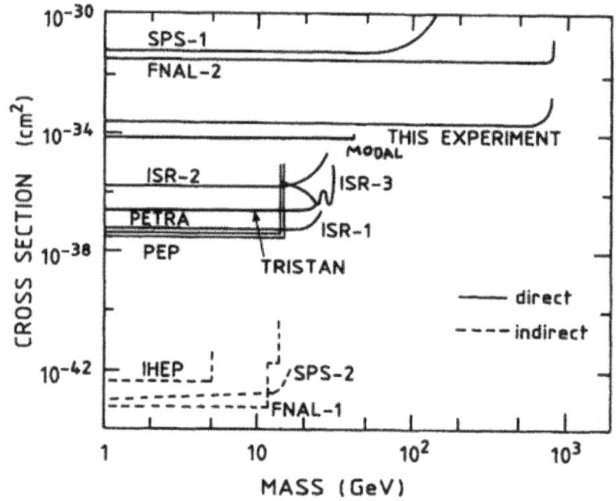

Fig. 1. Compilation of upper limits for Dirac magnetic monopole production in pN, $\bar{p}p$ and e^+e^- high energy collisions [4,7]. Solid and dashed lines refer to direct and indirect experiments.

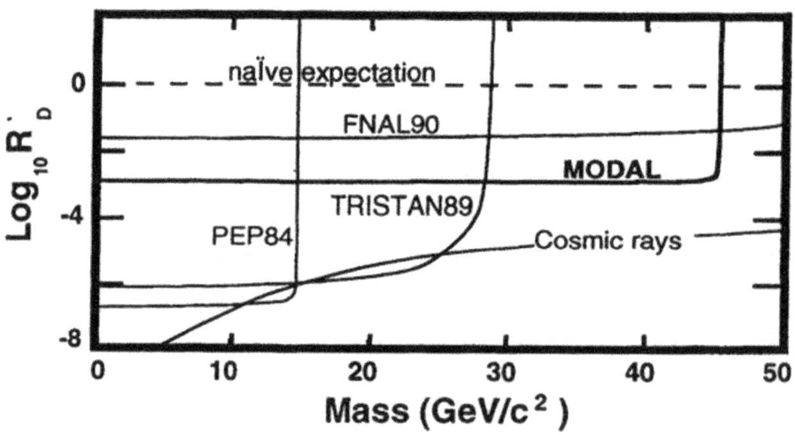

Fig. 2 Upper limits at 95% CL on $R_D = \sigma_{lim}(m)/\sigma_D(m)$ for isotropic exclusive production of monopole pairs with charge g_D [7]. The LEP and TRISTAN values include a phase space correction.

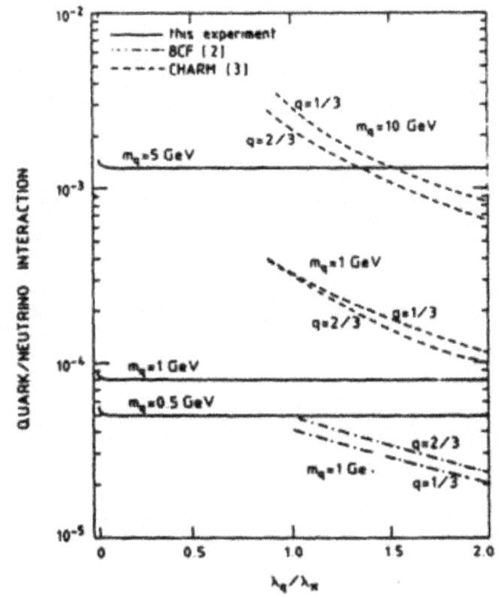

Fig. 3. Upper limits, at the 90% c.l., for the production of fractionally charged particles in (anti)neutrino-nucleon collisions as function of their mass and of their interaction length (solid lines are valid for both charge 1/3 and 2/3) [8].

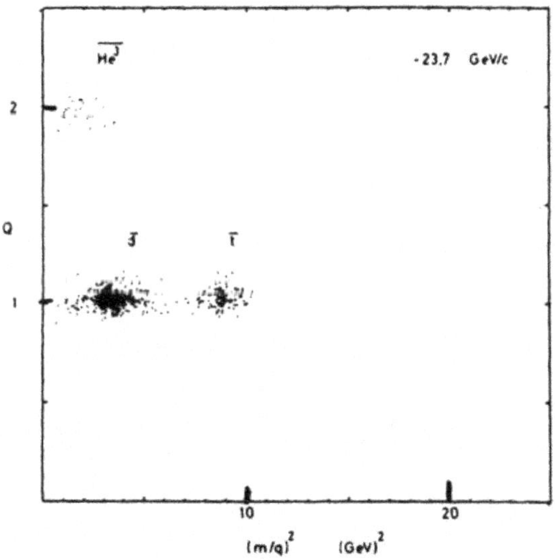

Fig. 4. Plot of the charge q versus $(m/q)^2$ for negative particles produced in p-Be collisions at the CERN-SPS [9]. One observes $\bar{d}$, $\bar{t}$ and $\bar{He}^3$ antinuclei; no other particle with charge -1/3, -2/3, -1, -4/3, -2 is observed.

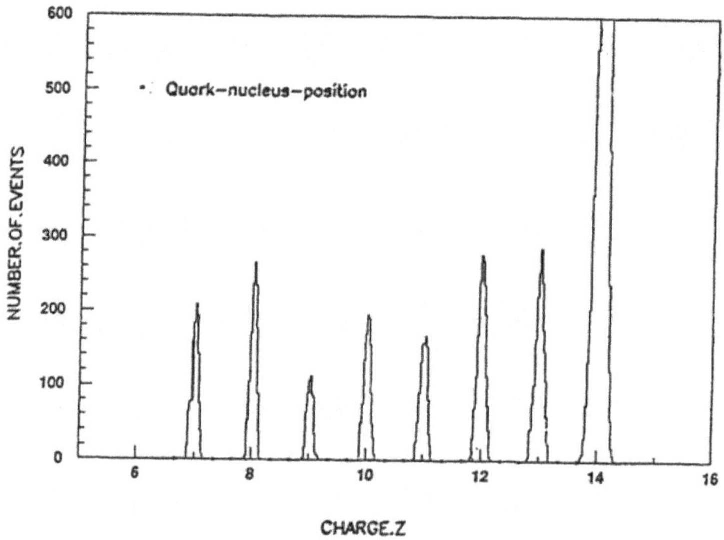

Fig. 5. Nuclear fragments produced in relativistic silicon-cupper collision detected in CR39 plastic [10]. The diameters of the etch pits left in CR39 were measured in 10 consecutives surfaces. No ion with fractional charge is observed.

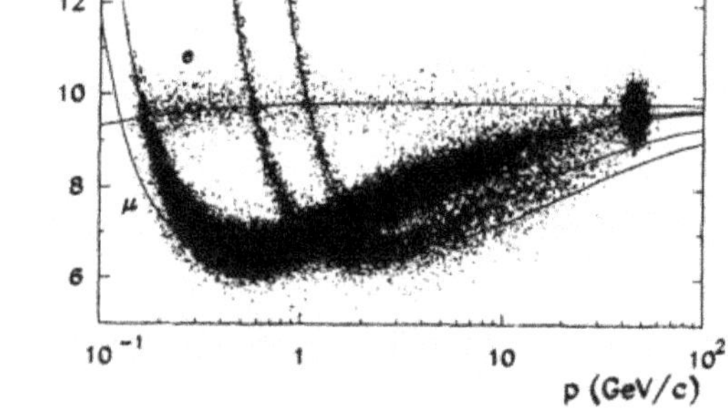

Fig. 6. Energy loss versus momentum for charged particles traversing the OPAL Central Detector [11]. Note the well defined bands due to the well known particles.

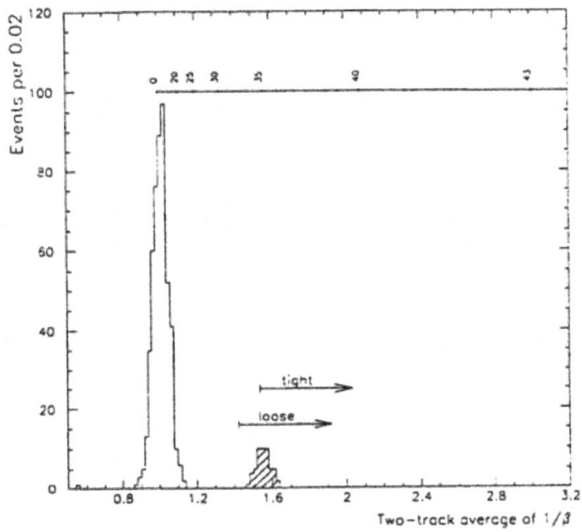

Fig. 7. Two-track average of $1/\beta$ for candidate events (open histogram), and for Monte Carlo events of 35 GeV stable singly charged heavy particles generated at $\sqrt{s} = 91.2$ GeV (hatched histogram) [13]. The MC normalisation is arbitrary. The loose and tight search zones are illustraded for a mass hypothesis of 35 GeV and $\sqrt{s} = 91.2$ GeV. Also shown is the mass (in GeV) corresponding to each value of $1/\beta$ at $\sqrt{s} = 91.2$ GeV.

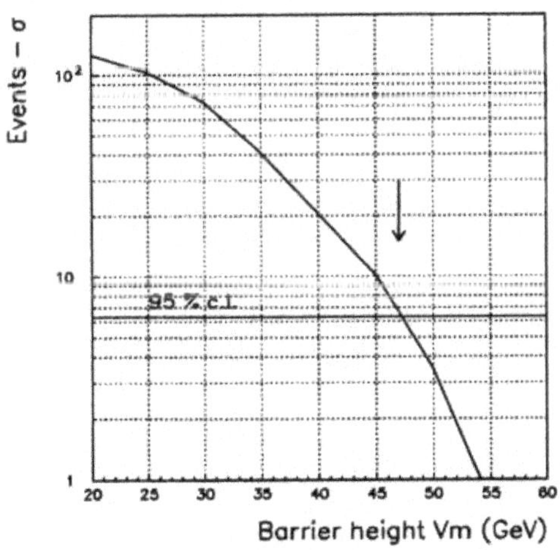

Fig. 8. The number of expected free gluon events subtracted by the systematic error on the selection efficiency as a function of the barrier height V_m [16]. The horizontal line corresponds to the 95% confidence level upper limit on the two observed (background) events. The arrow indicates the lower limit at 95% CL of 47 GeV on V_m.

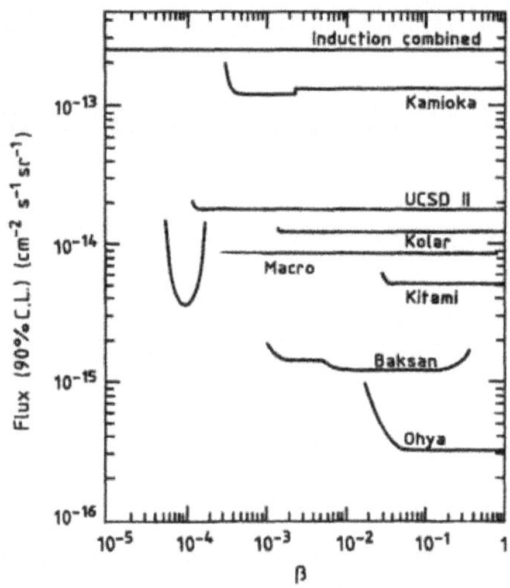

Fig. 9. Present upper limits on the flux of cosmic GUT magnetic monopoles of unit Dirac charge plotted versus monopole velocity $\beta = v/c$ for different types of detectors.

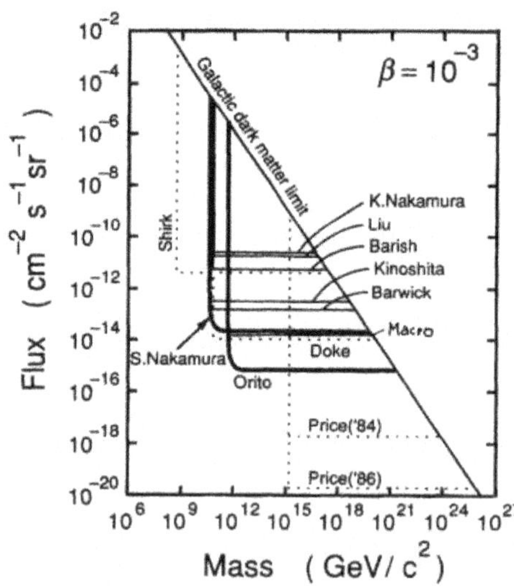

Fig. 10. The flux limit (90% confidence level) on strange matter with $\beta = 10^{-3}$ as a function of mass, assuming an isotropic flux [19–21].

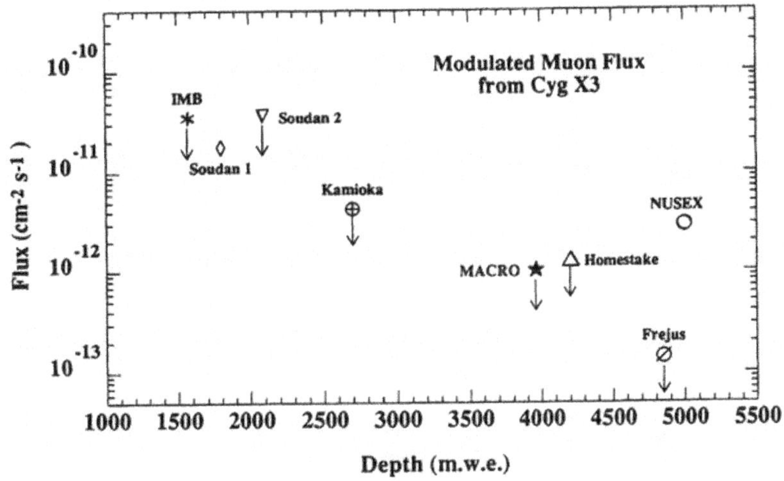

Fig. 11. The present limits and observations of the modulated muon flux from Cygnus X3 plotted versus depth (in meters of water equivalent) of underground detectors [23].

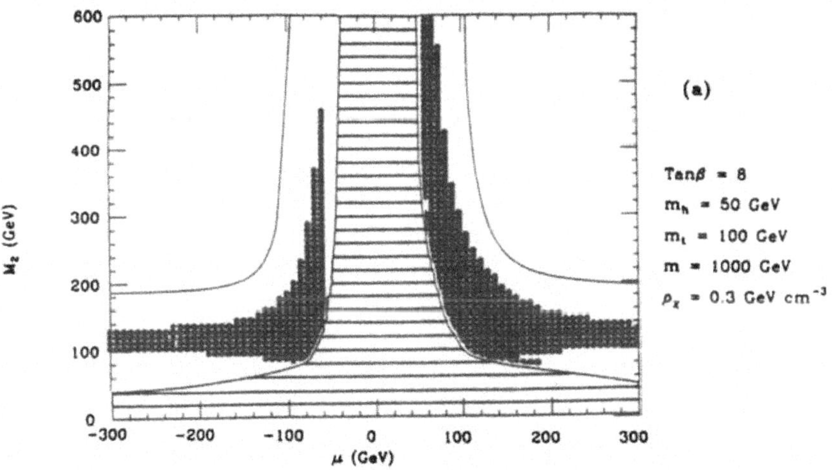

Fig. 12. Limits on the parameters M_2 and μ from direct and indirect neutralino searches for $\tan\beta = 8$, $m_h = 50$ GeV and radiative corrections with $m_t = 100$ GeV and $m_H = 1000$ GeV; Higgs couplings $g^h = 39$ MeV, $g^s = 354$ MeV [24]. Hatched regions are excluded by LEP, regions with stars are excluded by Kamiokande if $\rho_x = 0.3$ GeV cm^{-3}.

SEARCHES FOR NEW PARTICLES AT A TAU-CHARM FACTORY*

MARTIN L. PERL
Stanford Linear Accelerator Center
Stanford University, Stanford, California 94309

ABSTRACT

This paper calls attention to four topics in tau lepton physics which are outside our present areas of tau physics research: $\tau^+\tau^-$ atoms, τ^- nucleus atoms, photoproduction of τ's, and heavy production of τ's.

Table of Contents

* This work was supported by the Department of Energy, contract DE–AC03–76SF00515.

A. Introduction

The tau-charm factory consists of a high luminosity, two-ring, electron-positron collider and a detector with the following properties:

- Range of total energy = 3.0 to 5.0 GeV
- Design luminosity $\geq 10^{33}$ cm^{-2} s^{-1}
- High resolution, large acceptance detector specially designed for tau and charm physics

There is now an extensive literature on the design and physics potential of a tau-charm factory. The original descriptions are by Kirkby (1987, 1989) and Jowett (1987, 1988, 1989). Two international workshops have been devoted to the tau-charm factory Beers (1989), Kirkby and Quesada (1992). Studies of the design of the collider have been done by Brown *et al.* (1989), Gonichon *et al.* (1990), Barish *et al.* (1990), Danilov *et al.* (1990) and Baconnier *et al.* (1990). Papers on the physics and on detector designs include Schindler (1989, 1990a, 1990b), Perl (1991), Pich (1991), Wermes (1992), and Davier (1991). A recent review of the concept and potential physics of a tau-charm factory has been written by Kirkby and Rubio (1992).

In research at a tau-charm factory the experimenters will be able to carry out precise and probing studies of tau and charm physics through the use of the four basic elements of the tau-charm factory concept:

- Control of and direct measurement of backgrounds and contaminations.
- Very large statistics.
- Production of particles in known quantum mechanical states.
- Detector with very high quality particle identification, very good momentum and energy resolution, and close to 4π acceptance.

In experiments at the tau-charm factory the particles are produced at resonances or at energies just above production thresholds, Fig. 1.:

$$\text{Tau Pair}: \quad e^+ + e^- \to \tau^+ + \tau^- \tag{1a}$$

$$\begin{aligned} \text{Charmed meson}: \quad & e^+ + e^- \to D^+ + D^-, \quad \text{at } \Psi'' \\ & e^+ + e^- \to D^0 + \bar{D}^0, \quad \text{at } \Psi'' \\ & e^+ + e^- \to D^0 + \bar{D}^{*0} \\ & e^+ + e^- \to D_s^+ + D_s^- \end{aligned} \tag{1b}$$

$$\text{Charmed baryon}: \quad e^+ + e^- \to \Lambda_c + \bar{\Lambda}_c \tag{1c}$$

$$\begin{aligned} \text{Charmonium}: \quad & e^+ + e^- \to J/\Psi \\ & e^+ + e^- \to \Psi' \end{aligned} \tag{1d}$$

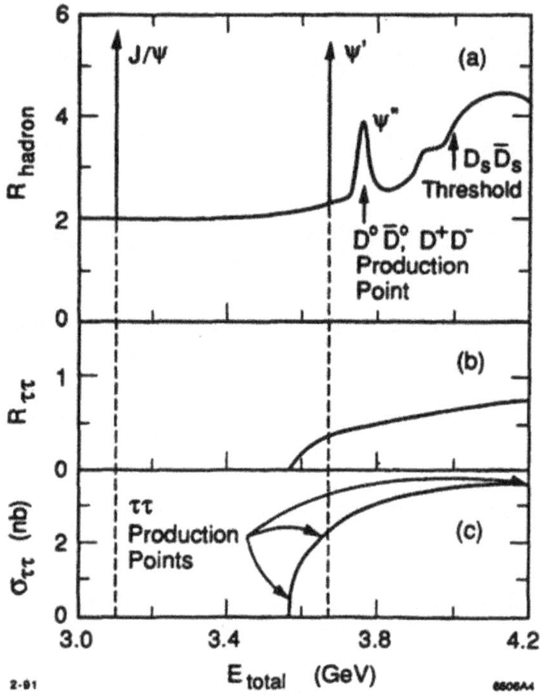

Figure 1.

For example, the primary operating energies for τ pair production are:

$$\begin{array}{ll} 3.57 \text{ GeV:} & \text{just above the } \tau \text{ threshold} \\ 3.67 \text{ GeV:} & \text{just below the } \Psi' \text{ threshold} \end{array} \tag{2}$$

At these energies there is no production of charm or bottom hadrons, the only background is from ordinary hadrons, and that background is almost constant between these operating energies and the energy just below τ pair threshold. Gomez-Cadenas *et al.* (1989) have discussed the value of studying some areas of τ physics just above the τ pair threshold where the produced τ's are almost at rest.

Table 1 gives the rates for particle production at the tau-charm factory.

Table 1. Particle production rates at the tau-charm factory year, based on 10 fb^{-1} per year*.

Particle	Events per year
D^0 (single)	5.8×10^7 at Ψ''
D^+ (single)	4.2×10^7 at Ψ''
D_s (single)	1.8×10^7 at 4.24 GeV
$\tau^+\tau^-$ (pairs)	0.5×10^7 at 3.57 GeV
$\tau^+\tau^-$ (pairs)	2.4×10^7 at 3.67 GeV
$\tau^+\tau^-$ (pairs)	3.5×10^7 at 4.25 GeV
J/Ψ (events)	1.7×10^{10}
Ψ' (events)	0.4×10^{10}

*Based on L = 10^{33} cm^{-2} s^{-1} for 10^7 s/year. As machine operation matures, the collider will operate for 2×10^7 s/year.

There are two basic principles in the new particle searches to be made at a tau-charm factory:

- The total energy and the particle masses are relatively small, therefore only relatively small mass new particles can be produced and directly detected.
- Precise studies of tau lepton and charm particle decays, or detection of unconventional decays, can lead to the indirect discovery of new particles. The new particle might occur as a virtual particle in an intermediate state in the decay process. Or, an unexplained property of a decay mode or set of decay modes might reveal the presence of a new particle coupled to the known particles in the decay.

B. Searches in Tau Physics

There are four powerful advantages in studying the physics of the tau lepton and tau neutrino at the tau-charm factory:

- The availability of large data sets collected over short times.
- The selection of a τ pair data sample by single-tagging of events. That is, only one τ decay need be identified in order to select the event.
- There are no backgrounds from D or B meson decays.

- The backgrounds from non-τ pair events are directly measured by moving below the τ pair threshold.

B.1 Unconventional τ Decays to Known Particles

Unconventional τ decays to known particles in general require violation of both τ lepton number conservation and e or μ lepton number conservation. Examples are

$$\begin{aligned} \tau^- &\rightarrow e^- + \gamma \\ \tau^- &\rightarrow \mu^- + \gamma \\ \tau^- &\rightarrow e^- + \pi^0 \\ \tau^- &\rightarrow e^- + e^+ + e^- \end{aligned} \tag{3}$$

If a proton or neutron is allowed in the decay then only τ lepton number conservation needs to be violated:

$$\tau^- \rightarrow \bar{p} + \pi^0 \tag{4}$$

Figure 2 shows two possibilities for how an unknown intermediate particle might lead to the type of unconventional τ decays in Eq. 3.

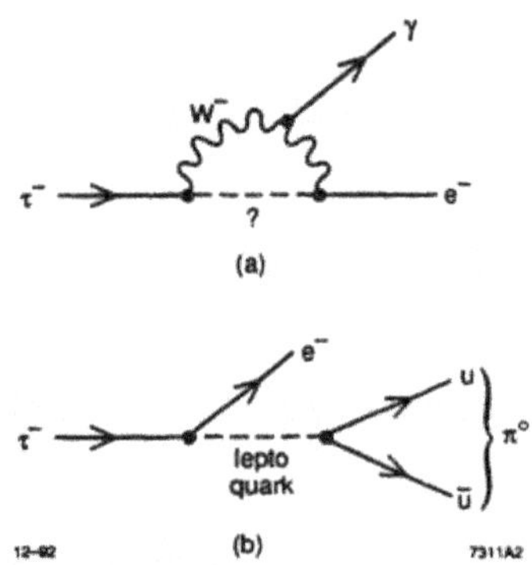

Figure 2.

It is relatively easy to look for such unconventional decays because all the particles in the final state can be detected and hence the mass of the τ can be reconstructed. If an experiment has n produced τ pairs whose unconventional decays can be detected, and if no unconventional decays are detected, then upper limits on the branching ratios of roughly

$$B\ (\text{unconventional decay}) \lesssim \text{ few/n} \tag{5}$$

will be set. Thus Albrecht *et al.* (1992) using about 10^5 τ pairs have set the upper limits in Table 2. The CLEO II experimenters with about 10^6 τ pairs have recently set the limit (Barish, 1992)

$$B(\tau^- \to e^-\gamma) < 4.2 \times 10^{-6} \tag{6}$$

Table 2. Upper limits on unconventional branching fractions from Albrecht *et al.* (1992).

		Upper Limits $[10^{-5}]$ (90% CL)				
Nr.	decay channel	MARK II	ARGUS 86	Crystal Ball	CLEO	ARGUS 91
1.	$\tau^- \to e^-e^+e^-$	40	3.8		2.7	1.3
2.	$\tau^- \to e^-\mu^+\mu^-$	33	3.3		2.7	1.9
3.	$\tau^- \to e^+\mu^-\mu^-$				1.6	1.8
4.	$\tau^- \to \mu^-e^+e^-$	44	3.3		2.7	1.4
5.	$\tau^- \to \mu^+e^-e^-$				1.6	1.4
6.	$\tau^- \to \mu^-\mu^+\mu^-$	49	2.9		1.7	1.9
7.	$\tau^- \to e^-\pi^+\pi^-$		4.2		6.0	2.7
8.	$\tau^- \to e^+\pi^-\pi^-$				1.7	1.8
9.	$\tau^- \to \mu^-\pi^+\pi^-$		4.0		3.9	3.6
10.	$\tau^- \to \mu^+\pi^-\pi^-$				3.9	6.3
11.	$\tau^- \to e^-\rho^0$	37	3.9			1.9
12.	$\tau^- \to \mu^-\rho^0$	44	3.8			2.9
13.	$\tau^- \to e^-\pi^+K^-$		4.2		5.8	2.9
14.	$\tau^- \to e^+\pi^-K^-$				4.9	2.0
15.	$\tau^- \to \mu^-\pi^+K^-$		12		7.7	11
16.	$\tau^- \to \mu^+\pi^-K^-$				4.0	5.8
17.	$\tau^- \to e^-K^{*0}$	130	5.4			3.8
18.	$\tau^- \to \mu^-K^{*0}$	100	5.9			4.5
19.	$\tau^- \to e^-\gamma$	64		20		12
20.	$\tau^- \to e^-\pi^0$	210		14		17
21.	$\tau^- \to \mu^-\gamma$	55				3.4
22.	$\tau^- \to \mu^-\pi^0$	82				4.4
23.	$\tau^- \to e^-\eta$			24		6.3
24.	$\tau^- \to \mu^-\eta$					7.3
25.	$\tau^- \to \bar{p}\gamma$					29.0
26.	$\tau^- \to \bar{p}\pi^0$					65.5
27.	$\tau^- \to \pi^-\gamma$					28
28.	$\tau^- \to \pi^-\pi^0$					37
29.	$\tau^- \to \bar{p}\eta$					129

The tau-charm factory experiment will have about 10^8 τ pairs and thus can look for unconventional decays of the types in Eqs. 3 and 4 with branching ratios as small as few $\times 10^{-8}$ The use of a tau-charm factory is necessary not only to achieve sample sizes of 10^8 τ pairs but also to carry out the precise measurements which are necessary to eliminate backgrounds. For example, the unconventional decays

$$\begin{aligned} \tau^- &\to e^- + e^+ + e^- \\ \tau^- &\to \mu^- + e^+ + e^- \end{aligned} \tag{7a}$$

can be faked by the conventional decays

$$\begin{aligned} \tau^- &\to e^- + \bar{\nu}_e + \nu_\tau + e^+ + e^- \\ \tau^- &\to \mu^- + \bar{\nu}_\mu + \nu_\tau + e^+ + e^- \end{aligned} \tag{7b}$$

if the total momentum of the neutrinos is close to zero.

Masiero (1991) has reviewed theories which can lead to the unconventional decays discussed in this section.

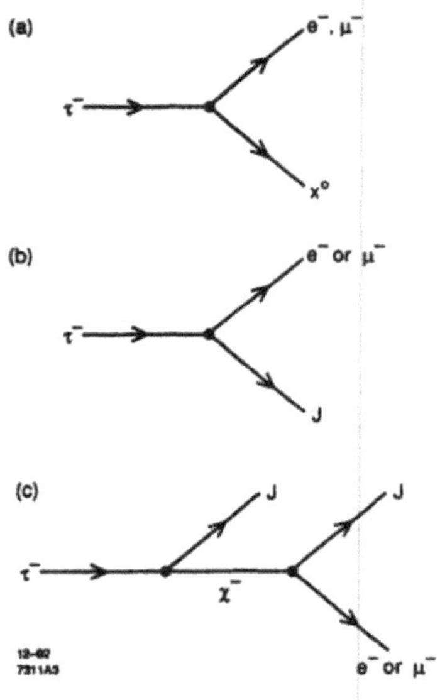

Figure 3.

B.2 Unconventional τ Decays to Unknown Particles

It is much more difficult to look for unconventional τ decays with an unknown, *weakly interacting* particle, x^0, in the final state. Examples are:

$$\begin{aligned}\tau^- &\to e^- + x^0 \\ \tau^- &\to \mu^- + x^0 \\ \tau^- &\to e^- + \text{hadrons} + x^0\end{aligned} \tag{8}$$

as shown in Fig. 3a. Since the x^0 has only weak interactions it will not be detected.

The only search so far made for such unconventional decays was by Baltrusaitis *et al.* (1985) looking for a light Goldstone boson in

$$\begin{aligned}\tau^- &\to e^- + G \\ \tau^- &\to \mu^- + G\end{aligned} \tag{9}$$

They set the 95% confidence level upper limits

$$\begin{aligned}B(\tau \to e\,G)/B(\tau \to e\nu\nu) &< 4.0\% \\ B(\tau \to \mu G)/B(\tau \to \mu\nu\nu) &< 12.0\%\end{aligned} \tag{10}$$

There are two difficulties in finding modes such as those in Eq. 8, or even in setting small upper limits on the branching fractions. First, the τ mass cannot be reconstructed because the x^0 cannot be detected. Second, there are severe background problems. For example, the backgrounds to the decay

$$\tau^- \to e^- + x^0 \tag{11a}$$

are

$$\tau^- \to \pi^- + \nu_\tau \tag{11b}$$

with the π misidentified as an e, and

$$\tau^- \to e^- + \bar{\nu}_e + \nu_\tau \tag{11c}$$

Indeed the decay in Eq. 11c always provides some background because the two neutrinos can always be reinterpreted as a particle of mass

$$m = \left[(E_\tau - E_e)^2 - (\vec{p}_\tau - \vec{p}_e)^2\right]^{\frac{1}{2}} \tag{12}$$

In a tau-charm factory experiment the uncertainty in m can be reduced by using data acquired near threshold, hence the significance of this background can be reduced.

Romão *et al.* (1991) have discussed a theoretical example of a τ decay of the type in Eq. 8.

$$\tau^- \rightarrow e^- \text{ or } \mu^- + J \tag{13}$$

where J is a majoron. They also discuss

$$\tau^- \rightarrow e^- \text{ or } \mu^- + J + J \tag{14}$$

B.3 Searches Using Precise Measurements of Calculable τ Decays

The decay widths, Γ_i, for five decay modes can be calculated from weak interaction theory and non$-\tau$ measurement (Perl 1992a) as outlined here:

$$\left.\begin{array}{l} \tau^- \rightarrow \nu_\tau + e^- + \bar{\nu}_e \longleftarrow \\ \tau^- \rightarrow \nu_\tau + \mu^- + \bar{\nu}_\mu \longleftarrow \end{array}\right[\begin{array}{l} \text{use weak interaction} \\ \text{theory} \end{array} \tag{15a}$$

$$\left.\begin{array}{l} \tau^- \rightarrow \nu_\tau + \pi^- \longleftarrow \\ \tau^- \rightarrow \nu_\tau + K^- \longleftarrow \end{array}\right[\begin{array}{l} \text{use weak interaction theory} \\ \text{theory and} \\ \pi^- \rightarrow \mu^- \bar{\nu}_\mu,\ K^- \rightarrow \mu^- \bar{\nu}_\mu \text{ widths} \end{array} \tag{15b}$$

$$\tau^- \rightarrow \nu_\tau + \rho^- \longleftarrow \left[\begin{array}{l} \text{use weak interaction} \\ \text{theory and} \\ e^+e^- \rightarrow \rho^0 \text{ cross section} \end{array}\right. \tag{15c}$$

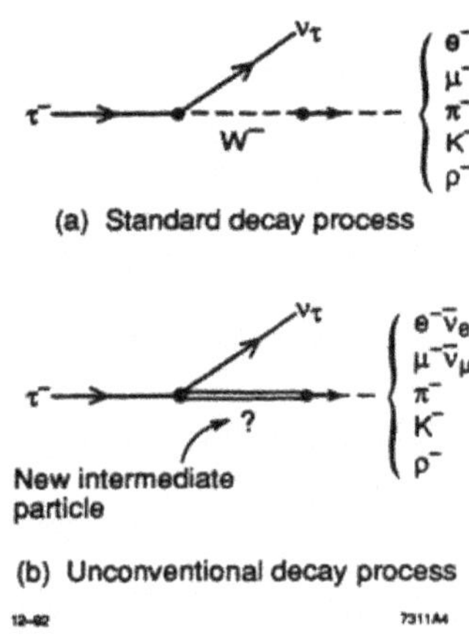

Figure 4.

As shown schematically in Fig. 4b, an unconventional decay process could occur through an unknown intermediate particle, and this would interfere with the conventional decay process in Fig. 4a. This could change the decay widths, or in the case of three body decay modes, the kinematic distributions could be altered.

Table 3 gives recent average values of the five decay modes in Eq. 15. For the first four modes the fraction errors in the branching ratios are in the range

$$1\% \lesssim \frac{\sigma_B}{B} \lesssim 3\% \tag{16}$$

(The $\tau^- \to K^-\nu_\tau$ mode has a much larger fractional error in B.) But the errors in Table 3 and in Eq. 16 may be underestimated. The errors depend upon the conventional but murky process of adding statistical and systematic errors in quadrature (Hayes and Perl 1988). In addition, the average values in Table 3 are obtained by averaging over many measurements, a calculation which has other uncertainties.

Table 3. Branching fractions for some decay modes whose decay widths can be calculated from weak interaction theory and non$-\tau$ measurements. The first four branching fractions are taken from or deduced from Galik (1992), the fifth is from Aguilar-Benitez *et al.* (1992).

Mode	Branching Fraction in %
$\nu_\tau e^- \bar{\nu}_e$	17.75 ± 0.15
$\nu_\tau \mu^- \bar{\nu}_\mu$	17.39 ± 0.17
$\nu_\tau \pi^-$	11.73 ± 0.35
$\nu_\tau \rho^-$	24.29 ± 0.24
$\nu_\tau k^-$	0.67 ± 0.23

But even if we take these errors as given, the comparison of branching fraction measurements with the calculation outlined in Eq. 15 has yet a larger error. We must compare the ratios

$$B_i/B_j = \Gamma_i/\Gamma_j \tag{17}$$

For example, weak interaction theory predicts (Perl 1992a)

$$(B_\mu/B_e)_{theory} = 0.973 \tag{18a}$$

and from Table 3

$$(B_\mu/B_e)_{measured} = 0.980 \pm 0.013 \tag{18b}$$

Thus the measured ratio has a fractional error of about 1.3%, but from the discussion in the last paragraph we might prefer assigning a 2% or 3% error.

Substantial reduction of these errors to say 0.1%, including more certainty about the systematic error, requires a single experiment with very high statistics and direct knowledge of backgrounds. This can only be achieved at a tau-charm factory.

B.4 Comparison of τ Lifetime with Leptonic Decays

From the first years of τ research in the 1970's it has been recognized that the τ lifetime, T_τ, is predicted by (Perl 1992a)

$$T_\tau = \frac{\hbar B(\tau^- \to \nu_\tau \ell^- \bar{\nu}_\ell)}{\Gamma(\tau^- \to \nu_\tau \ell^- \bar{\nu}_\ell)} \quad ; \ell = e \text{ or} \tag{19a}$$

$$\Gamma(\tau^- \to \nu_\tau \ell^- \bar{\nu}_\ell) = \frac{G_F^2 m_\tau^5}{192\pi^3} F_\ell(y) F_W F_{rad} \tag{19b}$$

$$F_\ell(y) = 1 - 8y + 8y^3 - y^4 - 12y^2 lny \tag{19c}$$

where

$$y = m_\ell^2 / m_\tau^2 \quad ;$$

and

$$F_W = 1.0003$$

is the correction for finite W mass and

$$F_{rad} = 0.9957$$

is the radiative correction.

Also, since the first measurements of T_τ seemed to be larger than the T_τ predicted by Eqs. 19 and measured values of B_e and B_μ, there has been speculation that (a) a second neutrino, N_τ, couples to the τ and (b) the mass of N_τ is larger than m_τ

$$m_{N_\tau} > m_\tau \tag{20}$$

Then the $\tau - W - \nu_\tau$ coupling would be given by

$$g_{\tau W \nu_\tau} = \cos\theta_\tau g_{eW\nu_e} \tag{21}$$

and the τ lifetime would be increased by $(\cos\theta_\tau)^{-2}$. A recent discussion has been given by Ma *et al.* (1992).

Thus if T_τ (measured) were larger than T_τ (calculated) from Eqs. 19 and measured values of B_e and B_μ, there would be evidence for new physics. That new physics might consist of a new particle N_τ. Or, the new physics might be that $e - \mu - \tau$ universality is violated, the $\tau\, W\, \nu_\tau$ coupling being different from the $e\, W\, \nu_e$ and $\mu\, W\, \nu_\mu$ couplings.

At present the experimental situation is as follows. From B_e and B_μ in Table 3 we predict

$$T_\tau \text{ (calculated from } B_e) = 288.8 \pm 2.4 \text{ fs} \tag{22a}$$
$$T_\tau \text{ (calculated from } B_\mu) = 290.9 \pm 2.8 \text{ fs} \tag{22b}$$

These calculations are not independent because some measured values of B_e and B_μ are correlated. Here I use the evaluation of Trischuk (1992) for

$$T_T(\text{measured}) = 295.7 \pm 3.2 \text{ fs} \tag{23}$$

Comparing Eq. 22a with Eq. 23, T_τ (measured) agrees with T_τ (calculated) to within 1.7 standard deviations with the former longer. Using Eq. 22b gives a similar result. At present we cannot claim that there is a discrepancy which requires new physics for an explanation. As I said in Perl, (1992a) it will take more precise measurements if B_e, B_μ and T_τ to decide whether or not there is a disengaging. The tau-charm factory will permit much more precise measurements of B_e and B_μ. The most difficult problem for future comparisons is how to substantially improve the precision of T_τ.

B.5 Summation Issues in τ Decays

Since the work of Gilman and Rhie (1985) and Truong (1984) the world of τ research has been faced with the question: Can we find and identify all the decay modes of the τ with branching fractions

$$B_i \gtrsim \text{ few } \times 0.1\% \tag{24a}$$

such that

$$\sum_i B_i = 100\% \quad ? \tag{24b}$$

On the face of it Eq. 24b is an identify; the fundamental question is: Are there some unknown and unconventional τ decays such that

$$\sum_i B_i \text{ (known and measured)} < 100\% \text{ ?} \tag{24c}$$

Historically (Perl 1992a) the question was first asked about decay modes with 1-charged particle, B_{1i}, since these made up most τ decays. The topological 1 and 3-charged particle branching fractions according to the Particle Data Group

(Aguilar-Benitez *et al.* 1992) are

$$\begin{aligned} B_1 &= (85.52 \pm 0.25)\% \\ B_3 &= (14.06 \pm 0.25)\% \quad ; \end{aligned} \tag{25a}$$

and in a more recent computation by Galik (1992)

$$\begin{aligned} B_1 &= (85.26 \pm 0.18)\% \\ B_3 &= (14.63 \pm 0.18)\% \quad ; \end{aligned} \tag{25b}$$

We usually break up the question in Eq. 24c into two questions. Does

$$\sum_i B_{1i} \text{ (known and measured) } = B_1 \tag{26a}$$

and does

$$\sum_i B_{3i} \text{ (known and measured) } = B_3 \tag{26b}$$

Some kinds of searches for new particles in τ physics are intimately related to these questions. If a τ decay mode contains a new particle, that mode might not be found. Or even if a τ decay mode contains known particles, it might be unconventional and not recognized. I'll give an example of such speculations later in this section, first I turn to the data.

In the past few years most of the new measurements on τ branching fractions have come from experiments at the CESR and DORIS II e^+e^- colliders with $E_{tot} \sim$ 10 GeV and from the four experiments at the LEP collider with $E_{tot} \sim 92$ GeV. In general, these experiments cannot distinguish $\pi^\pm$ from $K^\pm$, and hence cannot distinguish

$$\tau^- \to \nu_\tau + \pi^- + n\pi^0 \ , \ n \geq 0 \tag{27a}$$

from

$$\tau^- \to \nu_\tau + K^- + n\pi^0 \ , \ n \geq 0 \tag{27b}$$

Therefore, the two decay modes are measured as one with the designation.

$$\tau^- \to \nu_\tau + h^- + n\pi^0 \ , \ n \geq 0 \tag{27c}$$

A similar problem occurs with 3-charged particle. For approximate measurements the adding of the π and K modes is not important since

$$\frac{B(\tau^- \to \nu_\tau K^- n\pi^0)}{B\tau^- \to \nu_\tau \pi^- n\pi^0)} \sim 5\% \tag{28}$$

But the separation of the π and K modes is important for precise measurements of the separate modes, for understanding the physics of the K decay modes, and for the precise exploration of the summation questions in Eqs. 26.

Precise separation of the π and K modes requires production of τ pairs at low energy and a specially designed detector. The tau-charm factory is the ideal instrument.

Table 4. World average values of τ branching fractions in % from Aguilar-Benitez *et al.* (1992).

Decay Mode	B(%)
$e^-\bar{\nu}_e\nu_\tau$	17.85 ± 0.29
$\mu^-\bar{\nu}_\mu\nu_\tau$	17.45 ± 0.27
$h^-\nu_\tau$	12.47 ± 0.35
$h^-\pi^0\nu_\tau$	23.4 ± 0.6
$h^-2\pi^0\nu_\tau$	9.0 ± 0.6
$h^- \geq 3\pi^0\nu_\tau$	1.8 ± 0.6
$\sum_i B_{1i}$	82.0 ± 1.2
B_1	85.94 ± 0.23
$\Delta_1 = B_1 - \sum_i B_{1i}$	3.9 ± 1.2
$2h^-h^+\nu_\tau$	8.0 ± 0.3
$2h^-h^+ \geq 1\pi^0\nu_\tau$	5.2 ± 0.4
$\sum_i B_{3i}$	13.2 ± 0.5
B_3	14.06 ± 0.20
$\Delta_3 = B_3 - \sum_i B_{3i}$	0.9 ± 0.5
B_5	0.11 ± 0.03
$\sum_i B_i$	95.3 ± 1.3

Table 5. World average values of τ branching fractions in % from Galik (1992).

Decay Mode	B(%)
$e^-\bar{\nu}_e\nu_\tau$	17.75 ± 0.15
$\mu^-\bar{\nu}_\mu\nu_\tau$	17.39 ± 0.17
$h^-\nu_\tau$	12.40 ± 0.26
$h^-\pi^0\nu_\tau$	24.29 ± 0.24
$h^-2\pi^0\nu_\tau$	8.76 ± 0.33
$h^- \geq 3\pi^0\nu_\tau$	$1.26 \pm \pm 0.13$
$\pi^-\pi^0\eta\nu_\tau$	0.08 ± 0.03
$\sum_i B_{1i}$	81.93 ± 0.55
B_1	85.26 ± 0.18
$\Delta_1 = B_1 - \sum_i B_{1i}$	3.3 ± 0.6
$2h^-h^+\nu_\tau$	8.62 ± 0.19
$2h^-h^+ \geq 1\pi^0\nu_\tau$	5.45 ± 0.22
$\sum_i B_{3i}$	14.07 ± 0.29
B_3	14.63 ± 0.18
$\Delta_3 = B_3 - \sum_i B_{3i}$	0.6 ± 0.3
B_5	0.13 ± 0.03
$\sum_i B_i$	96.13 ± 0.62

Turning to the data, Tables 4 and 5 show two recent compilations from the Particle Data Group (Aguilar-Benitez *et al.* 1992) and from Galik (1992). Table 5 contains very recent data from the LEP experiments and from the CLEO II experiment as well as much of the data used in Table 4. The numbers in the tables are averages of measurements from several or even many experiments. To try to answer the questions in Eqs. 26 I give

$$\Delta_1 = B_1 - \sum B_{1i} \text{ (known and measured)} \tag{29a}$$

$$\Delta_3 = B_3 - \sum_1 B_{3i} \text{ (known and measured)} \tag{29b}$$

in Tables 4 and 5. Remember, these compilations have many data sets in common, they are not statistically independent. Hayes (1992) has given an important discussion of the problems in compiling such tables.

Understanding the true errors in these average values is very difficult, as I remarked in the previous section. Are the systematic errors underestimated? Have the proper corrections been made for modes which have a K^0_L? In a particular decay mode, do almost all experiments have the same bias, a bias which is not corrected? Therefore, at this time it is probably best to take the errors in $\sum B_{1i}$ and $\sum B_{3i}$ to be of the order of 1%, and to recognize that at present we do not know if there are missing decay modes, modes which are unconventional and hence not detected and not measured.

It has long been recognized that these questions would be best answered by a single experiment in which every B_{1i} and B_{3i} has high statistics. We do not yet have such an experiment with sufficient statistics to reduce errors to a few tenths of a per cent, the size of errors we would like. The closest we come at present to such an experiment is that carried out by the ALEPH experimenters at LEP (Davier 1992). They find (Davier 1992) Δ_1 and Δ_3 to be zero within errors, and

$$\sum_{u} B_{1i} + \sum_{i} B_{3i} + B_5 = (99.9 + 1.3)\% \tag{30}$$

Table 6 (Snow 1992) shows a measurement set from the ALEPH experimenters.

Table 6. A complete set of branching fraction measurements from the ALEPH experimenters (SNOW 1992). The third error on $\sum B_i$ is the normalization uncertainty.

Decay Mode	B(%)
$e^-\bar{\nu}_e\nu_\tau$	$18.23 \pm 0.30 \pm 0.22$
$\mu^-\bar{\nu}_\mu\nu_\tau$	$17.70 \pm 0.29 \pm 0.21$
$h^-\nu_\tau$	$12.63 \pm 0.28 \pm 0.24$
$h^-\pi^0\nu_\tau$	$26.04 \pm 0.57 \pm 0.63$
$h^-2\pi^0\nu_\tau$	$8.69 \pm 0.61 \pm 0.52$
$h^- \geq 3\pi^0\nu_\tau$	$1.65 \pm 0.41 \pm 0.41$
$2h^-h^+\nu_\tau$	$9.57 \pm 0.24 \pm 0.22$
$2h^-h^+ \geq 1\pi^0\nu_\tau$	$5.42 \pm 0.26 \pm 0.34$
$\sum B_i$	$99.93 \pm 0.83 \pm 0.72 \pm 0.67$

Finally the speculations, and they are speculations. One of the persistent problems in dealing with these summation questions is that no reasonable model has appeared as to what unconventional decay could explain a nonzero Δ_1 or Δ_3 in Eq. 29. We have only speculations.

As a first example, suppose there is some unknown spin 1/2 charge particle, $c^{\pm}$, which can be produced via

$$\tau^- \rightarrow c^- + \gamma \tag{31}$$

If the c mass were different from the e or μ mass, it would not have been found in the searches for $\tau^- \rightarrow e^- + \gamma$ and $\tau^- \rightarrow \mu^- + \gamma$, Table 2. Could such a decay mode have a branching fraction of several per cent? The mode would be counted in B_1. It would not be counted in any conventional B_{1i} but might slip into

$$\tau^- \rightarrow \nu_\tau + \rho^- \rightarrow \nu_\tau + \pi^- + \pi^0 \tag{32}$$

if both γ's from the π^0 overlapped in the electromagnetic shower detector and were counted as one γ. But the great objection to this speculation is why would not $c^{\pm}$ have been seen in many other experiments, in

$$e^+ + e^- \rightarrow c^+ + c^- \tag{33}$$

for example.

I sometimes speculate that in some τ decays a new particle may be produced, a new particle which in turn decays to several photons and other particles. The general form of the decay would be

$$\tau^- \rightarrow \text{normal decay products} + \text{extra } \gamma\text{'s} \tag{34}$$

Such a decay would be counted in measuring B_1 and B_3 but the extra photons, since they do not come from π^0's, might prevent these decays from being counted in any conventional τ decay mode. Or, if the decay in Eq. 34 were counted, it would appear incorrectly in the set of modes.

$$\tau^- \rightarrow \nu_\tau + h^- + n\pi^0 \quad , \ n \geq 2 \tag{35a}$$

$$\tau^- \rightarrow \nu_\tau + h^- + h^+ + h^- + n\pi^0 \quad , \ n \geq 2 \tag{35b}$$

Figure 5 shows two forms of this speculation. In Fig. 5a the new particle N is produced at the $\tau - W$ vertex, and N subsequently decays to ν_τ and several photons. In Fig. 5b an unknown particle couples to the $\tau - \nu_\tau$ vertex and subsequently decays to several photons and leptons or quarks.

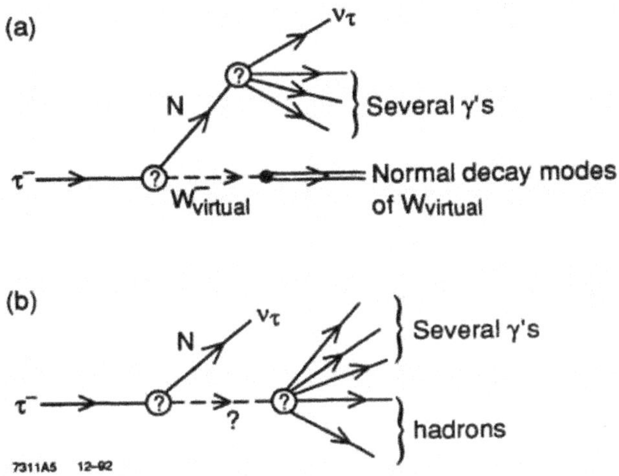

Figure 5.

There are certainly problems with this speculation. Can there be a new particle as required in Fig. 5a or 5b which would not have been found anywhere else in particle physics? And there is an experimental limitation, the ALEPH experimenters (Davier 1992) have placed an upper limit of 2.4% on the branching fraction of modes with photons not coming from π^0's.

It seems to me that at present it is impossible to know whether $\Delta_1 + \Delta_3$ in Eqs. 29 is consistent with zero, or whether there are unknown τ decay modes which are making $\Delta_1 + \Delta_3$ nonzero. Perhaps it is best to adopt the conservative attitude that there is no definite experimental evidence for a nonzero $\Delta_1 + \Delta_3$. Or perhaps it is best to be hopeful to continue to examine these summation issues.

In the end we need much more precise experiments with errors at the 0.1% level rather than the present 1 or 2% level. The tau-charm factory is the ideal machine for such experiments.

C. Searches in Charm Physics

There are six powerful advantages in studying the physics of the charm mesons at the tau-charm factory at the Ψ'' and at the D_s threshold energies:

- The ability to produce very large data sets over short collection times.

- The ability to cleanly select D and D_s by single-tagging of the events. That is, only one D or D_s decay is identified in order to select the pair in an event.
- The availability of kinematic constraints using the beam energy for the rejection of backgrounds.
- The production of the D meson pair in an initial state that is also a coherent quantum mechanical state, allowing further background control and also allowing certain unique physics studies to be performed.
- The absence of backgrounds from heavier meson decays, since production is at threshold.
- The ability to *directly* measure, as necessary, backgrounds from non-charm events by moving below the Ψ'' resonance or below the D_S threshold.

C.1 Unconventional D Decays to Known Particles

A D decay which violates lepton number conservation could be evidence for the existence of a new particle. Examples with present 90% CL upper limits on the branching fraction are:

$$\begin{array}{ll} Mode & B \\ D^0 \to e^\pm + \mu^\mp & < 1.0 \times 10^{-4} \\ D^+ \to \pi^+ + e^\pm + \mu^\mp & < 3.3 \times 10^{-3} \\ D^+ \to K^+ + e^\pm + \mu^\mp & < 3.4 \times 10^{-3} \\ D^+ \to \pi^- + e^+ + e^+ & < 4.8 \times 10^{-3} \\ D^+ \to \pi^- + e^+ + \mu^+ & < 3.7 \times 10^{-3} \end{array} \tag{36}$$

At a tau-charm factory the search for such decays can be carried down to branching fractions of 10^{-7} to 10^{-8}. Stockdale (1989) and Willy (1989) have discussed such searches.

C.2 D Decays Through a Flavor-Changing Neutral Weak Current

In the standard model, decays through a flavor-changing neutral weak current will not occur in lowest order. The general form of such decays is

$$D = \ell^+ + \ell^- + h \tag{37}$$

where

$$\begin{aligned} D &= D^0 \text{ or } D^\pm \\ \ell^+\ell^- &= e^+e^- \text{ or } \mu^+\mu^- \\ h &= \pi \text{ or } K \text{or } \rho \end{aligned}$$

But as shown in Fig. 6, these decays can occur in the standard model in higher order, and branching fractions of the order of

$$B \sim 10^{-7} \tag{38}$$

can be expected (Wiley 1989).

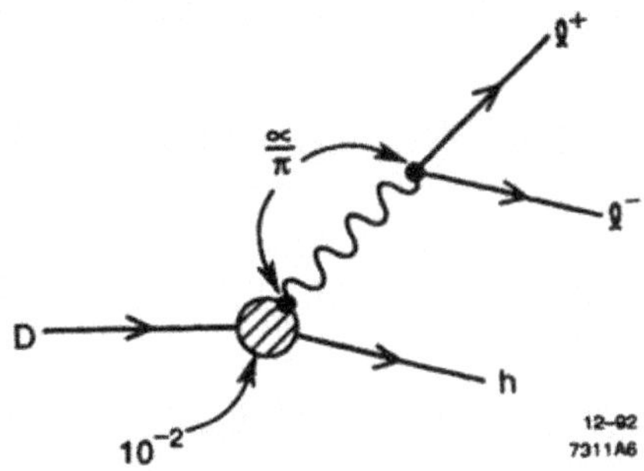

Figure 6.

However, if a D decay of the type in Eq. 37 occurs with a larger branching fraction, it could show the existence of a new particle intermediate in the decay process. Therefore, it is important to search for such decays down to branching fractions of 10^{-8}. This can be done with a tau-charm factory (Stockdale 1989).

It is interesting to look at some present 90% CL upper limits.

$$\begin{array}{ll} Mode & \quad B \\ D^0 \to e^+ + e^- & < \ 1.3 \times 10^{-4} \\ D^+ \to \mu^+ + \mu^- & < \ 1.1 \times 10^{-5} \\ D^+ \to \rho^0 + e^+ + e^- & < \ 4.5 \times 10^{-4} \\ D^+ \to \pi^+ + e^+ + e^- & < \ 2.5 \times 10^{-3} \\ D^+ \to \pi^+ + \mu^+ + \mu^- & < \ 2.9 \times 10^{-3} \end{array} \tag{39}$$

Finally a very interesting set of decays involves neutrinos:

$$\begin{array}{l} D^0 \to \nu + \bar{\nu} \\ D^+ \to h^+ + \nu + \bar{\nu} \end{array} \tag{40}$$

C.3 Charmonium Decays

At a tau-charm factory experimenters can produce $10^9 J/\psi$'s per month and $0.5\times 10^9\psi'$'s per month, thus an enormous number of J/ψ and ψ' decays can be acquired. As discussed by Toki (1989) and Burnett (1989), this allows powerful searches for rare and unconventional decays, decays which could lead to the discovery of new particles. I list some of these decays.

There are the weak decay

$$J/\psi \to D_s + \text{ hadrons} \tag{41a}$$
$$J/\psi \to D_s + \ell + \nu_\ell \tag{41b}$$

Note that the weak decay of a vector meson has never been seen.

A decay through an unknown neutral particle, x^0, might occur.

$$J/\psi \to \gamma + x^0 \ , \ x^0 \to \mu^+ + \mu^- \tag{42}$$

The x^0 might be a Higgs particle.

Once again there are the weak neutral current decays.

$$J/\psi \to \nu + \bar{\nu}$$

which could be detected through the sequence

$$\psi' \to \gamma + J/\psi \ , \ J/\psi \to \nu + \bar{\nu} \tag{43}$$

And finally, the enormous number of J/ψ's allows the search for all sorts of axion-like particles

$$J/\psi \to \gamma + a \tag{44}$$

I conclude this section and the paper with the basic motto of all new particle searches, a motto slightly modified from a fortune cookie:

She that seeks will find.

He that seeks will find.

References

Aguilar-Benitez M *et al* 1992 *Phys. Rev.* **D45** Part 2

Albrecht H *et al* 1992 *Z. Phys.* **C55** 25

Baconnier Y *et al* 1990 CERN/AC/90-07 unpublished

Baltrusaitis R M *et al* 1985 *Phys. Rev. Lett.* **55** 1842

Barish B C *et al* 1990 SLAC-PUB-5180 unpublished

Barish B 1992 *Proc. Second Workshop on Tau Lepton Physics* (Ohio State Univ., Columbus) to be published

Beers L, ed 1989 *Proc. Tau-Charm Factory Workshop* SLAC-REPORT-343 (SLAC, Stanford)

Brown K L *et al* 1989 *Proc. Tau-Charm Factory Workshop* ed L Beers SLAC-REPORT-343 (SLAC, Stanford) p 244

Burnett T H 1989 *Proc. Tau-Charm Factory Workshop* ed L Beers SLAC-REPORT-343 (SLAC, Stanford) p 733

Danilov M V *et al* 1990 ITEP-90-67 unpublished

Davier M 1991 *Les Rencontres de Physique de la Vallée d'Aoste* ed M Greco (Editions Frontiéres, Gif-sur-Yvette) p 595

Davier M 1992 *Proc. Second Workshop on Tau Lepton Physics* (Ohio State Univ., Columbus) to be published

Galik R 1992 *Proc. Second Workshop on Tau Lepton Physics* (Ohio State Univ., Columbus) to be published

Gilman F J and Rhie S H 1985 *Phys. Rev.* **D31** 1066

Gomez-Cadenas J J *et al* 1990 *Particle World* **1** 10

Gonichon J *et al* 1990 LAL/RT 90-02 unpublished

Hayes K G and Perl M L 1988 *Phys. Rev.* **D38** 274

Hayes, K G 1992 *Proc. Second Workshop on Tau Lepton Physics* (Ohio State Univ., Columbus) to be published

Jowett J M 1987 CERN LEP-TH/87-56 unpublished

Jowett J M 1988 *Proc. European Particle Accelerator Conf.* ed S Tazzari (World Scientific, Singapore) p 177

Jowett J M 1989 *Proc. Tau-Charm Factory Workshop* ed L Beers SLAC-REPORT-343, (SLAC, Stanford) p 7

Kirkby J 1987 *Spectroscopy of Light and Heavy Quarks* eds U Gastaldi, R Klapisch and F Close (Plenum, New York) p 401

Kirkby J 1989 *Particle World* **1** 27

Kirkby J and Quesada M 1992 *Proc. Mtg. on Tau-Charm Factory Detector and Machine* (Universidad de Sevilla, Andalucia) to be published

Kirkby J and Rubio J A 1992 *Particle World* **3** 77

Ma E *et al* 1992 *Particle World* **3** 27

Masiero A 1991 *Proc. Workshop on Tau Lepton Physics* (Editions Frontiéres, Gif-sur-Yvette) eds M Davier and B Jean-Marie

Perl M L 1991 *Proc. Lake Louise Sixth Winter Inst. on Particle Physics – The Factory Era* eds B A Campbell, A N Kamal, P Kitching and F C Khanna (World Scientific, Singapore) p 812

Perl M L 1992a *Rep. Prog. Phys.* **55** 653

Perl M L 1992 *Proc. 1992 SLAC Summer Institute on Particle Physics* (SLAC, Stanford) to be published; also SLAC-PUB-5900

Pich A 1991 in *Heavy Flavours* (World Scientific, Singapore) eds A J Burns and M Lindner

Romão J C *et al* 1991 *Nucl. Phys. B* **363** 369

Schindler R H 1989 *Les Rencontres de Physique de la Vallée d'Aoste* ed M Greco (Editions Frontiéres, Gif-sur-Yvette) p 89

Schindler R H 1990a *12th Int. Workshop on Weak Interactions and Neutrinos* eds P Singer and G Eilam (North Holland, Amsterdam) p 89

Schindler R H 1990b *Proc. Lake Louise Winter Inst. on the Standard Model and Beyond* ed A Astbury *et al* (World Scientific, Teaneck) p 217

Snow S 1992 *Proc. Second Workshop on Tau Lepton Physics* (Ohio State Univ., Columbus) to be published

Stockdale I E 1989 *Proc. Tau-Charm Factory Workshop* ed L Beers SLAC-REPORT-343 (SLAC, Stanford) p 724

Toki W H 1989 *Proc. Tau-Charm Factory Workshop* ed L Beers SLAC-REPORT-343 (SLAC, Stanford) p 207

Trischuk W 1992 *Proc. Second Workshop on Tau Lepton Physics* (Ohio State Univ., Columbus) to be published

Truong T N 1984 *Phys. Rev.* **D30** 1509

Wermes N 1992 *Sov. J. Nucl. Phys.* **55** 878

Wiley R S 1989 *Proc. Tau-Charm Factory Workshop* ed L Beers SLAC-REPORT-343 (SLAC, Stanford) p 196

SEARCHES FOR NEW PARTICLES AT A B-FACTORY

Peter C. Kim
Wilson Laboratory, Cornell University
Ithacs, New York 14853, U.S.A.

ABSTRACT

A review on the present experimental status of searches for rare decays of B, D, τ and Υ is given. The potential for observing new physics at a B factory is discussed. The importance of rare B decays, especially the radiative transiton $b \to s\gamma$, as an alternative to direct searches for new particles at higher energies is emphasized. We note that recent limits on penguin diagram decays from the CLEO experiment are within a factor of 2–3 of Standard Model predictions, and that a measurement might be just around the corner.

1. Introduction

The Standard Model (SM) has been very succesful in describing nature. There exists no known experimental fact concerning elementary particles and their interactions that is not consistent with the SM. However, two areas in the SM are not well understood: the Higgs sector and CP violation. Direct searches for the SM Higgs boson have not yielded any signal, and searches continue at machines of higher energies. CP violation has been observed only in the neutral kaon system.[1] Intense searches over the last three decades have not yet produced a single observation of CP violation in other systems. In the SM the phase of the CKM matrix, a unitary matrix constructed to relate the mass and weak eigenstates, is the sole cause of CP violation.[2] It is important that this explanation of the origin of CP violation be tested in a system other than kaons. The B system is the most likely candidate.

B mesons have already played an important role in the 1980's providing two vital pieces of information in our understanding of the Standard Model: the unexpectedly long B lifetime[3] and the large $B_d^0 - \overline{B_d^0}$ mixing rate.[4] A comprehensive study of the properties of B meson could provide a powerful test of the consistency of the Standard Model.

In the SM the unitarity of the CKM matrix implies

$$V_{ub}^* V_{ud} + V_{cb}^* V_{cd} + V_{tb}^* V_{td} = 0. \tag{1}$$

A geometrical representation of this relation in the complex plane is the so-called Unitary triangle (Fig. 1).[5] The three angles of this triangle, α, β and γ, can be

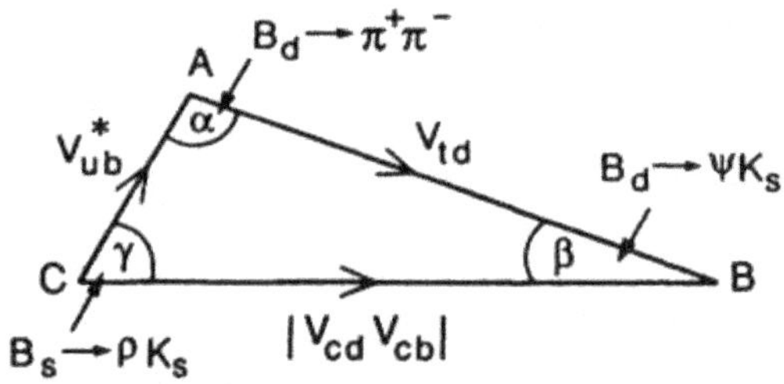

Figure 1. Representation in the complex plane of the triangle formed by the CKM matrix elements V^*_{ub}, V_{cd}, V^*_{cb}, and V_{td}.

determined by measurement of CP violation in the decay modes, $B^0_d \rightarrow \pi^+\pi^-$, $B^0_d \rightarrow \Psi K^0_s$, and $B^0_s \rightarrow \rho K^0_s$, respectively. Unfortunately, due to the small branching ratios of these CP eigenstates and low efficiency for flavor tagging, one needs in excess of 10^8 B mesons in order to observe a CP asymmetry in the B sector, a requirement which can only be fulfilled at a B factory.

It is also important to measure the lengths of the three sides of the triangle precisely. In conjuction with the measurements of CP asymmetries, these measurements will overconstrain the triangle and provide a severe test of the unitary condition. The values for $|V_{cd}|$ and $|V_{cb}|$ can be determined using semileptonic decays of charm and beauty mesons. The magnitude of V_{ub} has also been measured using $b \rightarrow u$ semileptonic decays with a promise of better precision in the near future.[6] The third leg of the triangle, V_{td}, is the most difficult one to measure. I will describe in later sections, how it might be determined from $B\overline{B}$ mixing or from radiative loop decays.

As we look for effects of new physics beyond the SM, there are two complementary options. One option is to build a new machine of sufficiently high energy that direct production of a new particle can occur. The present and future searches for the Higgs boson at LEP I/II and LHC/SSC are good examples of this approach. The other is to build a very high luminosity machine at low energy to look for indirect effects of new physics, probing new interactions at a much higher mass scale. If one pursues latter option, one can either measure a known quantity very precisely, or look for decays that are either strongly suppressed or forbidden. Searches for such rare decays of charm and beauty quarks and the τ can be performed at a B factory. Fig. 2 shows the measured hadronic cross section in the Υ energy region. The $\Upsilon(4S)$ cross section is about 1 nb, which corresponds to 1 million $B\overline{B}$ events per fb^{-1} of integrated luminosity. The underlying continuum produces $\sim 2.5 \times 10^6$ charm particles per fb^{-1}.

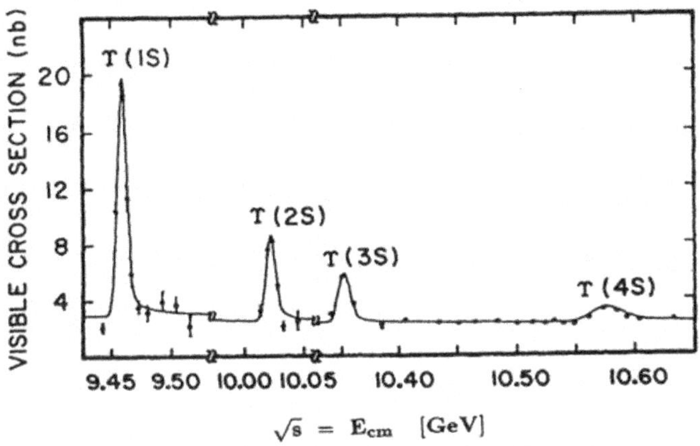

Figure 2. Visible hadronic cross section in the Υ energy region

The $(b\bar{b})$ bound states have large cross sections, especially the $\Upsilon(1S)$, which yields 3×10^7 events per month of running at a B factory. Reviews on the experimental status of rare decay searches in beauty, charm, and τ sectors will be given in the following sections.

2. B Factory Proposals

Proposals,[7] with the main goal of observing CP violation in B system, have been made for the construction of high luminosity double storage rings with asymmetric beam energies of either (8.0 GeV on 3.5 GeV) or (9.0 GeV on 3.1 GeV) with a design luminosity of $\mathcal{L} = 3 \times 10^{33}$ cm^{-1} sec^{-1}. The asymmetry in the beam energies is designed to produce B's with high momentum so that one can measure the time-dependence of their decays.

I will describe the plans for the Cornell B Factory here. The current CESR luminosity at the $\Upsilon(4S)$ is 2.5×10^{32} cm^{-1} sec^{-1}, the best ever achieved at e^+e^- storage rings. In 1991, a plan to upgrade CESR to an asymmetric B factory in phases was submitted to the U.S. National Science Foundation.[8]

Phase I (1991–1993) of the upgrade calls for the replacment of the rf cavities, which are currently limiting the beam currents in the machine. An increase of the luminosity can also be achieved by improving the tune shift, while the number of bunches stays at 7 per beam. In Phase II (1994–1995) beam currents will be increased by introducing a small crossing angle (±2.5 mr) between the colliding beams.[9] The number of bunches in the ring will increase to 27 per beam. In Phase III (1996) new superconducting cavities will help to push the beam current to the single beam tune shift limit with ∼ 40 bunches per beam. At this point, the CESR can be considered as a symmetric B factory with peak luminosity comparable to the original PSI proposal.[10] The final step, Phase IV (1996–1999), is upgrade to CESR–B; an asymmetric B factory. Construction of a second lower energy ring is planned. Further luminosity increases are expected with higher beam currents (∼ 164 bunches per beam) and better beam focusing at the interaction region with superconducting quadropole magnets.

Table 1. CESR upgrade schedule and luminosity goals

Phase	Date	$\mathcal{L}_{pk}$ goal ($cm^{-2}sec^{-1}$)	Total Integrated Luminosity
I	1991 – 1993	4.5×10^{32}	
II	1994 – 1995	6.0×10^{32}	5 fb^{-1} (1994)
III	1996	1.7×10^{33}	
IV	1996 – 1999	3.0×10^{33}	30 fb^{-1} (1998)
			100 fb^{-1} (> 2000)

Projections for integrated luminosity at CESR are shown in Table 1. By the end of 1998, the total integrated luminosity with the CLEO II detector will reach 30 fb^{-1} providing a superb opportunity to study the properties of B and D mesons, and the τ lepton. One expects significant advances will be made in searches for rare decays of these particles. A measurement of CP violation in the B meson system, however, will require a few years of running with an asymmetric machine.

3. Precision Measurements in B Decays

Semileptonic decay of the b quark is dominated by the transition, $b \to c\bar{l}\nu$. The theoretical width, including QCD corrections,[12] can be converted to a branching ratio using the measured B lifetime: $BR(b \to c\bar{l}\nu) = 12 - 14\%$, which is larger than the observed branching ratio of $(10.5 \pm 0.5)\%$.[11] The experimental value is determined from the yield of inclusive leptons in $\Upsilon(4S)$ decays and by assuming that the amount of B mesons in the data sample is known, *e.g.* the $\Upsilon(4S)$ is assumed to decay always to a $B\overline{B}$ pair. One way to avoid this potential pitfall is to measure the lepton yield in an event sample in which the other $\overline{B}$ is completely reconstructed. At a B factory,

$O(10^5)$ of tagged B events will be available per year. On a shorter time scale, the CLEO experiment is expected to have thousands of tagged events in the coming year.

At present, theoretical estimation of how the c quark hadronizes, which in turn affects the lepton momentum spectrum, is the main source of the systematic uncertainties in the measurement of $BR(b \to c\bar{l}\nu)$.[13, 14, 15] In the next few years, inputs from CLEO on the charm meson form factors and a better understanding of the semileptonic Dalitz distribution could reduce this model dependence.

Another issue regarding the semileptonic decay is that known exclusive decay modes, $B \to D\bar{l}\nu$, $D^*\bar{l}\nu$, and $D^{**}\bar{l}\nu$, do not seem to saturate the observed inclusive rate,[6, 11] reminding us of the 1-prong problem in the τ decay. It remains to be seen whether this discrepancy will persist with improved measurements of these branching ratios.

A lepton asymmetry measurement in $B \to D^*\bar{l}\nu$ can be used to test the V–A structure of the weak decay $b \to c$.[16] The left-handed (right-handed) helicity of the c quark in the V–A (V+A) currents is preserved in the D* system, and the chirality of the weak curent can be determined from the forward/backward asymmetry of the lepton direction in the $(\bar{l}\nu)$ rest frame. The current limit from CLEO[17] excludes a 30 % admixture of V+A (95% $C.L.$). At a B factory more precise tests of V+A components in the weak b decay can be performed.[18]

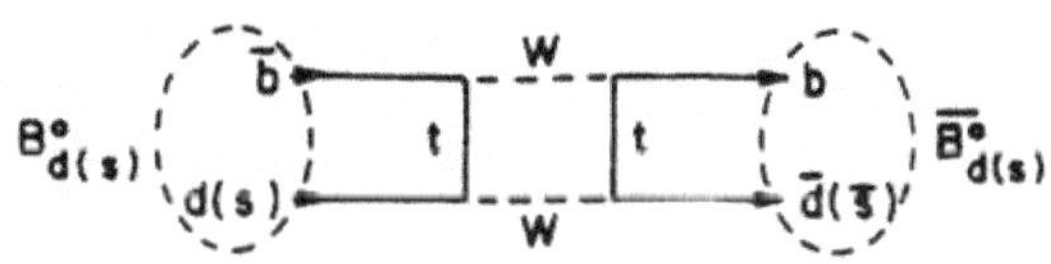

Figure 3. A box diagram contribution to $B^0\overline{B^0}$ mixing.

The next topic I would like to discuss is $B^0\overline{B^0}$ mixing. In the Standard Model, the box diagram with a virtual t quark (Fig. 3) is the dominant contribution to neutral B meson mixing. The mixing rate comes mainly from Δm, the mass difference between two mass eigenstates, rather than from $\Delta\Gamma$, the lifetime difference of two eigenstates. The quantity, $\Delta m/\Gamma$, can be written as

$$x_q \equiv \Delta m/\Gamma = \frac{G_F^2 m_{B_q}}{6\pi^2} m_t^2 \tau_{B_q} B_{B_q} f_{B_q}^2 |V_{tq}V_{tb}^*|^2 F(m_t^2/m_W^2)\eta, \tag{2}$$

where q could be a d or s quark, and B_{B_q}, f_{B_q}, and η are the B meson "bag" constant, B meson decay constant, and a QCD correction factor, respectively. All these factors make it difficult to extract useful information on fundamental Standard Model parameters such as m_t or V_{tq}. Most of these factors, however, cancel out in the ratio

$$\frac{x_d}{x_s} \approx \left|\frac{V_{td}}{V_{ts}}\right|^2. \tag{3}$$

Precise measurements of both B_d^0 and B_s^0 mixing could be the way to determine the CKM matrix element V_{td}, since V_{ts} is estimated to have a value close to V_{bc}.

One convenient mixing parameter experimentally accessible is:

$$\chi_q \equiv \frac{P(\overline{B_q} \to B_q \to lX)}{P(B_q \to lX)} = \frac{1}{2}\frac{x_q^2}{(1+x_q^2)}. \tag{4}$$

Unfortunately, due to the way x_q enters the equation, one can not extract a precise measurement of the x_q from a measurement of χ_q for large values of x_q. Combining all measurements from LEP, ARGUS and CLEO,[6] we have

$$x_s = \Delta m_{B_q}/\Gamma = 2.5^{+\infty}_{-2.5}. \tag{5}$$

It is unlikely that the situation will improve much through new measurements of the x_d and x_s.

A much better method is to use the time-dependence of $B^0\overline{B^0}$ mixing by reconstructing the B decay vertex. At LEP or SLC high momentum B's travel 1 *mm* or more before decaying. Feasibility studies at these machines have shown that when equipped with a high precision CCD or a silicon vertex detector, values of x_s up to 15 could be measured with a sample of 5×10^5 B mesons.[19] A time-dependent measurement of B_s mixing may also be possible at an asymmetric B factory. Even though the Lorenz boost factor, $\gamma\beta$, is near unity, one can partly compensate for the poorer vertex resolution with a larger B sample. Vertex position resolution is crucial here, and an x_s value of up to 15 might be reachable after a few years of running.[20]

4. Rare B Meson Decays

The B meson system is nearly ideal for the study of rare decays in the Standard Model for various reasons:

- The primary $b \to c$ transition is suppressed, $|V_{bc}|^2 \approx 2 \times 10^{-3}$, enhancing the rare decay branching ratios.

- B decays are theoretically clean, since they are short-distance interaction dominated.
- Large m_t and $|V_{tb}| \approx 1.0$ produce large loop effects. Comparison of the CKM matrix elements gives a factor of 10^5 enhancement over one-loop kaon decays.

Progress in the experimental area has been slow, however, mainly due to the small numbers of B's available in a background free environment.

4.1 *Electromagnetic Penguin Decays ($b \rightarrow s\gamma$) in the Standard Model*

The photon emission transition $b \rightarrow s\gamma$ (Fig. 4a and 4b) is an example of a one-loop Flavor Changing Neutral Currents (FCNC) process. In the Standard Model FCNC are forbidden at tree level,[21] but can be induced by higher order loop diagram processes. In the decay of B mesons, the loops with virtual top quarks and W bosons are the dominant contributions.

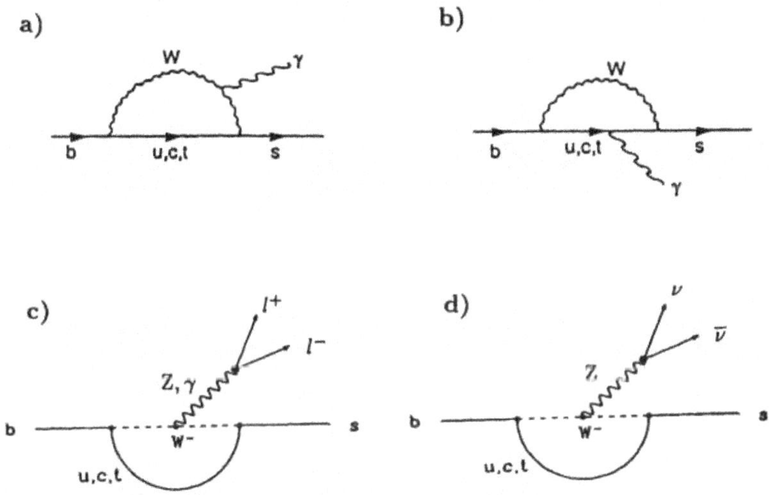

Figure 4. One-loop diagrams in the Standard Model:
a) and b) $b \rightarrow s\gamma$; c) $b \rightarrow sl^+ l^-$; d) $b \rightarrow s\nu\overline{\nu}$.

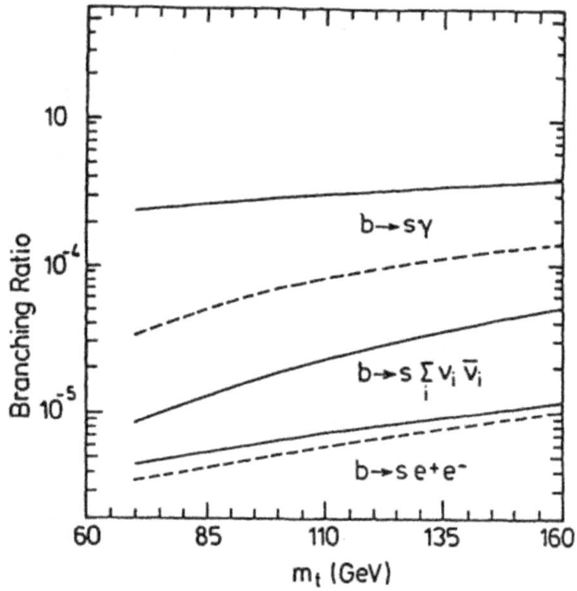

Figure 5. The SM prediction for $b \to s$ loop transitions, with (solid curve) or without (dashed curve) QCD corrections, are shown as a function of m_t.[22]

Two other types of loop diagrams ($b \to sl^+l^-$ and $b \to s\nu\bar{\nu}$) are shown in Fig. 4c and 4d. Their rates are much lower than the $b \to s\gamma$ rate, and do not gain much through QCD corrections (Fig. 5). Searches for $b \to s\nu\bar{\nu}$ are very challenging experimentally, and would require enormous tagged-B samples at a B factory.

It has been known for some time that the $b \to s\gamma$ transition, the so-called electromagnetic penguin, is sensitive to the value of the top quark mass.[23] Recently it has received more attention after the discovery of suprisingly large QCD corrections,[24] which modifies the usual GIM suppression in the amplitude from the form m_t^2/m_W^2 to a a softer form with a leading logarithmic term, $ln(m_t^2/m_c^2)$. This leads to an enhancement of the branching ratio by one order of magnitude at low m_t and by a factor of $3 - 4$ at $m_t = 150$ GeV. The branching ratio for $b \to s\gamma$ as a funtion of m_t before and after QCD corrections is shown in Fig. 5.[22] After correcting for the gluon bremsstrahlung contribution,[25] the branching ratio of $b \to s\gamma$ becomes 4×10^{-4} for $m_t = 150$ GeV.

Experimental searches for the $b \to s\gamma$ can be performed in two complementary ways. One approach is to fully reconstruct an exclusive decay mode such as $B \to K^*(892)\gamma$ or $B \to K_2^*(1430)\gamma$. The alternative is to look for the quasi-monochromatic photon signal inclusively. The exclusive search is rather straightforward, provided there are enough B's in the data sample and the combinatorial background is not too severe. On the other hand the theoretical interpretation of an exclusive measurement is not a very precise business. Predictions for the ratio

$$\frac{BR(B \to K^*(892)\gamma)}{BR(b \to s\gamma)} \tag{6}$$

range from 4.5% to 28%.[26, 27, 28, 29] Given such a wide range of theoretical values, the conversion of exclusive measurements into a decay width for $b \to s\gamma$ is quite model dependent. Heavy Quark Effective Theory (HQET)[30] with inputs from experimental form factors measured in charm semileptonic decays promises a more reliable determination of the $b \to s\gamma$ rate.[29, 31]

With an input branching ratio of 4×10^{-4} for $b \to s\gamma$, the range of $BR(B \to K^*(892)\gamma)$ becomes $(1.8 - 11.2) \times 10^{-5}$. The recent upper limit of 9.2×10^{-5} set by CLEO[32] is already within this theoretical range. With the large data sample being collected at CLEO, one might not have to wait until the completion of a B factory for the observation of a penguin.

One notes that if the s quark in Fig. 4a or 4b is replaced by a d quark, the amplitude is proportional to CKM matrix element V_{td} rather than V_{ts}. Hence, measurements of both the $B \to K^*\gamma$ and the $B \to \rho(\omega)\gamma$ branching ratios can be used to determine $|V_{td}|$,[33]

$$\frac{BR(B \to \rho(\omega)\gamma)}{BR(B \to K^*\gamma)} = \left|\frac{V_{td}}{V_{ts}}\right|^2. \tag{7}$$

As in the measurement of x_d and x_s from $B_d^0\overline{B_d^0}$ and $B_s^0\overline{B_s^0}$ mixing discussed earlier, most of the theoretical uncertainties from strong interaction effects cancel out. The process $B \to \rho\gamma$ has an estimated branching ratio of $(1-3) \times 10^{-6}$,[33] and this should be observable at a B factory.

We now return to the inclusive approach of looking for photons from $b \to s\gamma$ in the photon energy spectrum. When the s quark from b decay combines with the spectator quark $\overline{q}$, its hadronization is similar to semileptonic b decays, *i.e.* the lowest lying $(s\overline{q})$ states saturate the decay width. The inclusive photon energy spectrum from $b \to s\gamma$ decay in Fig. 6 shows a very hard peak near 2.5 GeV indicating the dominance of exclusive 2-body modes.[34] The advantage of this approach is that it is much less dependent on the details of the hadronization process compared to the exclusive approach, providing a direct test of Standard Model predicitons. The disadvantage is the presence of a large photon background from π^0 decays and initial state radiation, which make it experimentally difficult.

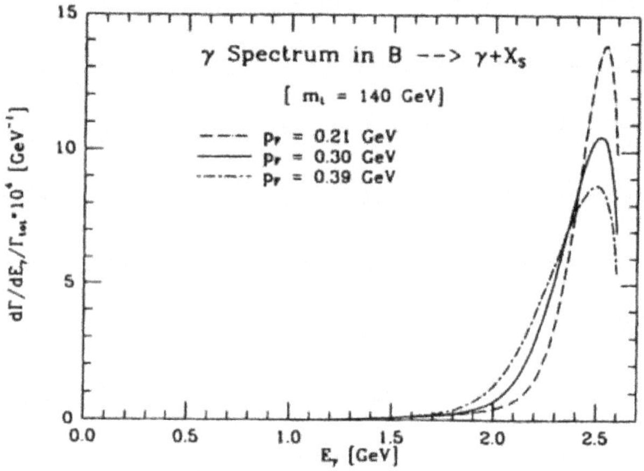

Figure 6. Inclusive photon energy spectrum for the process $B \to X_s + \gamma$. The parameter p_F is the Fermi momentum.[34]

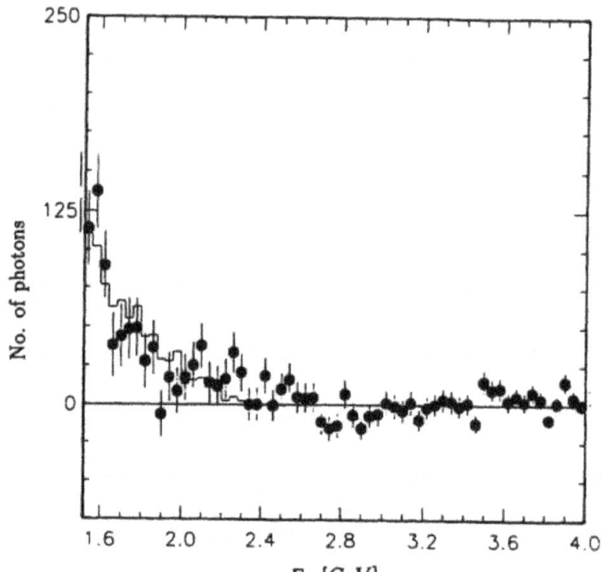

Figure 7. The inclusive photon energy spectrum from B decays after continuum subtraction.[32] The points are the data, and the solid histogram is the Monte Carlo predicted shape for photons from $b \to c$.

Inclusive searches for the $b \to s\gamma$ photons have been made at Crystal Ball[35] and CLEO[32] experiments. The inclusive photon energy spectrum from CLEO after background subtraction is shown in Fig. 7. No significant excess was observed in the $2.2 < E_\gamma < 2.7$ GeV signal region. An upper limit of $BR(b \to s\gamma) < 8.4 \times 10^{-4}$ (90% $C.L.$) is obtained, which is within a factor of two of the SM prediction. This 1991 analysis used only a subset of the current CLEO data sample, and a study with the full data set is now being carried out.

4.2 $b \to s\gamma$ *and New Physics*

Observation of $b \to s\gamma$ at a rate different from the Standard Model prediction could yield evidence for non-standard processes: a fourth generation, new scalars, new gauge bosons, or supersymmetric particles (Fig. 8). The question remains whether interference between the SM amplitude and amplitudes from new physics is constructive or destructive. In the latter case, the observation of a $b \to s\gamma$ branching ratio below the SM prediction would be an indication of new physics. I list below examples of four different types of models beyond the Standard Model.[36]

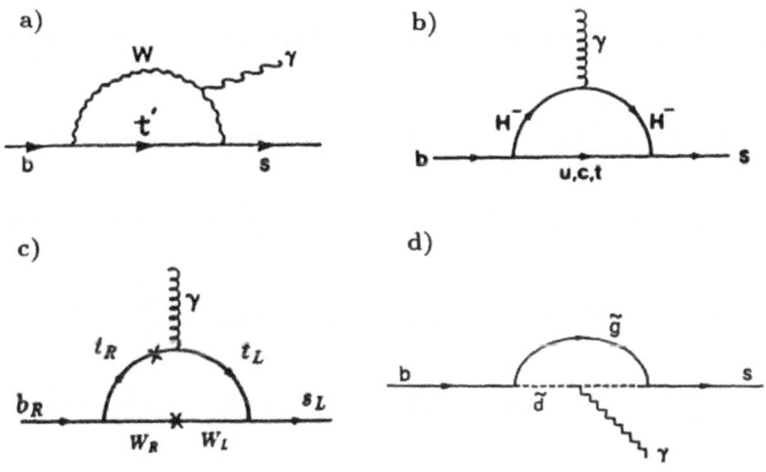

Figure 8. One-loop diagrams in non-Standard Model processes: a) new quark (Fourth Generation); b) new scalar (Higgs Doublet); c) new gauge boson (Left-Right Symmetry); d) Supersymmetric particle

A natural extension of the SM is the addition of a fourth family. Loop decays are enhanced, for example, via the exchange of a t' quark, the fourth family up-quark (Fig. 8a), which is expected to couple more strongly to the third family b quark than to lower generation quarks.[37] Interference between the SM loop and the t' quark loop could be either constructive or destructive, so that in the latter case the $b \to s\gamma$ rate would be below the SM prediction. The parameters that affect the $b \to s\gamma$ rate are the fourth family CKM matrix elements $V_{t'b}$, $V_{t's}$ and the t' quark mass. Predictions for the $b \to s\gamma$ branching ratios for various ranges of these parameters can be found in ref. 37. The precise measurements of the Z^0 width from LEP require that the neutrino of the fourth family have a mass greater than $M_Z/2$ and tightly constrain this class of models.

Left-Right Symmetric (LRS) theories have both V-A and V+A currents, with corresponding Intermediate Vector Bosons W_L and W_R.[38] Extra one-loop diagrams with $W_L - W_R$ mixing can contribute to the transition $b \to s\gamma$ (Fig. 8b).[39] The two main unknown parameters in these processes are: the W_R mass, the amount of mixing ξ between W_L and W_R. The CKM angles of the righthanded sector usually are taken to be the values of nominal CKM angles of the left-handed sector. Constraints on the unknown couplings and on the mass of W_R can be derived from kaon and muon decay. The most severe bound on the mass of W_R, $M_{W_R} > 1.4 - 2.5$ TeV, comes from the K_S-K_L mass difference.[40] Estimated limits on ξ range from 4×10^{-3} (derived from hadronic decays of strange mesons[41]) to 6×10^{-2} (obtained from direct observation of V-A dominance in μ decay[42]). A mixing rate of 10^{-2} would enhance the $b \to s\gamma$ branching ratio by an order of magnitude over the SM rate.[39] If the tighter limit on the mixing rate ξ holds, no significant enhancement of the $b \to s\gamma$ transition is expected.

In the simpliest non-minimal Two Higgs Doublet Model (2HDM), there are three neutral and one charged Higgs scalar. The charged Higgs loop contribution (Fig. 8c) could give a substantial increase in the $b \to s\gamma$ branching ratio, since it receives a large enhancement from QCD corrections and interferes constructively with the SM W loop amplitude.[43] A non-minimal Higgs sector arises naturally in many extensions of the Standard model,[44] including Left-Right Symmetric[38] and supersymmetric[45] models. In most 2HDMs tree level FCNC due to neutral Higgs bosons are avoided by requirements:

1) only one Higgs doublet is allowed to couple to both types of quarks (Type I)

or

2) one doublet is responsible to for the up-quark masses, the other for the down-quark masses (Type II).

It is the Type II model that has been found to have the largest effect on the $b \to s\gamma$ transition. Unkown parameters in 2HDM are m_{H^+} the charged Higgs mass and $tan\beta$ the ratio of two vacuum expectation values.

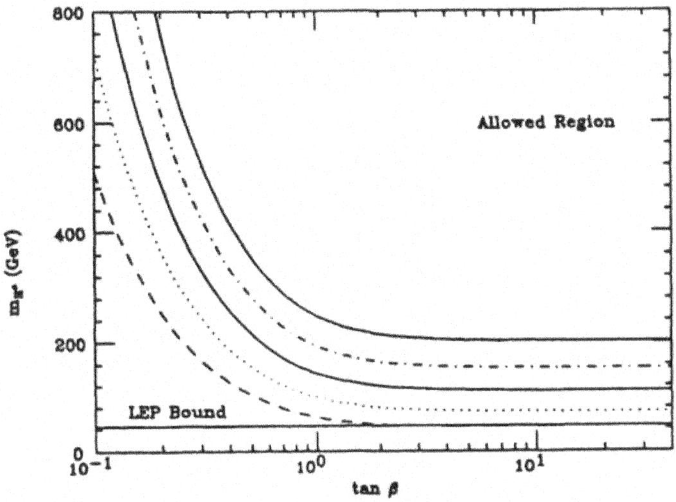

Figure 9. The excluded regions in the $m_{H^+} - tan\beta$ plane for various values of m_t coming from the CLEO limit on $BR(b \to s\gamma)$.[47] The curves, from top to bottom, correspond to m_t = 210, 180, 150, 120, and 90 GeV.

The CLEO upper limit of $BR(s \to s\gamma) < 8.4 \times 10^{-4}$ already implies severe constraints on the supersymmetric charged Higgs mass (Fig. 9), more stringent than those derived from the $B\overline{B}$ mixing rate.[46, 47] An improvement in this limit by a factor of 2–3 could rule out regions of the 2HDM parameter space that are not accessible to the searches at LEP I/II or even at LHS/SSC.[46]

Supersymmetric (SUSY) particles can contribute to the loop diagrams in many different ways.[48] One way is through the exchange of a charged Higgs particle as discussed above, another is to replace the particles in the SM loop diagram with their supersymmetric partners. For example, the up-quark can be replaced by the up-squark, W^- by the wino, and the charged Higgs by the higgsino. There are also new types of diagrams involving a gluino (Fig. 8d) for which there are no equivalents in SM loops. Hence, detailed studies of loop diagrams can provide tests of low energy supersymmetry. It has been found that contributions from diagrams with a charged Higgs and the chargino (a mixture of wino and higgsino) are the dominant SUSY contribution to $b \to s\gamma$, and that they enhance the $b \to s\gamma$ rate by interfering constructively with the SM prediction.[49] The CLEO limit on $b \to s\gamma$ has already been used to set constraints on the mass of gluino with a sensitivity comparable to direct searches at LEP I/II.[36]

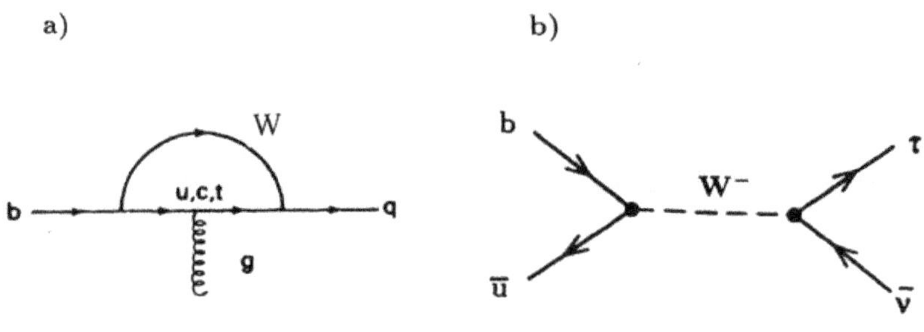

Figure 10. a) Hadronic penguin loop diagram; b) W-annihilation diagram: $B \to \tau\nu$.

4.3 *Hadronic Penguin Decays ($b \to s$ gluon)*

Penguin diagrams were first introduced to explain the $\Delta I = 1/2$ rule in the K meson decay,[50] and later identified as a possible source of direct CP violation in kaons.[51] The penguin diagram in B meson decay (Fig. 10a) is also expected to play the same role in CP asymmetries of the B system, which we will discuss further in section 5.

The total branching ratio of inclusive ($b \to s$ *gluon*) in the Standard Model after the QCD corrections is relatively large at the $O(10^{-3})$ level.[52] Each exclusive decay channel, however, has a branching ratio of only $O(10^{-5})$, since so many decay channels are available. Due to uncertainties in the hadronization process, theoretical calculations of branching ratios of exclusive modes are not very reliable. Two recent calculations[53] of branching ratios for exclusive two-body final states, using the factorization assumption, predict branching ratios of $BR(B^0 \to K^+\pi^-) = 1.1 \times 10^{-5}$ and 1.7×10^{-5}. The CLEO upper limit on this channel is 5.6×10^{-5} (90% *C.L.*).[54] With a larger data sample currently at hand, CLEO is expected to improve this limit in the near future.

New physics can enhance the ($b \to s$ *gluon*) rate via extra diagrams that contribute to the loop, similar to the ($b \to s\gamma$) enhancement. But again, due to hadronization uncertainties in the calculation of exclusive states, small deviations from SM predictions will be difficult to detect.

4.4 $B^- \to \tau^-\overline{\nu}$

In the Standard Model the decay $B^- \to \tau^-\overline{\nu}$ proceeds through annihilation of the b and $\overline{u}$ quarks (Fig. 10a). The width of this pure leptonic decay in the SM is dependent on the KM matrix element V_{ub} and the B meson decay constant f_B. With inputs of $|V_{ub}| = 0.0045$ and $f_B = 195\ MeV$, the predicted branching ratio is 1.0×10^{-4}. Using a reconstructed $B^{\pm}$ tag sample of 720 events, the CLEO experiment set the first limit of $BR(B^- \to \tau^-\overline{\nu}) < 1.3 \times 10^{-2}$ (90% $C.L.$).[55] With the expected two orders of magnitude increase in the tagged B sample, we could observe this decay at a B factory.

If a new charged scalar particle, which carries a new interaction different from the weak V–A interaction, replaces the W^- in Fig. 10b, the branching ratio of $B^- \to \tau^-\overline{\nu}$ or $B^- \to \mu^-\overline{\nu}$ could increase substantially above the SM prediction without the V–A helicity suppression.

5. CP Violation and New Physics

In the Standard Model there are two ways CP violation in the kaon system can arise:[51]

1) CP violation from K$\overline{\text{K}}$ mixing (box diagram) in the neutral K system
2) direct CP violation due to interference between two types of decay processes, tree diagrams (W-decay) and loop diagrams (penguin decay).

CP violation from mass mixing is parametrized by ϵ; while direct CP violation is parametrized by ϵ'. In both cases, the phase of the CKM matrix is responsible for CP violation. An alternative possibility is the superweak model of Wolfenstein.[56] In that model, mass mixing at tree level is the only source of CP violation, giving a non-zero value for ϵ, while ϵ' is identically zero. Hence, the observation of a non-zero value of ϵ' would be the first evidence for penguin diagrams and would rule out the superweak model. Results from direct CP violation search experiments are not yet conclusive.[6]

CP violation in the B system is analogous to the kaon system, and can either arise from mass mixing or from penguin decays. One difference is that the CP asymmetries in the B meson system are large, typically 10 – 20 %. However, due to the small branching ratios for decays to the CP eigenstates and the low flavor tagging efficiency, about 10^8 B mesons are required to observe an asymmetry. CP violation through CP eigenstates can be measured by studying the time-dependent correlated decay probabilities between the two B mesons at an asymmetric B factory. Direct CP violation can be detected from the charge asymmetry in the yields of B^+ and B^- mesons. Excellent review articles exist on the feasibility of observing CP violation at a B factory.[20]

If the superweak model is extended to the B system,[57] the asymmetry in $B \to \Psi K_s^0$ is smaller than the SM prediction, while the asymmetry for $B \to \pi^+\pi^-$ could be either smaller or larger. It is also possible that both sources of CP violation are present. In that case, ϵ has contributions from both, while ϵ' has a contribution only from the CKM matrix.

The effect of the fourth generation[58] would be the violation of the unitarity of the 3×3 CKM matrix. Other new physics, such as LRS,[59] 2HDM,[44] or SUSY[60] maintains the unitarity. Tests of the unitary condition which combine measurements of CP violation and the relevant CKM matrix elements in the unitary triangle, will provide insights on the nature of physics beyond the SM.[61]

New physics modifies Standard Model CP violation through extra contributions to box diagrams in CP violation through mass mixing, and or through extra contributions to loop diagrams in direct CP violation. These extra contributions should also be observed in rare B decays involving box diagrams and loop diagrams. This suggests that the effects of new physics will be more prominent in rare B decays. For example, we have seen in section 4.2 that the $b \to s\gamma$ decay rate is more sensitive to supersymmetric charged Higgs bosons than is the $B\overline{B}$ mixing rate. In addition, $b \to s\gamma$ is more sensitive to a supersymmetric Higgs than CP violation of B mesons through mass mixing. For the second type of CP violation, effects of supersymmetry enter through extra penguin diagrams, which again contribute more strongly to $b \to s\gamma$.

In other words, by the time we are able to observe non-Standard Model effects in CP asymmetries of the B system, it is quite likely that such effects will have been already observed in rare B decays.

6. Searches for New Particles in Charm Meson Decays

Table 2. Number of charm particles expected per year at a B factory running at $\mathcal{L} = 3 \times 10^{33}$ cm^{-2}sec^{-1}.

Particle Type	Events per year
charm quark	8.0×10^7
D^0	4.5×10^7
D^+	1.5×10^7
D_s^+	1.0×10^7
Λ_c^+	1.0×10^7

Charmed quarks will be produced in abundance at a high luminosity B factory, as shown in Table 2. We note that the yields are comparable to those at a tau-charm factory. At 10 GeV center-of-mass energy, charm particles have high momenta due to hard fragmentation, and hence any study that can utilize the displaced charm vertex can be performed at a B factory. Conversely, a tau-charm factory can use the double-tagging technique as well as the kinematical constraints of charm production near threhold.

Table 3. Upper limits on box or loop diagram decays of D mesons.[11]

Decay Mode	Upper Limit (90% $C.L.$)	Experiment
$D^0 \rightarrow \overline{D^0}$	3.7×10^{-3}	E691
	1.4×10^{-2}	ARGUS
$c \rightarrow e^+e^-X$	2.2×10^{-3}	CLEO
$c \rightarrow \mu^+\mu^-X$	2.2×10^{-3}	CLEO
$D^0 \rightarrow e^+e^-$	1.3×10^{-4}	Mark III
	1.7×10^{-4}	ARGUS
$D^0 \rightarrow \mu^+\mu^-$	1.1×10^{-5}	E615
	7.0×10^{-5}	ARGUS
$D^0 \rightarrow \mu^+e^-$	1.0×10^{-4}	ARGUS
	1.2×10^{-4}	Mark III
$D^0 \rightarrow \overline{K^0}e^+e^-$	1.7×10^{-3}	Mark III
$D^0 \rightarrow \rho^0e^+e^-$	4.5×10^{-4}	CLEO
$D^0 \rightarrow \rho^0\mu^+\mu^-$	8.1×10^{-4}	CLEO
$D^+ \rightarrow \pi^+e^+e^-$	2.5×10^{-3}	Mark II
	2.6×10^{-3}	CLEO
$D^+ \rightarrow \pi^+\mu^+\mu^-$	2.9×10^{-3}	CLEO
$D^+ \rightarrow \pi^+\mu^+e^-$	3.8×10^{-3}	CLEO

Rare charm decays via the box or loop diagrams are strongly suppressed in the Standard Model, since the heaviest quark in the loop, the b quark, has a much lower mass than the W boson. Signatures of new physics are more likely to show up first in B decays.

The interpretation of rare D decays is also complicated by long-distance effects (*e.g.* final state rescattering), because the energy region of $1.0 - 2.0$ GeV is rich with hadronic resonances. A good example is the decay $D^0 \to \rho\gamma$.[62] The contribution of loop diagram in the Standard Model gives a branching ratio of 10^{-14}, while the more conventional decay

$$D^0 \to \rho\rho^* \text{ or } \rho\phi^* \to \rho\gamma \tag{8}$$

is expected at a branching ratio of $10^{-6} - 10^{-7}$.

In Table 3, I listed experimental upper limits on rare charm decays that are promising at a B factory. Experiment has been limited by the small number of charm events available (at least for e^+e^- experiments) and the combinatorial background in the mass signal region. With the help of a silicon vertex detector, one can greatly improve the S/N ratio by reconstructing the secondary charm vertex. Having leptons in the final state also helps in reducing the background.

For most of the decay modes in Table 3, searches can be carried out down to a branching ratio of 10^{-7}. The $D^0\overline{D^0}$ mixing search is expected to reach mixing rates of $few \times 10^{-5}$ by utilizing the charm vertex information, which can separate mixed events from doubly cabibbo-suppressed decays which produce the same final state.[20]

7. Searches for New Particles in τ Decays

With a sample of 3×10^7 τ pairs per year, one can carry out precise measurements of many τ decay parameters. A very thorough description of the status of the τ and where new physics might be detected can be found in a recent review by Perl.[63] Here I will list a few of the topics that are of interest at a B factory.

There exist two interesting problems in present measurements of the τ branching ratios. The first is the $m_\tau - \tau_\tau - B_e$ controversy. The leptonic decay width of the τ is simply:

$$\Gamma(\tau^- \to e^-\overline{\nu}\nu) = \frac{G_F^2 m_\tau^5}{192\pi^3}(1 - 8y + 8y^3 - y^4 - 12y^2 lny) \cdot f_{rad}, \tag{9}$$

where y is m_e^2/m_τ^2 and f_{rad} is the radiative correction. Until recently, the τ lifetime derived from the above formula seemed to be in disagreement with the measured value at the 2σ level. New measurements of the tau mass at BES[64] and the tau lifetime from LEP[6] indicate that the significance of discrepancy is only 1 standard deviation. Improved lifetime measurents with a silicon vertex detector at a B factory, could further clarify the situation more and also provide a stringent test of the lepton universality. The CLEO experiment with a new Silicon Vertex Detector (scheduled for installation in 1994) could play a role here until an asymmetric B factory is ready.[65]

Another interesting problem in τ decay is the so-called One-prong Problem.[66] Over the years, it has been noted that the sum of exclusive decay channels with one charged track does not saturate the inclusive 1-prong measurements at the 5σ level. A compilation[6] using the most recent LEP and CLEO data, however, shows that the discrepancy is only at the 2σ level. The controversy is still not yet resolved, since the method of combining the systematic-dominated measurements is quite arbitrary. It is clearly desirable to measure all modes in a single experiment to a high precison. A state-of-the-art detector at a B factory could be a such device.

Rare decays of the τ has been discussed by Perl at this conference.[67] There are many common experimental requirements in rare τ decay searches at tau-charm and B factories. Both require good lepton identification, hermiticity of the detector and fine photon energy resolution. The tau-charm factory, though, has the advantage that continuum background can be greatly reduced by running below the $D\overline{D}$ threshold.

I will mention only one rare decay search which stresses the effectiveness of rare τ decays over rare μ decays. In the Standard Model, the transition $\tau^- \to \mu^-\gamma$ has a rate of 10^{-40}. It can be enhanced through the exchange of new particles such as Higgses, heavy lepton, or SUSY particles. Couplings to these particles are usually dependent on the masses of the leptons. For example, in a Superstring-inspired theory,[68] it is predicted:

$$BR(\tau^- \to \mu^-\gamma) = 2 \times 10^5 \cdot BR(\mu^- \to e^-\gamma). \tag{10}$$

Hence, in this model the recent limit of $BR(\tau^- \to \mu^-\gamma) = 1.1 \times 10^{-5}$ from CLEO[69] is comparable to a limit of $BR(\mu^- \to e^-\gamma) < 10^{-10}$.

8. Searches for New Particles in Υ Decays

At a B factory running at $\mathcal{L}_{peak} = 3 \times 10^{33}\mathrm{sec}^{-2}\mathrm{sec}^{-1}$ and a beam energy spread of $\sigma = 5MeV$, one could obtain 7×10^7 $\Upsilon(1S)$ events per month.[20] At this rate of operation, one expects an $O(10^2)$ increase over the integrated data samples of present experiments, in two months. The $(b\bar{b})$ bound states below the $B\overline{B}$ threshold, $\Upsilon(1S), \Upsilon(2S)$ and $\Upsilon(3S)$, have very narrow widths of 20–50 KeV due to OZI suppression of hadronic decays. Such small widths allow rare processes to have observable branching ratios. Another advantage in dealing with the b quark as compared to the J/Ψ meson is that particles such as axions or Higgs have couplings proportional to the quark masses. I list below a few examples of searches for direct production of exotic particles with masses less than m_Υ.

8.1 $\Upsilon(1S) \to \gamma H^0$

This reaction, where H^0 denotes the Higgs boson, is the well-known Wilczek mechanism.[70] Expected rates are:

$$BR(\Upsilon \to \gamma H^0) = BR(\Upsilon \to \mu\mu) \cdot \frac{G_F m_b^2}{\sqrt{2}\pi\alpha}\left(1 - \frac{m_H^2}{m_\Upsilon^2}\right) x^2 f_{rad}, \tag{11}$$

where $x = |v_2/v_1|$ is the ratio of vacuum expectation values and f_{rad} is the radiation correction factor.

Since the coupling of the Higgs boson is expected to be mass dependent, the H^0 should decay mainly into final states with heavy particles, i.e., $H^0 \to c\bar{c}$, $\tau\bar{\tau}$, if M_{H^0} is sufficient. Lighter Higgs Bosons are expected to decay to $\pi^+\pi^-$, $\mu^+\mu^-$, e^+e^-, and $\gamma\gamma$. Current experimental limits from CUSB[71] and ARGUS[72] rule out the mass range $0.21 < M_{H^0} < 5.0$ GeV.

8.2 $\Upsilon(1S) \to \gamma$ *"glueball"*

The observed rate of $BR(\Upsilon(1S) \to \gamma gg) = 3.0\%$ suggests the possibility of observing a bound state of two gluons. Searches for low mass glueball states are not competitive with experiments running at the J/Ψ energy at a tau-charm factory, but a much larger mass range can be scanned assuming that the widths of higher mass states are not too broad. Current upper limits on exclusive two-body decays, $\Upsilon \to \gamma$ *"light meson"*, are at the level of 10^{-4}.[11]

8.3 $\Upsilon(1S) \to \gamma$ *"Nothing"*

Here *"Nothing"* refers to any particle that escapes without detection, for example axions(a) or Higgs particles(h^0) that are long-lived.[73] We expect two orders of magnitude improvement in the the following limits set by Crystal Ball,[74] $BR(\Upsilon(1S) \to \gamma\, a) < 4.0 \times 10^{-5}$ and $BR(\Upsilon(1S) \to \gamma\, h^0) < 3.5 \times 10^{-5}$.

8.4 $\Upsilon(1S) \to$ *"Nothing"*

Searches for this type of event are difficult at the $\Upsilon(1S)$ energy, since the final state does not satisfy any of the usual detector triggers. An alternative is to utilize the cascade chain,

$$\Upsilon(3S) \to \Upsilon(1S)\pi^+\pi^-,\ \Upsilon(1S) \to \text{"Nothing"}. \tag{12}$$

A recent search with the CLEO data sample yielded an upper limit of 5.2×10^{-3}.[76] Interpreting the undetected particles as a cosmion-anticosmion pair, for example, can contrain the cosmion mass.[76, 75]

9. Conclusions

One of the main goals at a B factory is the study of CP violation. With $O(10^8)$ $B\overline{B}$ events, we should be able to:

- Observe CP violation in the B system, showing that it exists in a system other than the neutral kaons.
- Observe direct CP violation arising from interference between tree and loop diagrams.
- Test the CKM matrix origin of CP violation
- Compare the observed CP asymmetry to the Standard Model predictions and look for effects of new physics.

Rare B decay search is another important area of physics at a B factory. Observation of loop diagram (penguin) decays will be the first evidence of such a process that was proposed almost 20 years ago. Comparisons with the Standard Model predicitons could give us indirect evidence of new physics at a higher mass scale. An inclusive measurement of the electromagnetic penguin decay ($b \to s\gamma$) holds most promise, since it suffers little from hadronization uncertainties.

The CESR machine at Cornell will deliver a total integrated luminosity of ~ 30 fb^{-1} by 1998 according to the upgrade plan. With such a large data sample, we will be able to carry out comprehensive study of B physics before the arrival of an asymmetric B factory. Present limits on penguin decays from CLEO is already within a factor of 2–3 from the expected SM rates.

Searches for new physics in charm, τ, and Υ should not be neglected. Two orders of magnitude improvements in most of the present limits are expected at a B factory.

As a last note, I would like to remind you that the observation of CP violation in 1964 and the discovery of the *tau* lepton in 1975, as examples, were unexpected surprises that challenged the "Standard Model" of the day. I hope that the Standard Model as we now know it will encounter such a surprise in the near future.

10. Acknowledgements

I would like to thank Martin Perl and Gregor Herten for organizing this stimulating conference, and the ICTH for the great hospitality during my stay in Trieste. I thank Tom Browder for useful discussions and help during preparation of this paper.

REFERENCES

1. J. Christenson et al., *Phys. Rev. Lett.* **13** (1964) 138.
2. M. Kobayashi and T. Maskawa, *Prog. Theo. Phys.* **49** (1973) 652.
3. For a review of B meson lifetime measurements, see W.B. Atwood and J.A. Jaros, SLAC-PUB-5671 (1991), also in *B Decays*, ed. S. Stone, World Scientific, Singapore, 1992.
4. H. Albrecht et al. (ARGUS Collaboration), *Phys. Lett.* **192B** (1987) 245.
5. L.-L. Chau and W.-Y. Keung, *Phys. Rev. Lett.* **53** (1984) 1802;
   M. Gronau and J. Schechter, *Phys. Rev. Lett.* **54** (1985) 385;
   J.D. Bjorken and I. Dunietz, *Phys. Rev.* **D36** (1987) 2109.
6. P. Drell and J.R. Patterson, CLNS 92/1177, Plenary session report at the XXVI International Conference on High Energy Physics, Dallas, Aug. 1992.
7. "Feasibility Study for an Asymmetric B Factory based on PEP", SLAC-PUB-4476 (1987); "Task Force Report on Asymmetric B-Factory at KEK", KEK report (1990); "Novosibirsk B-Factory", Novosibirsk INP preprint 90-56 (1990); "Proposal for a B factory Upgrade for the Cornell Electron Storage Ring, CLNS 91-1041-43 (1991); "Helena: A Beauty Factory in Hamburg", DESY-92-041 (1992).
8. "Proposal for a B factory Upgrade for the Cornell Electron Storage Ring", CLNS 91-1041-43 (1991);
   K. Berkelman, *The CESR B Factory*, CLNS 92/1181 (1992).
9. K. Oide and K. Yokiya, *Phys. Rev.* **A40** (1989) 315.
10. "Proposal for an Electron Positron Collider for Heavy Flavour Particle Physics and Synchrotron Radiation", PSI-PR-88-09 (1989).
11. Particle Data Group, Review of Particle Properties, *Phys. Rev.* **D45** (1992).
12. G. Altarelli and S. Petraca, *Phys. Lett.* **261B** (1991) 303.
13. N. Isgur et al., *Phys. Rev.* **D39** (1989) 799.
14. J.G. Körner and G.A. Schuler, *Z. Phys.* **C38** (1988) 511.
15. M. Wirbel, B. Stech, and M. Bauer, *Z. Phys.* **C29** (1985) 637.
16. J.G. Körner and G. A. Schuler, *Phys. Lett.* **226B** (1989) 185.
17. S. Sanghera et al. (CLEO Collaboration), CLNS 92/1156 (1992).
18. M. Gronau and S. Wakaizumi, *Phys. Rev. Lett.* **68** (1992) 1814.
19. R. Schindler, in the *Proceedings of the SLD Physics Week*, Kirkwood, CA Jul. 1989; P. Roudeau, LAL-89-21 (1989).
20. The Physics Program of a High-Luminosity Asymmetric B Factory at SLAC,

SLAC-353 (1989); Workshop on Physics and Detector Issues for a High-Luminosity Asymmetric B Factory at SLAC, SLAC-373 (1991); CESR B Factory (Physics Rationale for a B-Factory), CLNS 91-1043 (1991).

21. S. Glashow, J. Iliopoulos and L. Maiani, *Phys. Rev.* **D2** (1970) 1285.
22. S. Bertolini, DESY 89-169 (1989).
23. B.A. Campbell and P.J. O'Donnell, *Phys. Rev.* **D25** (1982) 1989.
24. S. Bertolini, F. Borzumati and A. Masiero, *Phys. Rev. Lett.* **59** (1987) 180; N.G. Deshpande et al., *Phys. Rev. Lett.* **59** (1987) 183; B. Grinstein, R. Springer and M.B. Wise, *Nucl. Phys.* **B339** (1990) 269.
25. A. Ali, DESY 91-080 (1991).
26. T. Altomari, *Phys. Rev.* **D37** (1988) 677.
27. C.A. Dominguez, N. Paver and Riazuddin, *Phys. Lett.* **214B** (1988) 459.
28. N.G.Deshpande, P. Lo, and J. Trampetic, *Z. Phys.* **C40** (1988) 369.
29. A. Ali, T. Ohl and T. Mannel, DESY 92-113 (1992).
30. N. Isgur and M. Wise, *Phys. Lett.* **232B** (1989) 113; **237B** (1990) 527.
31. G. Burdman and J.F. Donoghue, *Phys. Lett.* **270B** (1991) 55; P.J. O'Donnell and H.K.K Tung, U. of Toronto report UTPT -92-12 (1992).
32. M. Battle et al. (CLEO Collaboration), Contributed paper to the Joint International Lepton-Photon Symposium and Europhysics Conference on High Energy Physics, Geneva (1991).
33. A. Ali and C. Greub, DESY 92-048 (1992).
34. A. Ali and C. Greub, *Phys. Lett.* **259B** (1991) 182.
35. T. Lesiak et al., *Z. Phys.* **C55** (1992) 33.
36. S. Bertolini, F. Borzumati and A. Masiero, MPI-ph/91-71 (1991), also in *B Decays*, ed. S. Stone, World Scientific, Singapore, 1992.
37. W.S. Hou, A. Soni and H. Steger, *Phys. Lett.* **192B** (1987) 441; J.L. Hewett, *Phys. Lett.* **193B** (1987) 327; N.G. Deshpande and J. Trampetic, *Phys. Rev.* **D40** (1989) 3773.
38. J.C. Pati and A. Salam, *Phys. Rev. Lett.* **31** (1973) 661; R.N. Mohapatra and J.C. Pati, *Phys. Rev.* **D11** (1975) 566,2558.
39. D. Cocolicchio et al, *Phys. Rev.* **D40** (1989) 1477.
40. G. Beall, M.Bander and A. Soni, *Phys. Rev. Lett.* **48** (1982) 848; G. Ecker and W. Grimus, *Nucl. Phys.* **B258** (1985) 258.
41. I.I. Bigi and J.M. Frere, *Phys. Lett.* **110B** (1982) 255 ; J. Donoghue and B. Holstein, *Phys. Lett.* **113B** (1982) 382.

42. J. Carr et al , *Phys. Rev. Lett.* **51** (1983) 627; *Erratum – ibid* **51** (1983) 1222.

43. B. Grinstein and M.B. Wise, *Phys. Lett.* **201B** (1988) 274;
T.G. Rizzo, *Phys. Rev.* **D38** (1988) 820;
W.S. Hou and R.S. Willey, *Phys. Lett.* **202B** (1988) 591;
V. Barger, J.L. Hewett and R.J.N. Phillips, *Phys. Rev.* **D41** (1990) 3421.

44. For a review, see *The Higgs Hunters Guide* by J.F. Gunion, H.E. Haber, G.L. Kane and S. Dawson, Addison Wesley, Menlo Park, CA 1989.

45. H.P. Nilles, *Phys. Rep.* **110** (1984) 1.

46. V.Barger, M.S. Berger and R.J.N. Phillips, UW Madison report, MAD/PH/730 (1992).

47. J.L. Hewett, U. of Oregon report, OITS-501 (1992).

48. S. Bertolini et al., *Nucl. Phys.* **B253** (1991) 591.

49. S. Bertolini, F. Borzumati and A. Masiero, *Phys. Lett.* **192B** (1987) 437; *Nucl. Phys.* **B294** (1987) 321.

50. M.A. Shifman, A.I. Vainshtein, and V.I. Zakharov, *Nucl. Phys.* **B120** (1977) 316.

51. For a review, see J. Rosner, *Heavy Quarks, Quark Mixing, and CP Violation*, EFI 90-63 (1990).

52. R. Grigjanis et al., *Phys. Lett.* **224B** (1989) 209.

53. N.G. Deshpande and J. Trampetic, *Phys. Rev.* **D41** (1990) 895;
L.-L. Chau et al., *Phys. Rev.* **D43** (1991) 2176.

54. S. Sanghera et al. (CLEO Collaboration), Contributed paper to the XXVI International Conference on High Energy Physics, Dallas, Aug. 1992.

55. M. Artuso et al. (CLEO Collaboration), Contributed Paper to the XXVI International Conference on High Energy Physics, Dallas, Aug. 1992.

56. L. Wolfenstein, *Phys. Rev. Lett.* **13** (1964) 562.

57. J. Liu and L. Wolfenstein, *Phys. Lett.* **197B** (1987) 536.

58. I.I. Bigi and S. Wakaizumi, *Phys. Lett.* **188B** (1987) 501;
M. Tanimoto, Y.Suetaka and K. Senba, *Z. Phys.* **C40** (1988) 539;
T. Hasuike et al., *Phys. Rev.* **D41** (1990) 1691;
D. London, *Phys. Lett.* **234B** (1990) 354.

59. G. Ecker and W. Grimus, *Z. Phys.* **C30** (1986) 293;
D. London and D. Wyler, *Phys. Lett.* **232B** (1989) 503.

60. I.I. Bigi and F. Gabbiani, *Nucl. Phys.***B352** (1991) 309.

61. C.O. Dib, D. London and Y. Nir, SLAC-PUB-5323 (1990);
Y. Nir and H.R. Quinn, SLAC-PUB-5643, also in *B Decays*, ed. S. Stone,

World Scientific, Singapore, 1992.

62. S. Egli et al., ETHZ-IMP PR/92-1 (1992).
63. M. Perl, *Rep. Progr. Phys.* **55** (1992) 1, and references therein.
64. J.Z. Bai et al., *Phys. Rev. Lett.* **69** (1992) 3021.
65. J.P. Alexander et al., CLNS 92/1141 (1992).
66. T.N. Truong, *Phys. Rev.* **D30** (1984) 1509;
F.J. Gilman and S.H. Rhie, *Phys. Rev.* **D31** (1985) 1066.
67. M. Perl, in these Proceedings.
68. R. Arnowitt and P. Nath, *Phys. Rev. Lett.* **66** (1991) 2708.
69. A. Bean et al. (CLEO Collaboration), CLNS 92/1172 (1992).
70. F. Wilczek, *Phys. Rev. Lett.* **39** (1977) 1304.
71. P. Franzini, *Phys. Rev.* **D35** (1987) 2883.
72. H. Albrecht, *Z. Phys.* **C42** (1989) 349.
73. F. Wilczek, *Phys. Rev. Lett.* **40** (1978) 279;
S. Weinberg, *Phys. Rev. Lett.* **40** (1978) 223.
74. D. Antreasyan et al., *Phys. Lett.* **251B** (1990) 204.
75. P. Fayet and J. Kaplan, *Phys. Lett.* **269B** (1991) 213.
76. D. Besson, Talk presented at the Spring Meeting of the American Physcial Society, April, 1992.

SSCL–599

The SSC: Programme and Searches for New Particles

Benjamín Grinstein†

Superconducting Super Collider Laboratory
2550 Beckleymeade Ave, Dallas, Texas 75237

Talk presented at the Trieste Workshop on the Search for New Elementary Particles: Status and Prospects.

10/92

† email: grinstein@sscvx1.bitnet, @sscvx1.ssc.gov

THE SSC: PROGRAMME AND SEARCHES FOR NEW PARTICLES

BENJAMIN GRINSTEIN
Superconducting Super Collider Laboratory
2550 Beckleymeade Ave, Dallas, Texas 75237

1. Introduction

The Superconducting Super Collider (SSC) Laboratory is under construction. Many critical components have been designed, small and full scale prototypes built and successfully tested, and in some cases samples have been produced by industry. The 'Project', as we call the task of building the SSC Lab, is well under way. In this talk I will describe the Project and give an idea of what its status is to date.

The physical potential of the SSC is extraordinary. Whatever is responsible for the spontaneous breaking of the electroweak symmetry, it must loom at the energies accessible at the SSC. It will produce top quarks copiously. And, most significantly, it will probe the shortest distance scales ever, opening up a new frontier in High Energy Physics. Also in this talk, I will touch on some of the physics of the SSC.

I end the introduction with a disclaimer. This talk is about subjects outside my field of expertise. Describing the Project would be better left to an administrator or manager, while a discussion of what could be found and how, could certainly be done more properly by one of the many brilliant experimenters involved in the design of the major detectors for the SSC. Not all is lost, though. You will get what I hope is an interesting and peculiar, if not illuminating, insider/outsider's view. Insider's because I am employed by the SSC, and so I am embedded in the day to day discussions about the Project. Outsider's because I am neither a manager nor an experimentalist.

Most of the material covered in this talk comes from conversations with my SSC colleagues and from their transparencies. I have also used some technical documents.[1,2]

2. The Project

2.1 The accelerator systems

The SSC is a facility for producing two counter-rotating high energy and high intensity proton-proton beams, and for studying their collisions. The design of the facility was made on the premise that interesting physics would be accessible to the laboratory if the center of mass energy were to be at least 20 TeV, and that integrated luminosities of $10^{40}\mathrm{cm}^{-2}$ would be needed to make discoveries possible. The design parameters of each of the two collider rings were chosen to meet these criteria; see table 1.

Table 1. SSC Parameters

Proton Energy	20 Tev
Circumference of Rings	87 Km
Protons per r.f. bunch	0.75×10^{10}
Bunch Spacing	5 m
Number of Bunches(per ring)	17,424
Total Particle Energy (per ring)	418×10^{6} J
Emittance (RMS)	1π mm-mrad
Interaction Region Focal Spot Size RMS Radius ($\beta^* = 0.5$m)	5×10^{-6} m
p-p collisin rate	60 MHz
Luminosity	1×10^{33} cm^{-2} sec^{-1}
Synchrotron Radiation Power	8.75 KW (per ring)

Four stages of injection will accelerate protons from zero to 2 TeV, for delivery into the collider rings. The parameters of these accelerators are given in table 2.

The first stage, the linac, is really an array of several systems that work in series. An Ion Source is followed by the Low Energy Beam Transport. The 2.2m long Radio Frequency Quadrupole accelerates the H^-s from 35 keV to 2.5 MeV, which are then accelerated to 70 MeV in the 23m Drift Tube Linac. The last system of the linac is the 120m Coupled Cavity Linac which boosts the ions to an energy of 600 MeV. Three days before this talk was given news came out of a $4.9 million general construction contract awarded to Sedalco, Inc. of Fort Worth, Texas, to begin construction of the facilities that will house the Linac.

Table 2. Injector and Collider Parameters.

	Linac	LEB	MEB	HEB	Collider
Kinetic Energy (GeV)	0.6	11.1	180–200	2,000	20,000
Circumference (Km)	(length)0.11	0.54	3.96	10.89	87.12
Cycle Time	0.1sec	0.1sec	3sec	2min	~24 hr
Protons per Cycle	—	10^{12}	8×10^{12}	2×10^{13}	1.3×10^{14}
Protons per Bunch	—	10^{10}	10^{10}	10^{10}	0.75×10^{10}
Normalized RMS Emittance (π mm-mrad)	< 0.5	0.6	0.7	0.8	1.0

The Low and Medium Energy Boosters use warm magnet systems, while the High Energy Booster and the Collider use cold magnets. The design of the High Energy Booster is both interesting and challenging: to provide protons to the two collider rings it must handle rotating and counter-rotating beams, and this is accomplished by quickly alternating the polarity of the dipoles.

It is interesting that the location chosen for the collider has rather favorable geological attributes. Most of the collider tunnel will be bore through a layer of

Austin Chalk, a type of limestone. Below this lays some Eagle Ford Shale. My knowledge of geology is very limited; my son reads to me from his book that 'limestone is made of shells all pressed together, sandstone is made of grains of sand all pressed together, shale is made of mud and clay all pressed together'. The point is that, under pressure, shale behaves like mud but limestone is more like concrete. Earlier this year a decision was made to locate the experimental halls for the large multipurpose detectors (see below) on the east side. With the main campus on the west side, this is certain to make it inconvenient to the users of those east side facilities. The reason for the decision is that experimental halls for large detectors on the west side would reach down into the shale, making it hard to stabilize a very heavy detector.

2.2 Magnet Systems

The success of this enterprise lies heavily on the ability to produce industrially high quality superconducting dipoles. So not only must dipoles be designed and constructed in laboratories that will conform to demanding specifications, but the technology has to be transfered to industry.

Table 3 displays the characteristics of the collider's dipoles. The performance of dipoles constructed both by FNAL and BNL, and by technicians from industry working at these labs is superb. Typical quench sequences (at 4.35°K) involve at most one quench below the operating current of 6500 Amps (on the first ramp up), with every subsequent quench occurring always at currents exceeding 7000 Amps.

Table 3. Parameters of SSC Superconducting Dipoles (and Quadrupoles)

Operating Field	6.6 T
(Operating Gradient)	206 T/m
Operating Current	6.5 kAmps
Stored Energy	1.58 MJ (1.32 MJ)
Length	15.8 m (13.3 m)
Inner Coil Aperture	50 mm
Cold Mass	12,700 kg (10,800 kg)
Conductor	NbTi
J_c (5T, 4.2°K)	2750 Amps/mm
Filament Diameter	6 μm
Operating Temperature	4.35°K
Dipoles (Quads) per Ring	4230 (832)

The Accelerator Systems String Test (ASST) has as objective operating a half cell of five collider dipole magnets, one quadrupole magnet, and two spool pieces at the design current of 6500 Amps. It is housed in a long structure built to reproduce precisely the (almost imperceptible) actual curvature of the tunnel. A half cell is the minimum reproducible unit of which the collider arcs are built. Meeting the AAST objectives successfully is necessary to demonstrate that the many different systems that go into a minimal collider unit can work together.

2.3 Other systems and facilities. Schedule

The SSC Laboratory will be a complex array of very many systems. It is impossible to properly describe them here, so I won't. Just as an example of what is involved, the cryogenic systems for the collider will collectively perform as the biggest cooling plant in the world. There are also satellite facilities with important functions. For example, in collaboration with Southwest Medical Center a medical applications station has been planned. It will use H^- beams from the linac for cancer therapy, among other applications.

Table 4. SSC Baseline Schedule: Major Project Milestones.

Description	*Schedule*
Baseline Validation Complete	July 1990
A-E/CM Letter Contract & NTP	August 1990
CDM Authorization to Incur Costs	November 1990
SEIS Record of Decision (ROD)	February 1991
Begin Conceptual Design for Detectors	February 1991
Start SSC Civil Construction	February 1991
Accelerator String Test Complete	**September 1992**
Notice to Proceed Experiment Halls	June 1993
Full-rate Production Decision on Magnets	April 1994
Start First Half Sector – CDM Delivery	April 1994
First Collider Half Sector – Start Installation	June 1994
LINAC Start Commissioning (600 MeV)	October 1994
First Collider Half Sector – Start Cooldown	March 1995
LEB Start Commissioning	October 1995
Beneficial Occupancy of Large Experimental Halls	January 1996
MEB Start Commissioning	June 1996
HEB Start Installation	August 1996
MEB Test Beams Available	January 1997
HEB Start Commissioning	September 1998
West Detectors – Start Commissioning	March 1999
Collider – Start Commissioning	March 1999
Beam to Experiments (End of Project/Begin Ops.)	September 1999

At the beginning of time (1990) a document, the 'SSC Baseline' was created. (As far as the Project is concerned, the beginning of time *is* the baseline). In varying degrees of detail, the baseline specs out the project, its timetable and cost. Table 4 lists the major project milestones together with their baseline schedule. Of note is that the Accelerator Systems String Test (see 2.2, above) met its objectives six weeks ahead of schedule (at 11:39 AM, on Friday, August 14). The cooldown for the key test began in June and the current was slowly raised in the string, testing the quench protection system along the way. The test demonstrated the quality of the industrially assembled magnets and the associated power, cooling, and control equipment to operate together successfully as a system.

Also of note, beam should be delivered to experiment by September 1999. Of course, it will take some additional time before the beam to experiment achieves design luminosity. Studies of operation procedures are currently under way.

3. Experimental Program

3.1 Generalities

The initial experimental program of the SSC should match the physics potential and the investment in such a discovery machine. It was therefore decided, upon recommendation of the advisory panels, that the initial program should involve:

- ▷ Two major, complementary detectors aimed at the physics opportunities opened at multi-TeV energies; and
- ▷ Several smaller experiments, adding diversity and coverage of phenomena at low transverse momentum and B-physics.

The experimental program of the SSC is proposal driven. Expressions of Interest were called for starting in May 1990. By the time this talk was given 21 Expressions of Interest had been submitted, involving some 2300 collaborators from about 350 institutions worldwide. They include large all purpose detectors and smaller specialized detectors. The latter include, among others, a proposal for study of "Low p_T Physics at the SSC", four for studies of B decays (and CP violation), one proposal for "A Very Long Baseline Neutrino Oscillation Experiment" and one for "A Full Acceptance Detector for SSC Physics at Low and Intermediate Mass Scales".

(If I understand correctly) Decisions on the experimental program are made by the Director of the Laboratory upon recommendation from advisory panels (the Scientific Policy Committee and the Program Advisory Committee), constituted from renowned international experts.

3.2 Status of the Two Major All Purpose Detectors

The Program Advisory Committee stated (July 1990) that "A healthy initial program requires two detectors with complementary as well as overlapping strengths that address the physics at high p_T". That their strengths be complimentary has the obvious effect of increasing the total capabilities and discovery reach. Overlapping strengths allow for cross checks of discoveries. And, of course, that there are two experiments with some overlapping capabilities introduces that element of competitiveness that brings the best out of scientists!

Out of the many Expressions of Interest for major detectors eventually two were selected to proceed forward with Letters of Intent and then Technical Design Reports:

GEM Its design Goals emphasize identification and precision measurement of gammas, electrons and muons, with capabilities for higher luminosity. The Letter of Intent was submitted on November 29, 1991, and in January 1992 was given approval to proceed towards a Technical Design Report. At present detector R & D and Engineering is under way, with some major decisions for calorimetry to be made by the fall of 1992.

SDC Its design Goals emphasize charged particle tracking, hermetic calorimetry, lepton energy measurement and identification and vertex detection. Letter of Intent was submitted on November 30, 1990, and in January 1991 was given approval to proceed towards a Technical Design Report. This was submitted on April 1, 1992. Detector R & D and Engineering is under way.

3.3 Physics

As an example of the physics that can be studied at the SSC, we focus on discovery of a standard model higgs. By 1999 LEPII will have either discovered the higgs or set a lower bound on its mass of about 80 GeV. Theoretical upper bounds on the standard model higgs mass are in the 600 GeV to 800 GeV range. Therefore a strategy for higgs discovery at SSC needs only cover the 80 GeV to 800 GeV mass range.

The main problem for almost any SSC experiment is the enormous two jet cross section. This translates into large, often unmanageable detector backgrounds. For example, the rate for higgs production anywhere in the above mass range is at least five orders of magnitude smaller than the rate of two jet events. Thus discovery in (or study of) purely hadronic decay channels, such as $H \to b\bar{b}$ is nearly impossible.

Discovery strategies are driven by the higgs decay fractions. For masses above $2M_Z$ the dominant decay modes are into pairs of weak vector bosons. The cleanest channel is higgs $\to ZZ \to \ell^+\ell^-\ell^+\ell^-$, but it is statistically limited. Not as clean is higgs $\to ZZ \to \ell^+\ell^-\nu\bar{\nu}$. Much more abundant are the processes higgs $\to ZZ \to \ell^+\ell^- +$ 2jets and higgs $\to W^+W^- \to \ell\nu +$ 2jets, which are respectively 20 and 150 times more frequent than the four charged leptons mode. They are less clean, as they involve backgrounds from $W/Z +$ jets and $t\bar{t}$, so in the end they give smaller signal to background ratios. The main backgrounds to higgs $\to ZZ \to \ell^+\ell^-\ell^+\ell^-$ are from $q\bar{q}$ or $gg \to ZZ$, $Z + Q\bar{Q}$ or $Q\bar{Q}$. An additional complication comes from the fact that the width of the higgs boson increases from 1.4 to 30 to 270 GeV as the mass increases from 200 to 400 to 800 GeV. Except at the upper end of this mass range, simple cuts seem to ensure discovery of the higgs boson in the four lepton channel in 1 SSC year (1 year of running at design luminosity, i.e., 10^{40}cm^{-2}). At the upper end of the mass range either the inclusion of other channels, or additional integrated luminosity, or both, would be necessary for discovery.

Below $2M_Z$ decay into two bottom quarks quickly becomes dominant, but the decay into ZZ^*, i.e., into Zl^+l^-, $Z\nu\bar{\nu}$ and $Zq\bar{q}$, is still significant and can be used for discovery in the range 120 GeV $\leq M_h < 2M_Z$. (The branching fraction into ZZ^* dips to a minimum at $\sim 2M_Z$, reaching a maximum at $\sim$ 150 GeV). The most important backgrounds for higgs $\to ZZ^* \to e^+e^-e^+e^-$ come from production of $t\bar{t}$, $Zb\bar{b}$ and $Zt\bar{t}$; it seems possible to bring these under control through simple transverse momentum and rapidity cuts. The situation is similar for other channels. The combined signal for $4e$, 4μ and $2e2\mu$ seems very significant after 1 SSC year.

In the lowest end of the mass range, 80 GeV $< M_h <$ 130 GeV, the decay mode of choice is into two photons. Although its branching fraction is small (approximately constant and $\sim 10^{-3}$ in the range 80 Gev$< M_h <$ 150 Gev), it is a relatively clean mode. Directly produced higgs with subsequent decay into two photons seems hard, if not impossible, at SDC and GEM; the irreducible backgrounds from $q\bar{q} \to \gamma\gamma$ and $gg \to \gamma\gamma$ are several orders of magnitude larger than the signal. If the higgs is produced in association with another particle which may be used to tag the event the prospects of discovery seem brighter. Studies by both SDC and GEM indicate that the combined $W +$ higgs and $t\bar{t}+$higgs signals may yield a 4-sigma discovery limit in this mass range within 1 SSC year. The mass resolution in this range clearly depends critically on the specific choices made for electromagnetic calorimetry, and is estimated to be somewhere between 0.5 GeV and 2.5 GeV.

Space limitations preclude us from discussing the many other discovery potentials of the SSC experiments. The list is long (Supersymmetric partners of standard particles, technicolor resonances, new gauge bosons, multiple higsses, and on and on), *but hopefully incomplete!* What makes the SSC so unique among present day accelerators is that it is designed to probe the dynamics underlying the breaking of the electroweak symmetry, the stuff of which matter is made. The mechanism responsible for this could be completely new and unanticipated!

4. Concluding Remarks

The odyssey has begun. Tunnels are being dug, magnets are being fabricated, systems are being tested. Two gargantuan experimental collaborations have been formed and are quickly transiting from the design into the development stage. The SSC counts with over 2200 employees, and $517 million in funding for Fiscal Year 1993 has been signed into law. If everything proceeds according to plan (and so far it has), we may very well have a first look at a completely new repertoire of high energy phenomena by 2001.

5. References

1. Technical Design Report, Solenoidal Detector Collaboration, 1 April 1992, SDC-92-201
2. Letter of Intent, GEM Collaboration, November 30, 1991, GEM-92-49

The Experimental Potential of HERA for the Discovery of New Particles

Ch. Berger
1. Physikalisches Institut
RWTH Aachen

Abstract

We discuss the production of leptoquarks, leptogluons, new heavy fermions (mostly neutrinos) and excited leptons at HERA. Production cross sections and signatures for a search for these particles are presented. Exclusion limits for an integrated luminosity of 100 pb^{-1} are given.

1 Introduction

The recently commissioned electron-proton storage ring HERA at the DESY laboratory in Hamburg, Germany, collides electrons of 30 GeV with protons of 820 GeV. It thus reaches a center of mass (*cms*) energy of 314 GeV in the *ep* system. An important task of HERA will be to search for traces of new physics beyond the Standard Model (SM). The cleanest signature would clearly be manifestations of new particles and hence this aspect received central attention in preparing for the experimental programme. In two workshops (1987 and 1991) it was concluded that with respect to new particles HERA is very well suited for a search for leptoquarks, leptogluons, excited leptons and (with some reservations) new heavy fermions. We therefore concentrate in the following on these items and refer the reader to [1] for more details.

2 Leptoquarks and Leptogluons

2.1 Cross Section

A large variety of theoretical models (a summary is given in ref [2]) motivate the existence of particles which carry lepton and baryon quantum numbers (L, B) and interact with the known fermions via Yukawa couplings. For leptoquarks[1], Buchmüller,

[1] The discussion mainly deals with leptoquarks but is equally valid for leptogluons, since the experimental aspects are very similar for both particle types (see below).

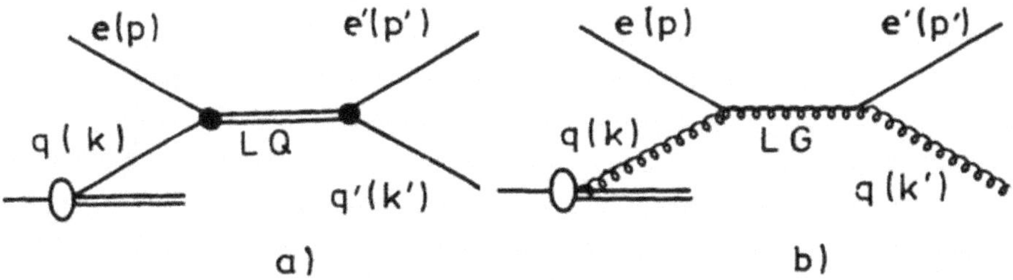

Figure 1: Feynman diagrams for direct channel production of leptoquarks (a) and leptogluons (b). The symbols in brackets denote the four-vectors of the particles.

Rückl and Wyler [3] have given a model independent description of all possible dimensionless, B and L conserving couplings, which are invariant with respect to the standard model transformations. Present bounds on these couplings are summarized in detail in ref. [2]. Here only those limits which are relevant for masses below the *cms* energy of HERA are mentioned:

- Limits for flavour changing neutral currents demand that family violating couplings are suppressed. Therefore, assuming B and L conservation, leptoquarks produced in *ep*– collisions will decay into eq or $\nu q'$.
- Bounds from rare pion decays forbid for each leptoquark species the existence of both left *and* right handed couplings. Therefore only chiral couplings are used in the following.
- The measurements of the Fermi constants in μ and β decay constrain (in certain models) the strength of some of the lefthanded Yukawa couplings λ_L. This limit may be parameterized as
$$\lambda_L < \frac{M_{LQ}}{1.7 \text{ TeV}}$$
- In $p\bar{p}$ collisions at CERN, scalar leptoquarks have been excluded [4] independent of their Yukawa couplings for masses up to 72 GeV if only decays into electrons are allowed and up to 67 GeV if the branching ratio into electrons and neutrinos are equally high. Limits from CDF are expected to exceed 100 GeV, but are not yet published.

At HERA the process with the largest production cross section for leptoquarks and leptogluons is the fusion of the incoming electron and a parton forming an s-channel resonance (fig.1). The final state consists of the spectator jet from the proton remnant together with a lepton and a jet from the decay of the resonance, which are balancing each other in p_T. This signature is identical to the neutral and charged current modes in deep inelastic scattering (DIS). This means that both processes will interfere and

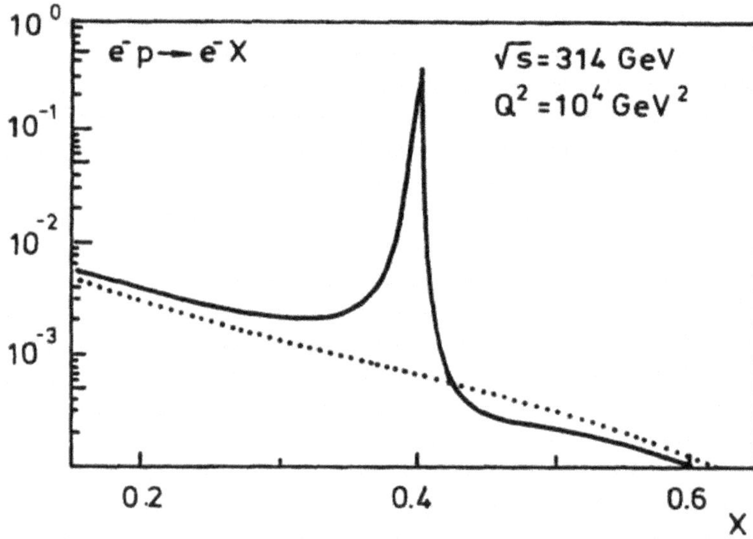

Figure 2: Production cross section for leptoquarks with mass 200 GeV as tunction of x [3] .

that they can only be separated by statistical and not by topological arguments. In fact, most of the kinematic considerations given here can be applied to DIS analysis and vice versa.

The mass of the leptoquark or leptogluon M ($M = M_{LQ}, M_{LG}$) would become visible as a peak in the Bjorken-x distribution (fig.2) at

$$x_0 s \approx (p+k)^2 = M^2 \tag{1}$$

where $\sqrt{s} = 314$ GeV is the center of mass energy and p, k denote the four-vectors of the beam electron and the incoming quark (or gluon).

For all leptoquark species an overview of the cross sections is given in ref. [5]. The production cross sections can be expressed as

$$\sigma_{ep\to LQ} = \frac{\pi}{4s}\lambda^2 q(x_0, Q^2) \times 1(2) \tag{2}$$

with the factor 1(2) for the scalar (vector) leptoquarks. λ is a dimensionless coupling constant and $q(x_0, Q^2)$ is the quark density in the proton for the quark flavour(s) which couple to the leptoquark species. This formula applies only if the natural width of the particle is very small. For the small couplings detectable at HERA, this is a valid approximation because the partial decay width

$$\Gamma_{LQ} = \lambda^2 M_{LQ} \times \frac{1}{16\pi(24\pi)} \tag{3}$$

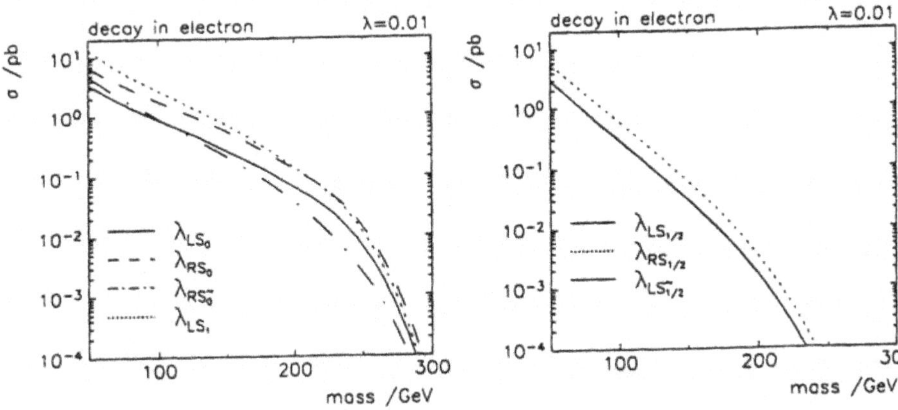

Figure 3: Leptoquark production cross section in ep collisons at HERA energies. L and R denote left and righthanded couplings, S refers to scalar leptoquarks with an index giving the weak isopsin. S and $\tilde{S}$ can be distinguished by their charge. The structure functions are taken from ref. [8] at a scale Q^2.

(16π for scalar, 24π for vector leptoquarks) is typically of the order $O(100\,\mathrm{MeV})$. The spin of the leptoquarks also determines the y dependence of the cross sections, where y is as usual defined by

$$y = \frac{(p-p')\cdot k}{p\cdot k} \quad . \tag{4}$$

In their rest frame, scalar particles decay isotropically. Since the lepton scattering angle with respect to the proton direction, Θ, is related to the squared four momentum transfer Q^2 like $1+\cos\Theta$, and $Q^2 = x_0 ys$, the cross section is flat in y. Vector particles however are produced with a y-dependence $\sim (1-y)^2$. The production cross section at HERA for scalar leptoquarks decaying into electrons and quarks respectively antiquarks is shown in fig. 3

The coupling of gluons to leptons in the production of leptogluons is of the magnetic type [7] like in e^* production discussed below. It is therefore described by a parameter Λ with dimension of a mass which is related to the scale of the underlying interaction. For leptogluons with spin 1/2, similar arguments like given above result in a shape of the cross section proportional to $(1-y)$. The leptogluon decay width depends strongly on the ratio M_{LG}/Λ,

$$\Gamma_{LG} = \frac{\alpha_s}{4}\frac{M_{LG}^3}{\Lambda^2}. \tag{5}$$

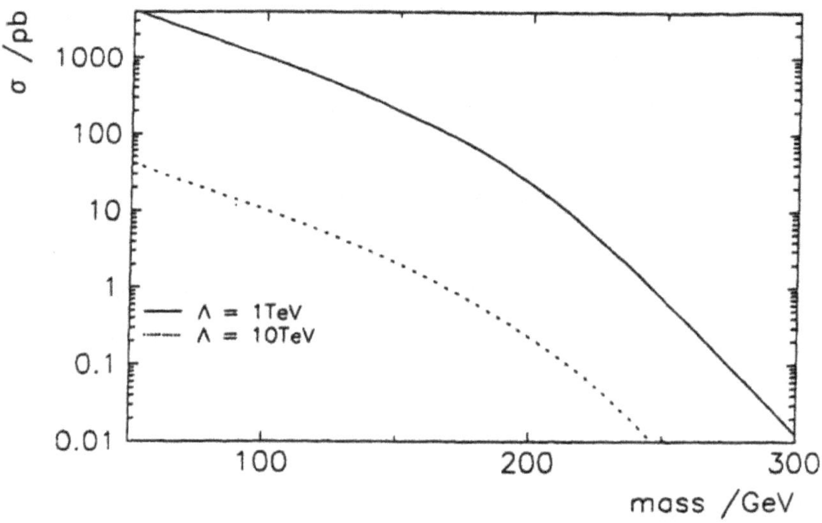

Figure 4: Leptogluon production in ep collisions. The structure function is taken from [8] at a scale Q^2.

Here α_s is the strong coupling constant. For small widths ($M_{LG} \ll \Lambda$) the total cross section is

$$\sigma_{ep\to LG} \approx \frac{2\pi^2\alpha_s}{s}\left(\frac{M_{LG}}{\Lambda}\right)^2 g(x_0, Q^2) \tag{6}$$

where $g(x_0, Q^2)$ is the gluon density in the proton. In fig. 4 the cross section is calculated for $\Lambda = 1$ and 10 TeV.

2.2 Limits

Since the topologies of leptoquark events and of deep inelastic scattering events are indistinguishable[2] and the natural width of leptoquarks is very small, an optimum reconstruction of the leptoquark mass is of greatest importance for the suppression of background.

There are basically two different approaches to compute the mass. It is possible to use the decay products of the leptoquarks, the electron and the quark (antiquark). This method has been tried out using a Jade algorithm [9] in order to define the quark jet. The main deficiency of this method is the imperfection of the jet reconstruction

[2]This also means that, without modifications, the standard trigger set-up for deep inelastic events can be used.

giving rather moderate mass resolutions (10– 20 GeV) with long tails to lower and higher masses.

A much more efficient approach is to use the relation $M_{LQ} = \sqrt{x_0 s}$. However, in contrast to the previous method, this introduces effects of radiative corrections which depend on the reconstruction scheme used for x and y or M_{LQ} and y respectively. For neutral current events the kinematic variables can be reconstructed from the electron alone or from the hadrons in the detector using the Jacquet-Blondel method [10]. The mass resolution for both methods is of course different and depends in a different way on the value of y. This can be studied in figs. 5a and b where leptoquark events with a mass of 100 GeV have been reconstructed in the $y - M_{LQ}$ plane including initial state bremsstrahlung.

For neutral current events the kinematics are overdetermined and many reconstruction algorithms are possible using combinations of the electron and the hadron measurements. All these combinations vary in their systematic behaviour in the presence of bremsstrahlung and, since the electron and the hadrons are measured with different precision, the resolutions for x, y, and Q^2 depend strongly on the kinematic region considered.

For overdetermined systems, the standard method promising optimum resolutions is a constrained kinematic fit. This is independent of the considered kinematic region because in any case all event information, including measurement errors, are taken into account. This idea has been studied by P. Schleper for the analysis of leptoquarks [5],[6]. The results of the fit are an improved reconstruction of the electron and hadron vectors and the radiated photon, which was not seen in the detector, is also reconstructed. The result of the fitting procedure is clearly a big improvement of the mass resolution (fig 5c). Also the tails due to initial state radiation are recovered. The experimental width (≈ 2 GeV for a typical detector like H1 at HERA) has become almost independent of y which very much simplifies a further separation from deep inelastic scattering background.

Neutral and charged current events are characterized by just two kinematic variables if the effect of initial state bremsstrahlung is corrected for. Since the mass resolution and y are fairly uncorrelated (fig. 5) these variables are chosen to separate leptoquark events from the deep inelastic scattering events. The separation is performed with independent upper and lower cuts for M_{LQ} and y. In an optimization procedure these four cuts are varied until best limits are obtained. Because of the varying y-shapes of background and signal, this has been carried out separately for neutral and charged current events of scalar and vector leptoquarks and for leptogluons. Only events at very high and very low y are excluded ($0.05 < y < 0.95$) since in these regions radiative corrections are expected to become important for the DIS cross section. The much better mass resolution, obtained by the fit in comparison to other methods, allows to extend the measurement to much lower y. As result of the

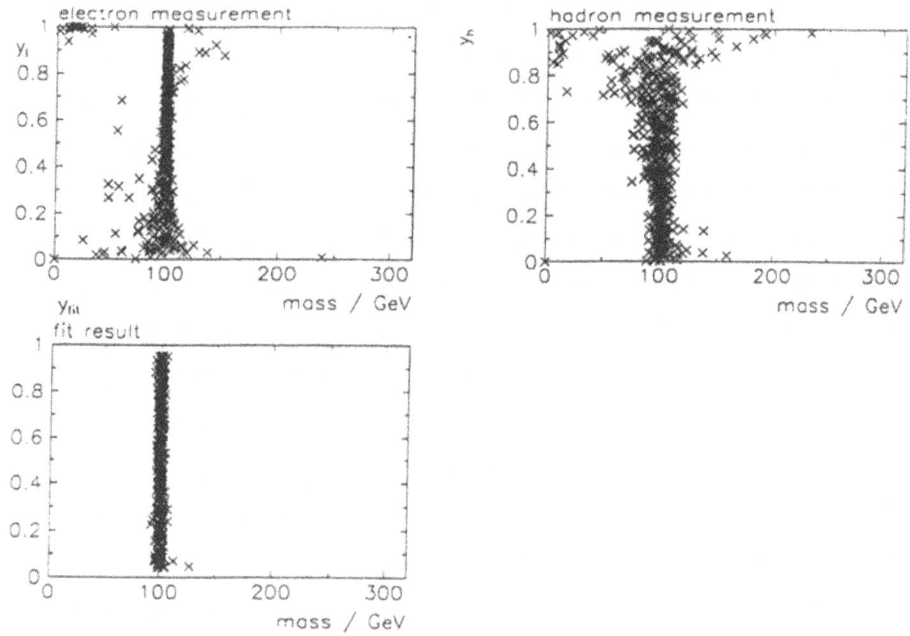

Figure 5: Reconstruction of leptoquark events in the $y - M_{LQ}$ plane with different methods, a) from electron alone, b) from hadrons with the Jacquet-Blondel method, c) using the constrained fit.

optimization procedure, leptoquarks with high masses can be separated from background in almost the whole y range. At lower masses the DIS background becomes more and more significant. For scalar leptoquarks decaying into an electron, the lower y cut has to be raised to 0.4 for M_{LQ} = 200 GeV and to 0.6 for M_{LQ} = 100 GeV. For vector particles the cross section drops at high y making the low y region even more important. This also holds for decays into neutrinos because of the decreasing mass resolution obtained at high y with the hadrons alone.

Now it is straightforward to obtain limits in terms of masses and coupling constants for any specific model if the branching ratios into electron or neutrino are known. We again consider leptoquarks in the framework of B and L conserving, dimensionless, $SU(3)_c \times SU(2)_L \times U(1)_Y$ invariant couplings of ref. [3]. Here the branching ratios are well defined and limits in the mass versus coupling constant plane are obtained. In fig. 6 results are shown for the leptoquark types of fig. 4. Naturally the limits are much better for leptoquarks with couplings to quarks than to anti-quarks. It would, however, be the opposite situation if a positron beam becomes

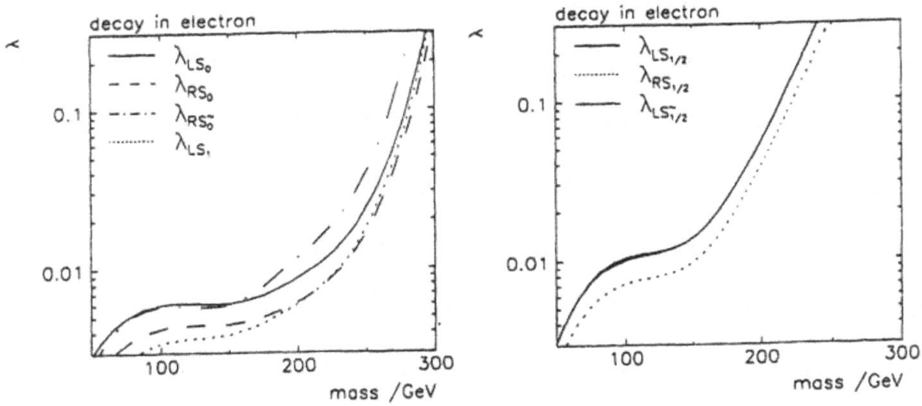

Figure 6: Exclusion limits for coupling constants and masses of the leptoquarks of fig. 4 at 95% confidence level for $L_{int} = 100\ \text{pb}^{-1}$.

available at HERA. For a leptoquark discovery at HERA with a significance of 5 standard deviations the limits in fig. 6 have to be scaled up by roughly a factor 2.

From this very general description, also the limits for more specific models can be obtained. As an example, a recent suggestion [11] to 'save' $SU(5)$ from disagreement with experimental data can be tested at HERA. In this model leptoquarks of the type $\tilde{S}_{1/2}$ are introduced. The corresponding limit in fig. 6 is directly valid for this model.

As discussed above, the most restrictive bounds from other experiments are $M_{LQ} >$ 72 GeV and $\lambda_L < M_{LQ}/1.7$ TeV. Obviously HERA has a great potential to improve these bounds or even establish the existence of such exotic particles, for masses up to almost the kinematic limit and for Yukawa couplings at least one order of magnitude below the region observable up to now. The Yukawa couplings envisaged are already much weaker than for instance the electromagnetic gauge interaction. As an example for a scalar leptoquark decaying with equal probability into electron and neutrino, couplings of $\lambda^2/4\pi = 4 \cdot 10^{-4} \cdot \alpha_{em}$ at $M_{LQ} = 100$ GeV ($\lambda^2/4\pi = 4 \cdot 10^{-3} \cdot \alpha_{em}$ at $M_{LQ} = 250$ GeV) might be excluded.

The remarkable sensitivity of the leptogluon process for the scale Λ is shown in figure 7. The high production cross section will allow to explore values between $\Lambda = 80$ TeV for low masses and $\Lambda = 1$ TeV close to the kinematic limit.

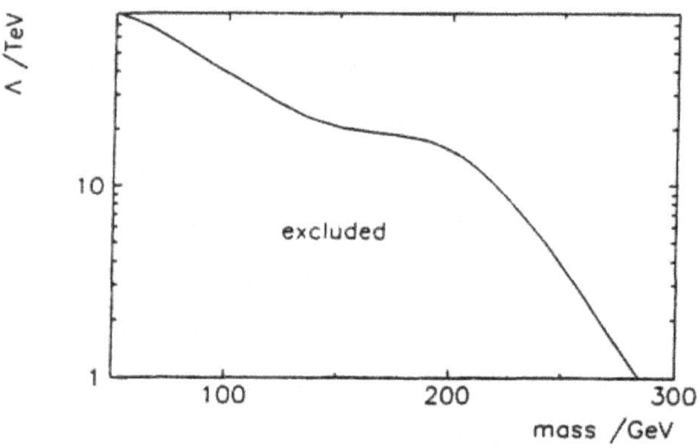

Figure 7: Exclusion limit for the scale Λ from leptogluon production at 95% confidence level.

3 Heavy Fermions

3.1 Cross section

Heavy neutrinos which for example decay into $e\mu\nu_\mu$ lead to $e\mu$ events with a large missing transverse momentum and thus form a typical exotic signature at HERA. Given an ideal detector with 100% efficiency and perfect background suppression the nonobservation of a signal constitutes an exclusion limit at 95% CL of

$$\sigma BR \leq \frac{3}{L_{int}} \tag{7}$$

which results in $\sigma BR = 3 \cdot 10^{-2}$ pb for an integrated luminosity L_{int} of 100 pb^{-1}. Because this product of cross section and branching ratio (BR) is in the range of a standard model weak interaction process like electroproduction of W-bosons the discovery potential of HERA is not very high. On the other hand the reactions in question are of fundamental importance and certainly deserve a close look in preparing for the experimental programme.

An extension of the standard model with right handed neutrinos has been recently studied by Buchmüller and Greub [12]. In this model heavy Majorana-type neutrinos N are predicted with a charged current given by

$$J^{\mu}_{CC} = \bar{e}\xi\gamma^{\mu}\frac{1-\gamma^5}{2}N \ , \tag{8}$$

where e describes the electron spinor and ξ is a mixing parameter, which is experimentally bound to values $|\xi| \leq 0.1$. In contrast to the definition of ref.[12] the effect of an additional leptonic Kobayashi-Maskawa matrix is already included in the ξ parameter. The production and decay of these neutrinos in electron proton interactions is described by the Feynman diagram of fig. 8. We define (neglecting the masses of electrons and quarks)

$$\mu = \frac{m_N}{\sqrt{\hat{s}}} \tag{9}$$

the mass of the heavy neutrino in units of the center of mass energy in the subprocess ($\hat{s} = 2p \cdot k$). The *cms* scattering angle Θ can then be expressed in terms of y and μ via

$$\cos\Theta = 1 - \frac{2y}{1-\mu^2} \ . \tag{10}$$

On the parton level the calculation of the production cross section is straightforward. The relevant helicity amplitudes are contained in table 1. Using these amplitudes we obtain the helicity averaged production cross section

$$\frac{d\sigma}{dy}(eq \rightarrow Nq') = \frac{G_F^2}{2\pi}|\xi|^2 F_W^2 \hat{s}(1-\mu^2) \tag{11}$$

off quarks and

$$\frac{d\sigma}{dy}(e\bar{q} \rightarrow N\bar{q}') = \frac{G_F^2}{2\pi}|\xi|^2 F_W^2 \hat{s}(1-y)(1-y-\mu^2) \tag{12}$$

off antiquarks. The corresponding cross sections for electroproduction off protons are readily calculated via the usual folding with the quark distribution functions inside the proton.

As an example we show in fig. 9 the cross section for the production of heavy Majorana neutrinos as function of m_N with $|\xi| = 0.1$. The scale for the parton distribution functions (Duke Owens set 1 [13]) is given by $\hat{s}$. At HERA already for $m_N = 160$ GeV the cross section decreases down to 0.1 pb. At LEP/LHC this limit is reached for $m_N = 720$ GeV! Our results agree with those given in ref.[12].

3.2 Limits

For the study of heavy fermion production these expressions are however not sufficient. The heavy neutrino decays with a very short lifetime ($\approx 10^{-22}$ s [12]) into 3 fermions, which have to be followed through the detector. The general 8-fold differential cross section becomes very complicated e.g. due to interference effects of the amplitudes in table 1. Fortunately in real life things are not so difficult. Note from table 1 that the neutrinos produced off quarks are purely lefthanded polarized. Because the distribution function for antiquarks suppresses the production of heavy objects ($m_N \geq 100$ GeV) the whole process can be simulated to a very good approximation

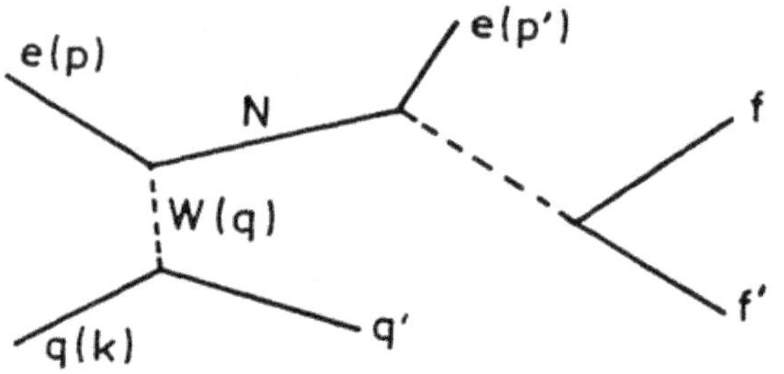

Figure 8: Feynman diagram for production and decay of heavy neutrinos. The symbols in brackets denote the four-vectors of the particles.

$e+q \to N+q'$	$T_{fi}/(2\sqrt{2}G_F F_W \lvert\xi\rvert\sqrt{1-\mu^2})$
$L+L \to L+L$	$2\hat{s}$
$L+L \to R+L$	0
$L+\bar{R} \to L+\bar{R}$	$(1+\cos\Theta)\hat{s}$
$L+\bar{R} \to R+\bar{R}$	$\sin\Theta\sqrt{\hat{s}}m_N$

Table 1: Helicity amplitudes for the production of heavy neutrinos's. The left column contains the helicities of the particles (and antiparticles) involved. All other amplitudes vanish. The expressions in the right column have to be multiplied by the factor given in the headline in order to give the usual scattering amplitude. G_F is the Fermi constant and $F_W = 1/(1+|q|^2/M_W^2)$ a factor which accounts for the propagator of the exchanged weak boson. All other symbols are explained in the text.

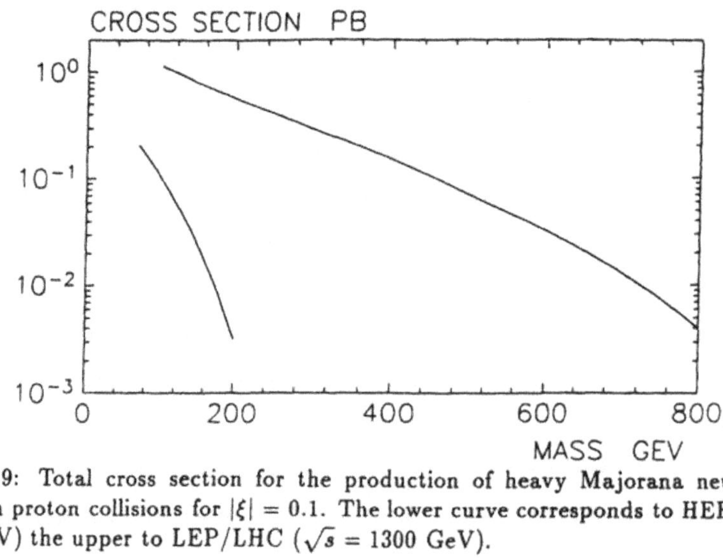

Figure 9: Total cross section for the production of heavy Majorana neutrinos in electron proton collisions for $|\xi| = 0.1$. The lower curve corresponds to HERA ($\sqrt{s} = 314$ GeV) the upper to LEP/LHC ($\sqrt{s} = 1300$ GeV).

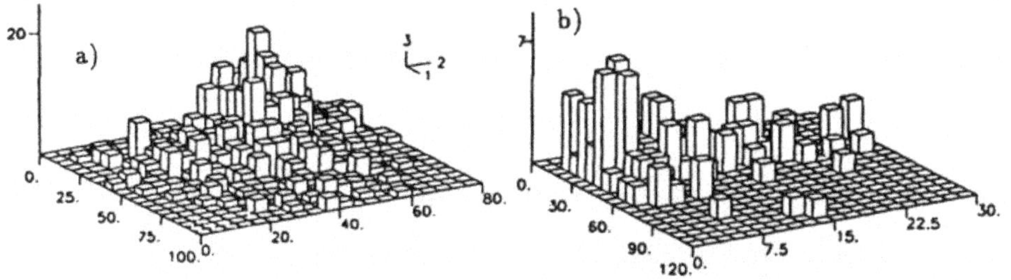

Figure 10: Event distribution in the $p_{T,\mu}, p_{T,e}$ plane. The transverse momentum of the electron is along the 2-axis. The left figure (a) corresponds to the signal the right one (b) to the background from W-production.

in a sequential Monte Carlo generator (e.g. [14]). In the first step heavy neutrinos are produced according to (11) and (12) whereas in the second step *lefthanded* heavy neutrinos decay into 3 fermions, eg. $e\mu\nu_\mu$.

Here we only look at the decay channel $N \rightarrow e\mu\nu_\mu$ due to the clear signature and probably small background. In fig. 10 a LEGO plot of $p_{T,\mu}$ versus $p_{T,e}$ is shown. The transverse momenta of both leptons can reach rather high values. This is in contrast to possible background reactions. Take for example electroproduction of W-bosons where the W decays to $\mu\nu$. The total cross section for W production is in the order of 1 pb, the scattered electrons are however strongly peaked at small p_T. For a quantitative study we used a generator [15] which is based on the paper of Baur and Zeppenfeld [16]. The corresponding $p_{T,\mu}, p_{T,e}$ distribution is shown in fig. 10b. With a cut at $p_{T,e} = 20$ GeV we expect 0 background events for an integrated luminosity of 100 pb^{-1} (the plot corresponds to 10000 pb^{-1}). At the same time we still have an efficiency between 50% and 65% for the signal.

Other sources of background have not been studied, so that the exclusion limits given below have to be taken with caution. They are given at the 95% CL in fig. 11 for $L_{int} = 100$ pb^{-1}. It is clear that for masses $m_N \geq 100$ GeV HERA does not seem to be very attractive, taking into account that the branching ratio of this and similar type neutrinos into $e\mu\nu_\mu$ will be less than 10%.

The results are not restricted to heavy Majorana neutrinos. New heavy fermions (Dirac and Majorana type) are e.g. also discussed in superstring inspired $E6$ models [17] and in the context of mirror fermions, which have been investigated recently by Csikor and Montvay [18]. Our results can be translated to limits for these classes of fermions if we restrict ourselves to only *left*handed couplings of the new fermions because in this case the charged current has the same structure as in the model discussed previously.

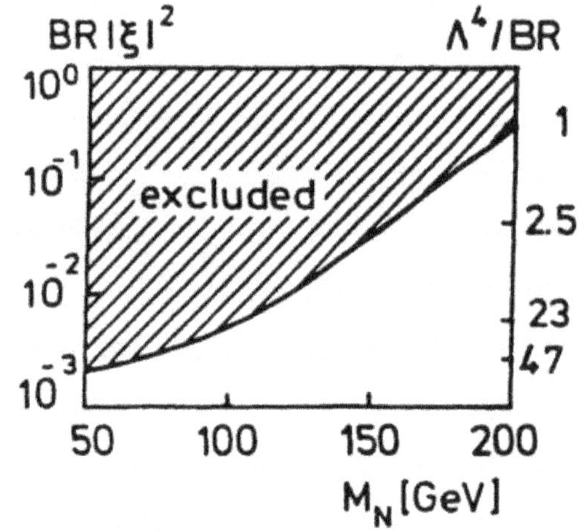

Figure 11: Exclusion limits ($CL = 95\%$) for heavy Majorana neutrinos (left scale) or heavy neutrinos produced via contact interactions (right scale). The unit of Λ is TeV.

Heavy fermions (not only heavy neutrinos) have also been looked at in the more general approach of compositeness and contact interaction models [19]. In these models usually a chiral coupling is assumed in order to avoid conflict with the precision measurements of $(g-2)$. One can therefore use the same programs for the simulation of contact interactions by simply putting $F_W = 1$ and replacing $4G_F\xi/\sqrt{2}$ by $4\pi/\Lambda^2$. The result is valid for the contact interaction of either lefthanded or righthanded currents. This is due to the fact that the structure of the cross sections (11) and (12) does not change if lefthanded couplings are replaced by righthanded couplings. In the latter case the new heavy fermions are of course righthanded polarized.

The propagator effects certainly don't influence the cross section very much, so that after proper rescaling fig. 11 also gives the cross section for heavy neutrino production via contact interactions. With the assumption that the detection efficiency after cuts in the additional decay channel $N \to e\mu\nu_\mu$ via *contact* interactions is the same as in the above example the limits in $|\xi|^2 BR$ can be easily converted to limits in Λ^4/BR. This is done at the right hand scale of fig. 11.

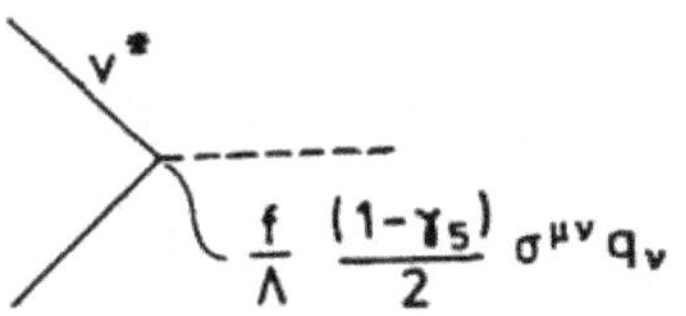

Figure 12: Feynman diagram for the production of excited heavy fermions via magnetic coupling.

4 Excited Leptons

4.1 Cross Section

In composite models of quarks and leptons it is natural to expect new heavy particles, which can be interpreted as excitations of the ground state. Because the electromagnetic current is conserved these states can in electromagnetic interactions only be reached via magnetic type coupling [7], fig. 12. In electron proton interactions one will first search for excited electrons, e^*. It has been shown [20], that about half of the cross section is in the elastic channel

$$e + p \rightarrow e^* + p \tag{13}$$

with a branching ratio of $\approx 28\%$ for a decay of the e^* with a mass of more than 150 GeV into $e\gamma$ [19]. This reaction has an extremely clear signature, a wide angle electron photon pair and nothing else in the detector! The cross section for e^* production is shown in fig. 13a. The magnetic coupling can also be applied for other gauge interactions. Excited neutrinos can for example be produced in charged current reactions. The relevant helicity amplitudes are listed in table 2. Note that this time excited neutrinos produced off antiquarks are completely (*righthanded*) polarized. The helicity averaged cross sections on the parton level are given by

$$\frac{d\sigma}{dy}(eq \rightarrow \nu^* q') = \frac{G_F}{4\pi\sqrt{2}M_W^2}\frac{f^2}{\Lambda^2}F_W^2 \hat{s}^2 y(1-y) \tag{14}$$

and

$$\frac{d\sigma}{dy}(e\bar{q} \rightarrow \nu^* \bar{q}') = \frac{G_F}{4\pi\sqrt{2}M_W^2}\frac{f^2}{\Lambda^2}F_W^2(1-\mu^2)\hat{s}^2 y(1-y-\mu^2) \; . \tag{15}$$

The cross section for excited neutrino production at HERA is shown in fig. 14, with the structure functions taken from ref.[8] (set 1) at a scale $\hat{s}$.

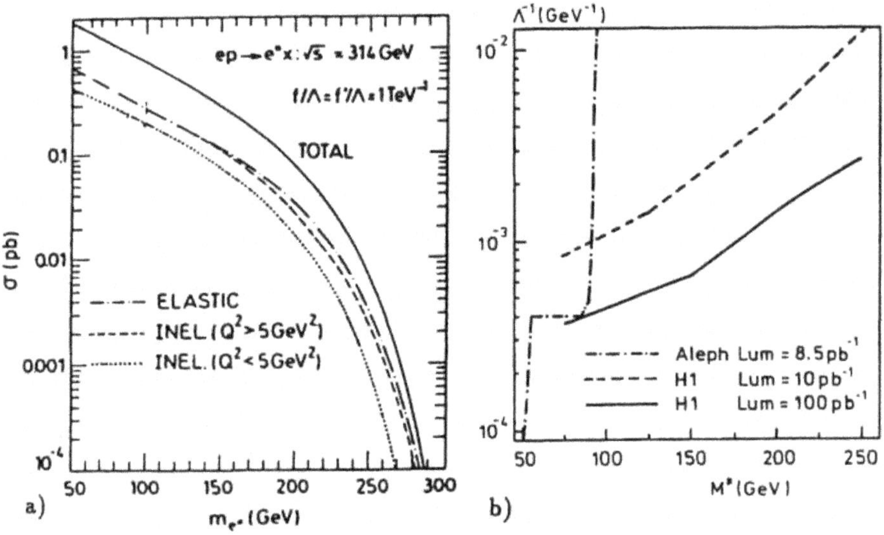

Figure 13: a) Cross section for e^* production in ep collisions,[20]. b) Exclusion Limit at 95% CL for the scale paramer Λ in excited electron production. Only the decay channel $e^* \to e + \gamma$ is considered.[21]

$e + q \to \nu^* + q'$	$T_{fi}/(\frac{fgF_W}{\sqrt{2}\Lambda M_W^2})$
$L + L \to L + L$	$2m^* \hat{s}\sqrt{1-\mu^2}\sin^2\frac{\Theta}{2}$
$L + L \to R + L$	$-2\sqrt{\hat{s}}\hat{s}\sqrt{1-\mu^2}\sin\Theta$
$L + \bar{R} \to L + \bar{R}$	0
$L + \bar{R} \to R + \bar{R}$	$-2\sqrt{\hat{s}}\hat{s}(1-\mu^2)^{\frac{3}{2}}\sin\Theta$

Table 2: Helicity amplitudes for the production of excited neutrinos via magnetic coupling in charged current interactions. The left column contains the helicities of the particles (and antiparticles) involved. All other amplitudes vanish. The expressions in the right column have to be multiplied by the factor given in the headline in order to give the usual scattering amplitude. g is the weak coupling constant, f the dimensionless coupling which together with the scale parameter Λ determines the strength of the interaction. $F_W = 1/(1 + |q|^2/M_W^2)$ is a factor which accounts for the propagator of the exchanged weak boson. All other symbols are explained in the text.

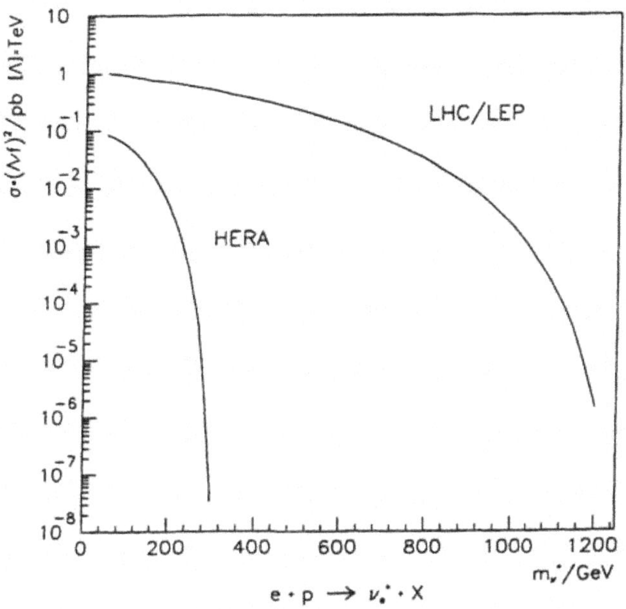

Figure 14: Cross section for the production of excited neutrinos in *ep* collisions.

4.2 Limits

The generic background to e^* production is given by wide angle bremsstrahlung. It is very important to have a precise knowledge of this process. One has to take into account elastic bremsstrahlung

$$e + p \to e + \gamma + p \tag{16}$$

off a real proton (described by magnetic and electric form factors) and inelastic bremsstrahlung

$$e + p \to e + \gamma + X \tag{17}$$

which also has a high cross section. For small masses of X (the resonance region) it cannot be distinguished from the elastic channel, because the decay products of the resonances will not escape the beam pipe. These processes have been investigated intensively by different groups,[22].

In fig. 13b the exclusion limits for e^* production are given. The branching ratio into the channel $e + \gamma$ has been taken into account. For comparison results from a direct search in electron positron annihilation are also given. It is evident that HERA can extend our knowledge on excited electron production substantially.

Excited neutrinos decay with a branching ratio of $\approx 50\%$ into electron and W. The mass of the excited neutrino can be reconstructed from the missing transverse momentum in the event and from the hadronic variables using e.g. the Jacquet-Blondel method [10]. The underlying assumption is, that the missing transverse momentum is only carried away by the decay neutrino of the ν^*. A detailed study along these lines is performed at the moment, final results will be available soon.

Conclusions

HERA will allow us to search for physics beyond the standard model in a completely new kinematical range of *ep* collisions. It would of course be extremely exciting to find one of the states discussed in this paper. If nature does not provide new particles in the HERA energy and luminosity range we will extend our present knowledge on exclusion limits substantially.

It is a pleasure to thank the organizers of the workshop G. Herten and M. Perl for a very fine meeting. The hospitality of the ICTP Trieste is gratefully acknowledged.

References

[1] Physics at HERA, Proceedings of the workshop, Ed. W. Buchmüller and G. Ingelman, DESY 1991
Proceedings of the HERA workshop, Ed. R.D. Peccei, DESY 1987.

[2] B. Schrempp, Physics at HERA, Proceedings of the workshop, Ed. W. Buchmüller and G. Ingelman, Vol.2, DESY 1991

[3] W. Buchmüller, R. Rückl, D. Wyler, Phys. Lett. B191 (1987) 442

[4] UA2 Collab., J. Alitti et al., Phys. Lett. B247 (1992) 507

[5] P. Schleper, Thesis RWTH Aachen, PITHA 92/29.

[6] P. Schleper, Physics at HERA, Proceedings of the workshop, Ed. W. Buchmüller and G. Ingelman, Vol.2, DESY 1991

[7] J.A. Bagger and M.E. Peskin Phys. Rev. D31 (1985) 2211

[8] E.Eichten, I.Hinchliffe, K.Lane, Ch.Quigg: Rev. Mod. Phys. 56 (1984) 579

[9] W.Bartel et. al., JADE Collaboration, Z.Phys. C33 (1986) 23

[10] A. Blondel and F. Jacquet, Proceedings of the study of an *ep* facility for Europe, Ed. U. Amaldi, DESY 79/48 (1979) 391

[11] H. Murayama, T. Yanagida, Tohoku University, TU–370, 1991

[12] W. Buchmüller and C. Greub, Nucl.Phys. B363 (1991) 345.

[13] D. Duke and J.F. Owens, Phys. Rev. D30 (1984) 49.

[14] K. Rosenbauer Physics at HERA, Proceedings of the workshop, Ed. W. Buchmüller and G. Ingelman, Vol.3, DESY 1991.

[15] E. Theuer, Physics at HERA, Proceedings of the workshop, Ed. W. Buchmüller and G. Ingelman, Vol.3, DESY 1991.

[16] U. Baur and D. Zeppenfeld, Nucl. Phys. B325 (1989) 253

[17] T.G. Rizzo and L. Hewett, Phys. Rep. 183 (1989) 193
A. Djouadi et al., Physics at HERA, Proceedings of the workshop, Ed. W. Buchmüller and G. Ingelman, Vol.2, DESY 1991.

[18] F. Csikor and I. Montvay, Phys.Lett. B231 (1989) 503
F. Csikor, Z.Phys. C49 (1991) 129

[19] U. Baur, M.Spira and P. Zerwas, Phys. Rev. D42 (1990) 815.

[20] K. Hagiwara, S. Kommamiya and D. Zeppenfeld, Z. Phys. C29 (1985) 115.

[21] T. Carli et al., Physics at HERA, Proceedings of the workshop, Ed. W. Buchmüller and G. Ingelman, Vol.2, DESY 1991

[22] F. Raupach, Physics at HERA, Proceedings of the workshop, Ed. W. Buchmüller and G. Ingelman, Vol.3, DESY 1991
T. Carli et al., Physics at HERA, Proceedings of the workshop, Ed. W. Buchmüller and G. Ingelman, Vol. 3, DESY 1991.

e^+e^- Linear Colliders and New Particle Searches*

TIMOTHY L. BARKLOW

Stanford Linear Accelerator Center
Stanford University, Stanford, California 94309

ABSTRACT

We discuss future e^+e^- linear colliders and new particle searches that can be done with them. In the discussion of new particle searches we examine the following topics: searches for gauge boson structure, searches for a strongly interacting Higgs sector, top quark studies, Higgs searches, supersymmetric particle searches and measurements of soft supersymmetry breaking parameters.

1. Introduction

Physicists from many institutions around the world are currently in the process of defining the physics, machine parameters and machine design of a future e^+e^- linear collider. We will first discuss ideas for future e^+e^- linear collider design and say a few words about the experimental environment at an e^+e^- linear collider. We will begin the discussion of physics topics with the process $e^+e^- \rightarrow W^+W^-$. This is an important topic in its own right, but it will also serve to illustrate many of the techniques used in analyzing final states at e^+e^- linear colliders. We will then briefly discuss the physics of top quarks. The remainder of the paper will be devoted to Higgs physics and supersymmetry (SUSY).

Fig. 1 is a schematic of a future e^+e^- linear collider. We shall refer to such a collider as the next linear collider or NLC. All NLC designs share the schematic shown in Fig. 1. Looking at Fig. 1 we see that there are damping rings to reduce the beam emittance, compressors to shorten the length of the e^+e^- bunches, two main accelerators, and two final focus sections.

The initial center-of-mass energy (E_{cm}) of all NLC designs is 500 GeV. Of course the E_{cm} can be lowered to study, for example, the $t\bar{t}$ threshold region. The design luminosity at $E_{cm} = 500$ GeV is about $6 \times 10^{33}\ cm^{-2}s^{-1}$. It will hopefully become clear as we discuss various physics topics why the initial E_{cm} is a good choice and why the luminosity is a good match for this E_{cm}.

The E_{cm} is upgraded by lengthening the accelerator sections, increasing the

⋆ This work was supported by the Department of Energy, contract DE–AC03–76SF00515.

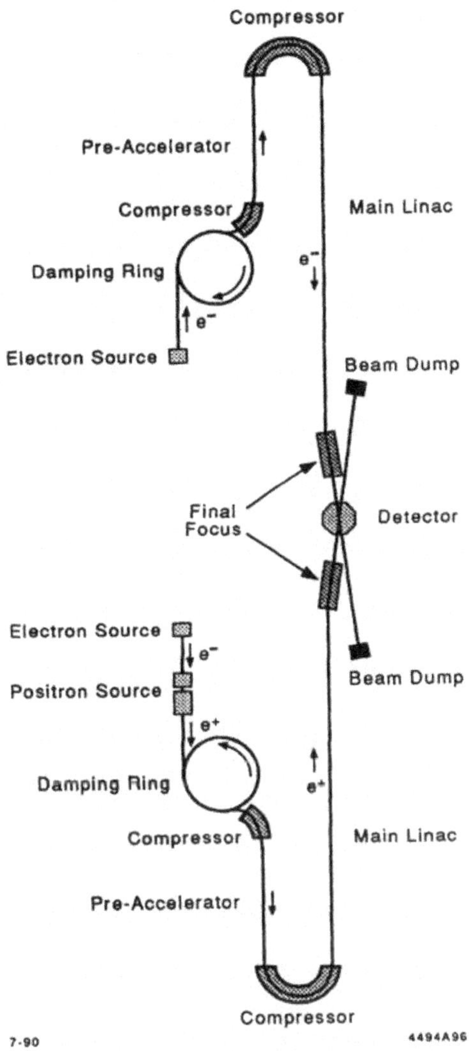

Fig. 1. Schematic diagram of an NLC.

accelerating gradient, or both. The only collider parts which must be moved when lengthening the main accelerators are the bunch compressors at the top and bottom of Fig. 1. These compressors move outward so that the collider is lengthened like a trombone. Most designs envisage an energy upgrade from $E_{cm} = 500$ GeV to $E_{cm} = 1500$ GeV. This factor of 3 increase in NLC energy can be compared to the factor of 2.5 increase in the SLAC energy during its first 20 years of operation, where the increase at SLAC was accomplished solely by upgrading the accelerating gradient.

Vigorous research and development programs in e^+e^- linear collider design have been under way for many years at SLAC, KEK, DESY and other institutions. With the exception of the main accelerator, there is general agreement on all aspects of NLC design. Three approaches to e^+e^- acceleration have emerged.

SLAC, KEK, and Novosibirsk are proposing[1,2] to extend the Stanford Linear Accelerator technology from the present RF frequency of 3 GHz (S-Band) to an RF frequency of 11.4 GHz (X-Band). When operating at a center-of-mass energy of 500 GeV the X-Band Linac would have an accelerating gradient of 50 MV/m which is to be compared to an accelerating gradient of 17 MV/m for the present S-Band Linac. It is thought that the gradient could be increased to 100 MV/m for operation of the collider at center-of-mass energies of 1.0 to 1.5 TeV The SLAC X-Band design has a repetition rate of 180 Hz with 0.65×10^{10} electrons per bunch and 90 bunches per RF pulse.

A group at DESY is proposing[3] to use the current S-Band technology. The DESY design would therefore have an accelerating gradient of 17 MV/m. The DESY S-Band repetition rate is 50 Hz and there are 0.70×10^{10} electrons per bunch and 172 bunches per RF pulse.

TESLA, a collaboration of physicists from Cornell, CEBAF, DESY, and other institutions, is proposing[4] to use superconducting RF technology. The highest gradient attainable today in a working superconducting RF accelerator is 7 MV/m. The TESLA collabortion believes they can achieve 25 MV/m eventually. The TESLA acclerator would operate at a frequency of 1.5 GHz (L-Band) so that the accelerator structure would be quite large. A larger structure reduces wakefield effects and is therefore an advantage. The TESLA collider has a repetition rate of 10 Hz with 5×10^{10} electrons per bunch and 800 bunches per RF pulse.

The luminosity of all three designs is about $6 \times 10^{33}\ cm^{-2}\ s^{-1}$ at $E_{cm} = 500$ GeV. At 1 TeV the TESLA and X-Band designs have a luminosity of 1 to $2 \times 10^{34}\ cm^{-2}\ s^{-1}$ but the S-Band design is having trouble keeping up at a luminosity of $0.3 \times 10^{34}\ cm^{-2}\ s^{-1}$.

Most of the analyses which we present assume $E_{cm} = 500$ GeV and an integrated luminosity of 10 fb^{-1}. Such an integrated luminosity would correspond to a 10^7 second run at 16% of the design differential luminosity at $E_{cm} = 500$ GeV. We note that the SLAC linear collider (SLC), an add-on to a 30 year old accelerator, currently operates at 15% of its design luminosity.

We will not talk about detectors at future e^+e^- linear colliders. For most physics analyses the features and resolution of existing e^+e^- collider detectors, such as SLD and the LEP detectors, are sufficient. Indeed, the resolution of SLD/LEP detectors is assumed for all the analyses described in this article. That is not to say that these detectors can be transplanted without modification to the interaction point of an NLC. There are new sources of backgrounds at an NLC, such as muons created upstream of the interaction point, and a large number of e^+e^- pairs produced each bunch crossing due to the large luminosity per bunch crossing. However, even with these and other constraints on detector design, it is felt that all of the features of SLD/LEP detectors can be retained and that the existing resolution specifications can be met or exceeded[5]

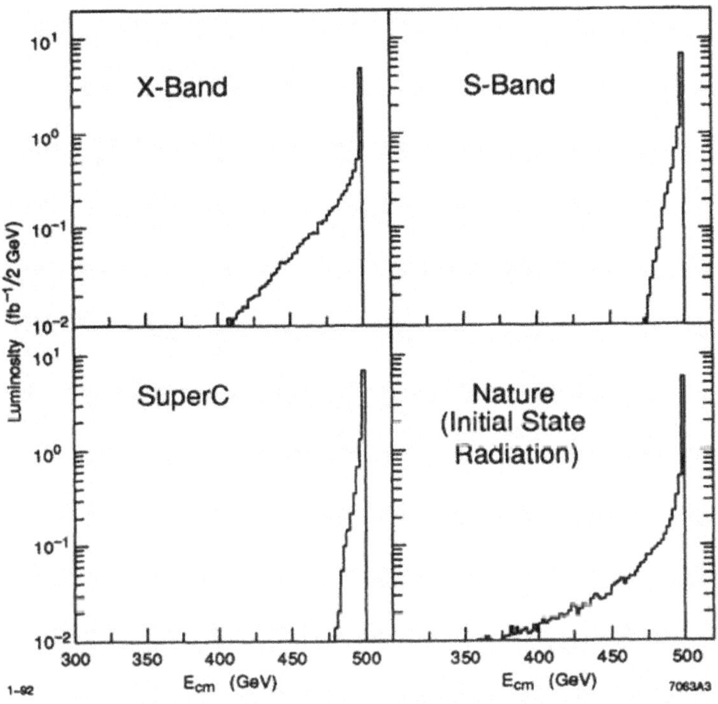

Fig. 2. e^+e^- luminosity spectra.

One important difference between the experimental environments at e^+e^- storage rings and future e^+e^- linear colliders is the presence of beamstrahlung at linear colliders. Beamstrahlung is the radiation by particles of one bunch as they bend in the intense collective magnetic field of the opposing bunch. Fig. 2 shows the one-dimensional luminosity-weighted e^+e^- E_{cm} distributions[6] for the three types of accelerators. The extent to which each distribution deviates from a delta function reflects the amount of beamstrahlung radiation emitted during a bunch crossing. Also shown is the e^+e^- E_{cm} distribution for a machine without beamstrahlung but with the effects of initial state radiation included. We see that the magnitude of the beamstrahlung radiation is equal to or less than the bremsstrahlung radiation we are accustomed to. Beamstrahlung radiation modifies the irreducible bremsstrahlung radiation spectrum but does not introduce anything new.

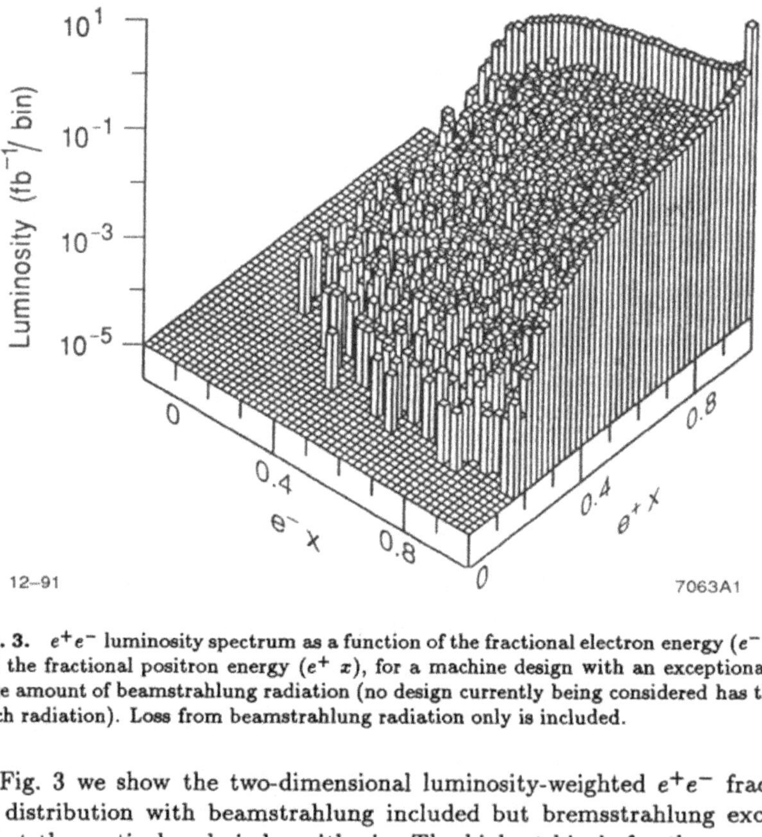

Fig. 3. e^+e^- luminosity spectrum as a function of the fractional electron energy (e^- x) and the fractional positron energy (e^+ x), for a machine design with an exceptionally large amount of beamstrahlung radiation (no design currently being considered has this much radiation). Loss from beamstrahlung radiation only is included.

In Fig. 3 we show the two-dimensional luminosity-weighted e^+e^- fractional energy distribution with beamstrahlung included but bremsstrahlung excluded. Note that the vertical scale is logarithmic. The highest bin is for the case where neither e^+ or e^- radiates a beamstrahlung photon. The second most likely case is that one, but not both, of the e^+ and e^- radiate a photon. The least likely case is when both the e^+ and e^- radiate a beamstrahlung photon. This means, for example, that it is a pretty good approximation to assume that neither the e^+

or e^- has radiated when performing beam energy–momemtum constraints on the final state momenta, that it is a better approximation to assume that only one of the e^+e^- has radiated, and that it is only necessary for very high precision analyses to allow for radiation by both the e^+ and e^-.

2. Gauge Boson Compositeness

Remarkably little is known about the interactions of the electroweak gauge bosons with each other. There are two distinct aspects of these interactions which are of interest. First, the triple and quartic gauge boson vertices need to be tested for any structure beyond what is produced by Standard Model radiative corrections. Such anomalous structure would appear, for example, if the W and Z bosons were themselves composite objects. Second, gauge boson scattering processes such as $W^+W^- \to W^+W^-$ can help us understand the mechanism responsible for electroweak symmetry breaking. W^+W^- scattering processes can be studied at an e^+e^- collider either by radiating virtual W's off of the e^- and e^+, or by studying W^+W^- rescattering in $e^+e^- \to W^+W^-$.

2.1 *Parameterization of Three Gauge Boson Vertex*

We use the formalism of Ref. 7 to describe the general W^+W^-V vertex, where $V = \gamma, Z$. The effective Lagrangian for the general W^+W^-V vertex as given in Eq.(2.1) of Ref. 7 is:

$$
\begin{aligned}
\mathcal{L}_{WWV}/g_{WWV} = & \, i g_1^V (W^\dagger_{\mu\nu} W^\mu V^\nu - W^\dagger_\mu V_\nu W^{\mu\nu}) + i\kappa_V W^\dagger_\mu W_\nu V^{\mu\nu} \\
& + \frac{i\lambda_V}{m_W^2} W^\dagger_{\lambda\mu} W^\mu{}_\nu V^{\nu\lambda} - g_4^V W^\dagger_\mu W_\nu (\partial^\mu V^\nu + \partial^\nu V^\mu) \\
& + g_5^V \varepsilon^{\mu\nu\rho\sigma} (W^\dagger_\mu \vec{\partial}_\rho W_\nu) V_\sigma + i\tilde{\kappa_V} W^\dagger_\mu W_\nu \tilde{V}^{\mu\nu} \\
& + \frac{i\tilde{\lambda}_V}{m_W^2} W^\dagger_{\lambda\mu} W^\mu{}_\nu \tilde{V}^{\nu\lambda} \quad .
\end{aligned}
\tag{1}
$$

The terms with the coupling parameters g_1^V, κ_V, λ_V separately conserve C, P, and T, while the others violate C or P, or both. We will concentrate in this article on measurements of the coupling parameters κ_γ, λ_γ, κ_Z, and λ_Z.

In the standard model, at tree-level, $g_1^\gamma = g_1^Z = 1$, $\kappa_\gamma = \kappa_Z = 1$ and all other coupling parameters in Eq. (1) are 0. There is currently much discussion in the literature[8,9] regarding the extent to which present day electroweak measurements, especially precision LEP measurements, constrain κ_V and λ_V. In order to set the stage for our discussion of the limits obtainable with the NLC, we quote the results of Kane *et al.*[8] who considered loop effects and unitarity constraints:

$$|\lambda_\gamma| \leq 0.6 \quad |\kappa_\gamma - 1| \leq 1.0$$
$$|\lambda_Z| \leq 0.6 \quad -0.8 \leq \kappa_Z - 1 \leq 0.0$$

where $\kappa_V - 1 \equiv \kappa_V - 1$. These authors also estimated that the SSC should achieve limits on λ_γ and κ_γ of

$$|\lambda_\gamma| \leq 0.02 \quad |\kappa_\gamma - 1| \leq 0.1$$

with 10 fb^{-1}.

2.2 *Final State Helicity Analysis of* $e^+e^- \to W^+W^-$

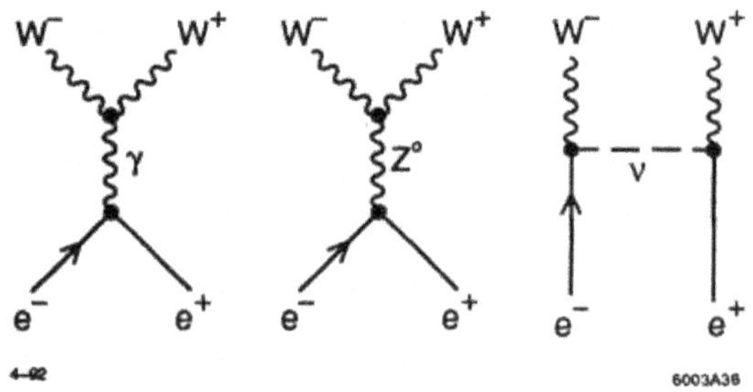

Fig. 4. The three tree-level Feynman diagrams for the process $e^+e^- \to W^+W^-$.

We start our discussion of $e^+e^- \to W^+W^-$ with some definitions. Following Ref. 7 we define our z-axis to be the direction of the W^- in the e^+e^- rest frame. The W^- production angle Θ is defined to be the angle between the initial state e^- and the W^- in the e^+e^- rest frame. The W^- and W^+ decay according to $W^- \to f_1\overline{f}_2$ and $W^+ \to f_3\overline{f}_4$. We define θ and ϕ to be the polar and azimuthal angles, respectively, of the fermion f_1 in the W^- rest frame, while $\overline{\theta}$ and $\overline{\phi}$ are the polar and azimuthal angles, respectively, of the anti-fermion $\overline{f}_4$ in the W^+ rest frame. We assume that the e^- beam has the polarization P_e where $P_e = 1$ is 100% right-handed and $P_e = -1$ is 100% left-handed.

$e^+e^- \to W^+W^-$ proceeds through the three Feynman diagrams shown in Fig. 4. The differential cross-section $d\sigma/\cos\Theta$ is shown in Fig. 5, with the cross-section decomposed into different W^+W^- helicity combinations. Our notation is such that $(\lambda,\overline{\lambda})$ indicates that the W^- has helicity λ, the W^+ has helicity $\overline{\lambda}$,

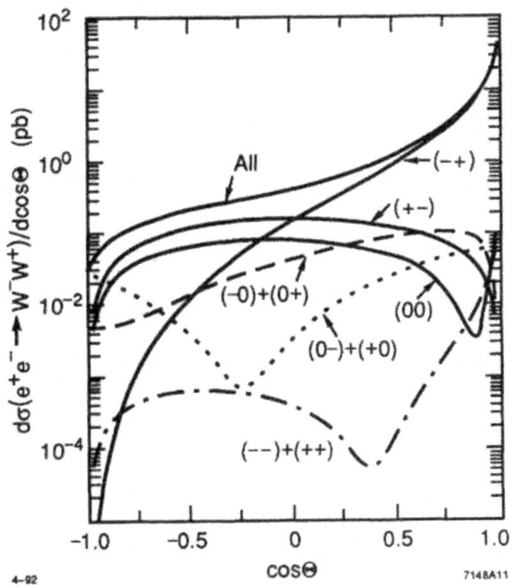

Fig. 5. The differential cross-section $d\sigma/\cos\Theta$ for $e^+e^- \to W^+W^-$ with the cross-section decomposed into different W^+W^- helicity combinations.

and that $\lambda = +, -, 0$ refers to right-handed transverse, left-handed transverse, and logitudinal W's, respectively. At forward angles the t-channel neutrino exchange diagram dominates and the cross-section consists almost entirely of left-handed transverse W^- and right-handed transverse W^+ gauge bosons. Transverse W's continue to dominate the cross-section at central angles. Only when $\cos\Theta$ drops below -0.9 do helicity combinations with at least one longitudinal gauge boson become important. The production angle θ therefore provides only limited information about individual W^+W^- helicity combinations.

The decay angles θ and ϕ can be used to improve the W^+W^- helicity analysis. To understand in detail how the decay angles reflect the helicity of the parent W we must look at the multi-differential cross-section for W^+W^- production and decay. The expression for the multi-differential cross-section depends on the event topology we are analyzing. Throughout this chapter we will concentrate on events where one W decays leptonically and the other decays hadronically. This topology has good statistics, can be separated from background with a high degree of purity, is rich in helicity information, and allows us to reconstruct all final state variables on an event-by-event basis. For events where the W^- decays leptonically and the W^+ decays hadronically (*topology A* events) the multi-differential cross section is given by:

$$\frac{\mathrm{d}\sigma(\kappa_\gamma,\lambda_\gamma,\kappa_Z,\lambda_Z;\cos\Theta,\cos\theta,\phi,\cos\overline{\theta},\overline{\phi},P_e)}{\mathrm{d}\cos\Theta\ \mathrm{d}\cos\theta\ \mathrm{d}\phi\ \mathrm{d}\cos\overline{\theta}\ \mathrm{d}\overline{\phi}} =$$

$$\frac{9\beta B_{lh}}{8192\pi^3 s}\sum_{\lambda,\lambda',\overline{\lambda},\overline{\lambda}'}\mathcal{Q}^{\lambda\overline{\lambda}}_{\lambda'\overline{\lambda}'}(\kappa_\gamma,\lambda_\gamma,\kappa_Z,\lambda_Z;\cos\Theta,P_e)\mathcal{D}^{\lambda}_{\lambda'}(\cos\theta,\phi)\overline{\mathcal{H}}^{\overline{\lambda}}_{\overline{\lambda}'}(\cos\overline{\theta},\overline{\phi}) \tag{2}$$

where

$$\begin{aligned}\mathcal{Q}^{\lambda\overline{\lambda}}_{\lambda'\overline{\lambda}'} \equiv&(1+P_e)\sum_{\overline{\sigma}}\mathcal{M}_1(+,\overline{\sigma},\lambda,\overline{\lambda};\Theta)\mathcal{M}_1^*(+,\overline{\sigma},\lambda',\overline{\lambda}';\Theta)\\ &+(1-P_e)\sum_{\overline{\sigma}}\mathcal{M}_1(-,\overline{\sigma},\lambda,\overline{\lambda};\Theta)\mathcal{M}_1^*(-,\overline{\sigma},\lambda',\overline{\lambda}';\Theta)\quad .\end{aligned}$$

and

$$\overline{\mathcal{H}}^{\overline{\lambda}}_{\overline{\lambda}'}(\cos\overline{\theta},\overline{\phi}) \equiv \left[\overline{\mathcal{D}}^{\overline{\lambda}}_{\overline{\lambda}'}(\cos\overline{\theta},\overline{\phi}) + \overline{\mathcal{D}}^{\overline{\lambda}}_{\overline{\lambda}'}(-\cos\overline{\theta},\overline{\phi}+\pi)\right]\quad .$$

The W^+W^- production amplitude $\mathcal{M}_1$, the W^- decay tensor $\mathcal{D}^{\lambda}_{\lambda'}$ and the W^+ decay tensor $\overline{\mathcal{D}}^{\overline{\lambda}}_{\overline{\lambda}'}$ are defined in Ref. 7. We note that

$$\mathcal{P}^{\lambda\overline{\lambda}}_{\lambda'\overline{\lambda}'} = \mathcal{Q}^{\lambda\overline{\lambda}}_{\lambda'\overline{\lambda}'}\,|_{P_e=0}$$

where $\mathcal{P}^{\lambda\overline{\lambda}}_{\lambda'\overline{\lambda}'}$ is the W^+W^- production tensor defined in Eq. 4.11 of Ref. 7. In Eq. (2), $\sqrt{s}$ is the e^+e^- center-of-mass energy, $\beta = \sqrt{1-4m_W^2/s}$ and B_{lh} is the product of the leptonic and hadronic branching ratios for the W boson. The sum $\overline{\mathcal{H}}^{\overline{\lambda}}_{\overline{\lambda}'}(\cos\overline{\theta},\overline{\phi})$ appears in Eq. (2), instead of the single term $\overline{\mathcal{D}}^{\overline{\lambda}}_{\overline{\lambda}'}(\cos\overline{\theta},\overline{\phi})$, because we assume that we cannot tag the flavor of the the quark jets in the decay of the W^+. When the W^+ decays leptonically and the W^- decays hadronically (*topology B*) we use the cross-section

$$\frac{\mathrm{d}\overline{\sigma}(\kappa_\gamma,\lambda_\gamma,\kappa_Z,\lambda_Z;\cos\Theta,\cos\overline{\theta},\overline{\phi},\cos\theta,\phi,P_e)}{\mathrm{d}\cos\Theta\ \mathrm{d}\cos\theta\ \mathrm{d}\phi\ \mathrm{d}\cos\overline{\theta}\ \mathrm{d}\overline{\phi}} =$$

$$\frac{9\beta B_{lh}}{8192\pi^3 s}\sum_{\lambda,\lambda',\overline{\lambda},\overline{\lambda}'}\mathcal{Q}^{\lambda\overline{\lambda}}_{\lambda'\overline{\lambda}'}(\kappa_\gamma,\lambda_\gamma,\kappa_Z,\lambda_Z;\cos\Theta,P_e)\mathcal{H}^{\lambda}_{\lambda'}(\cos\overline{\theta},\overline{\phi})\overline{\mathcal{D}}^{\overline{\lambda}}_{\overline{\lambda}'}(\cos\overline{\theta},\overline{\phi}) \tag{3}$$

where

$$\mathcal{H}^{\lambda}_{\lambda'}(\cos\overline{\theta},\overline{\phi}) \equiv \left[\mathcal{D}^{\lambda}_{\lambda'}(\cos\theta,\phi) + \mathcal{D}^{\lambda}_{\lambda'}(-\cos\theta,\phi+\pi)\right]\quad .$$

The W^- decay tensor $\mathcal{D}^{\lambda}_{\lambda'}(\cos\theta,\phi)$ is given explicitly by

$$\begin{aligned}
\mathcal{D}^{+}_{+} &= \frac{1}{2}(1-\cos\theta)^2, \qquad \mathcal{D}^{-}_{-} = \frac{1}{2}(1+\cos\theta)^2, \qquad \mathcal{D}^{0}_{0} = \sin^2\theta,\\
\mathcal{D}^{+}_{0} &= \frac{1}{\sqrt{2}}(1-\cos\theta)\sin\theta e^{i\phi},\\
\mathcal{D}^{-}_{0} &= -\frac{1}{\sqrt{2}}(1+\cos\theta)\sin\theta e^{-i\phi},\\
\mathcal{D}^{+}_{-} &= \frac{1}{2}\sin^2\theta e^{2i\phi},
\end{aligned} \tag{4}$$

and $\mathcal{D}^{\lambda'}_{\lambda} = (\mathcal{D}^{\lambda}_{\lambda'})^*$. We also show explicitly the expression for the W^- hadronic decay tensor $\mathcal{H}^{\lambda}_{\lambda'}(\cos\theta,\phi)$:

$$\begin{aligned}
\mathcal{H}^{+}_{+} &= 1+\cos^2\theta, \qquad \mathcal{H}^{-}_{-} = 1+\cos^2\theta, \qquad \mathcal{H}^{0}_{0} = 2\sin^2\theta,\\
\mathcal{H}^{+}_{0} &= \sqrt{2}\cos\theta\sin\theta e^{i\phi}\\
\mathcal{H}^{-}_{0} &= -\sqrt{2}\cos\theta\sin\theta e^{-i\phi}\\
\mathcal{H}^{+}_{-} &= \sin^2\theta e^{2i\phi} \quad .
\end{aligned} \tag{5}$$

Although the W^- hadronic decay tensor $\mathcal{H}^{\lambda}_{\lambda'}(\cos\theta,\phi)$ does not contain as much helicity information as the leptonic decay tensor $\mathcal{D}^{\lambda}_{\lambda'}(\cos\theta,\phi)$, its information content is nevertheless non-negligible.

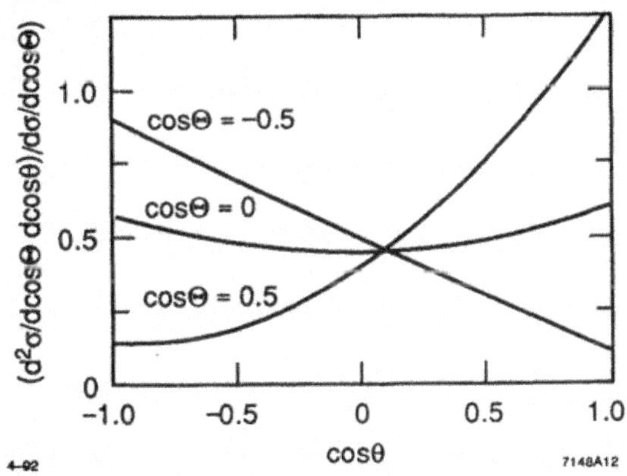

Fig. 6. Distributions in $d\sigma/\cos\theta$ for various values of $\cos\Theta$.

The decay tensor components $\mathcal{D}^{+}_{+}$, $\mathcal{D}^{-}_{-}$, and $\mathcal{D}^{0}_{0}$ for pure left-handed transverse, pure right-handed transverse, and pure longitudinal W^- gauge bosons, respectively, have distinctive polar angle distributions, and do not exhibit an azimuthal angle dependence. Distributions in $d\sigma/\cos\theta$ are shown in Fig. 6 for various values of $\cos\Theta$. At $\cos\Theta = +0.5$ we can see from the $\cos\theta$ distribution that the W^- is almost entirely left-handed. On the other hand, at $\cos\Theta = 0$ and $\cos\Theta = -0.5$ the W^- is a mixture of various polarizations.

The interference terms in Eq. (4) do depend on the decay azimuthal angle. For example, the decay tensor component $\mathcal{D}^0_0$ for the pure longitudinal W^- boson is indistinguishable from the component $\mathcal{D}^+_-$ for the interference between left-handed and right-handed transverse W^- bosons, except for the azimuthal angle factor $e^{2i\phi}$. Our set of experimental observables therefore includes not only the cosines of the polar angles, $\cos\Theta$, $\cos\theta$, and $\cos\overline{\theta}$, but also the azimuthal angles ϕ and $\overline{\phi}$. Our full helicity analysis then consists of a maximum likelihood fit of the form

$$\ln L(\kappa_\gamma, \lambda_\gamma, \kappa_Z, \lambda_Z; \mathcal{L}_k) = \sum_{i_1=1}^{N_\Theta}\sum_{i_2=1}^{N_\theta}\sum_{i_3=1}^{N_\phi}\sum_{i_4=1}^{N_{\overline{\theta}}}\sum_{i_5=1}^{N_{\overline{\phi}}}\sum_{k=1}^{N_{P_e}} \Big[r_{i_1i_2i_3i_4i_5k}\ln(\mu_{i_1i_2i_3i_4i_5k}) - \mu_{i_1i_2i_3i_4i_5k} + \overline{r}_{i_1i_2i_3i_4i_5k}\ln(\overline{\mu}_{i_1i_2i_3i_4i_5k}) - \overline{\mu}_{i_1i_2i_3i_4i_5k}\Big] \quad (6)$$

where $r_{i_1i_2i_3i_4i_5k}$ and $\mu_{i_1i_2i_3i_4i_5k}$ are the measured number of topology A events and the expected number of topology A events, respectively, in the bin centered at $\cos\Theta_{i_1}$, $\cos\theta_{i_2}$, ϕ_{i_3}, $\cos\overline{\theta}_{i_4}$, $\overline{\phi}_{i_5}$, when the e^- polarization is P_{ek}. $\overline{r}_{i_1i_2i_3i_4i_5k}$ and $\overline{\mu}_{i_1i_2i_3i_4i_5k}$ are the measured and expected number of topology B events. The volume of a bin is

$$\Delta\Omega = \frac{16\pi^2}{N_\Theta N_\theta N_\phi N_{\overline{\theta}} N_{\overline{\phi}}} \quad .$$

We define $\mu_{i_1i_2i_3i_4i_5k}$ as follows:

$$\mu_{i_1i_2i_3i_4i_5k}(\kappa_\gamma, \lambda_\gamma, \kappa_Z, \lambda_Z, \mathcal{L}_k) \equiv \mathcal{L}_k \int_{\eta^-_{i_1}}^{\eta^+_{i_1}} d\cos\Theta \int_{\eta^-_{i_2}}^{\eta^+_{i_2}} d\cos\theta \int_{\eta^-_{i_3}}^{\eta^+_{i_3}} d\phi \int_{\eta^-_{i_4}}^{\eta^+_{i_4}} d\cos\overline{\theta} \int_{\eta^-_{i_5}}^{\eta^+_{i_5}} d\overline{\phi}$$

$$\Bigg\{ \frac{d\xi(\cos\Theta, \cos\theta, \phi, \cos\overline{\theta}, \overline{\phi}, P_{ek})}{d\cos\Theta\ d\cos\theta\ d\phi\ d\cos\overline{\theta}\ d\overline{\phi}} + \Bigg[\int_{-1}^{1} d\cos\Theta' \int_{-1}^{1} d\cos\theta' \int_{0}^{2\pi} d\phi' \int_{0}^{1} d\cos\overline{\theta}' \int_{0}^{2\pi} d\overline{\phi}'$$

$$G(\cos\Theta, \cos\theta, \phi, \cos\overline{\theta}, \overline{\phi}; \cos\Theta', \cos\theta', \phi', \cos\overline{\theta}', \overline{\phi}')$$

$$\cdot \frac{d\sigma(\kappa_\gamma, \lambda_\gamma, \kappa_Z, \lambda_Z; \cos\Theta', \cos\theta', \phi', \cos\overline{\theta}', \overline{\phi}', P_{ek})}{d\cos\Theta'\ d\cos\theta'\ d\phi'\ d\cos\overline{\theta}'\ d\overline{\phi}'} \Bigg] \Bigg\}$$

where

$$\eta^{\pm}_{i_1} = \cos\Theta_{i_1} \pm \frac{1}{N_\Theta}; \qquad \eta^{\pm}_{i_2} = \cos\theta_{i_2} \pm \frac{1}{N_\theta}; \qquad \eta^{\pm}_{i_3} = \phi_{i_3} \pm \frac{\pi}{N_\phi}$$

$$\eta^{\pm}_{i_4} = \cos\overline{\theta}_{i_4} \pm \frac{1}{2N_{\overline{\theta}}}; \qquad \eta^{\pm}_{i_5} = \overline{\phi}_{i_5} \pm \frac{\pi}{N_{\overline{\phi}}}$$

and

$$\frac{\mathrm{d}\xi(\cos\Theta, \cos\theta, \phi, \cos\overline{\theta}, \overline{\phi}, P_e)}{\mathrm{d}\cos\Theta\ \mathrm{d}\cos\theta\ \mathrm{d}\phi\ \mathrm{d}\cos\overline{\theta}\ \mathrm{d}\overline{\phi}}$$

is the multi-differential cross-section for background processes that pass all the W^+W^- analysis cuts. The function

$$G(\cos\Theta, \cos\theta, \phi, \cos\overline{\theta}, \overline{\phi}; \cos\Theta', \cos\theta', \phi', \cos\overline{\theta}', \overline{\phi}')$$

is the resolution function for our $e^+e^- \to W^+W^-$ analysis. It is defined so that $G\mathrm{d}\Omega$ is the probability that a W^+W^- pair produced with the variables

$$\cos\Theta', \cos\theta', \phi', \cos\overline{\theta}', \text{ and } \overline{\phi}'$$

passes all the W^+W^- analysis cuts and is placed in a bin of volume $\mathrm{d}\Omega$ centered at

$$\cos\Theta, \cos\theta, \phi, \cos\overline{\theta}, \overline{\phi}\ .$$

We shall assume that the W^+W^- final state can be isolated sufficiently well that

$$\frac{\mathrm{d}\xi(\cos\Theta, \cos\theta, \phi, \cos\overline{\theta}, \overline{\phi}, P_e)}{\mathrm{d}\cos\Theta\ \mathrm{d}\cos\theta\ \mathrm{d}\phi\ \mathrm{d}\cos\overline{\theta}\ \mathrm{d}\overline{\phi}} = 0,$$

is a good approximation. Although we are confident that this indeed is a good approximation, it remains an important topic of future study to actually prove that this will be the case.

For most of the results that are presented in this chapter we assume that there is no initial state radiation, beamstrahlung or intrinsic linac energy spread. Furthermore, we assume that we have a perfect detector that can measure the variables $\cos\Theta, \cos\theta, \phi, \cos\overline{\theta}, \overline{\phi}$ with 100% efficiency and infinite accuracy for topology A and B events with $|\cos\Theta| < 0.8$:

$$G(\Omega; \Omega') = \begin{cases} \delta^5(\Omega; \Omega'), & \text{if } |\cos\Theta| < 0.8; \\ 0, & \text{otherwise,} \end{cases}$$

where we have used Ω to denote the set of variabes $\cos\Theta$, $\cos\theta$, ϕ, $\cos\overline{\theta}$, and $\overline{\phi}$. We shall also assume for most of this chapter that $G(\Omega; \Omega') = 0$ for topology A and B events in which the W decays via $W \to \tau\nu$. The only source of error remaining after we make these assumptions is the statistical error associated with each bin count. At the end of the chapter we shall discuss how initial state radiation and beamstrahlung might be dealt with, and we investigate how our results are changed once detector resolution and $W \to \tau\nu$ decays are incorporated into a realistic resolution function.

We first fix κ_Z and λ_Z at their standard model values and do a three parameter log-likelihood fit of κ_γ, λ_γ, and $\mathcal{L}$. Here, and throughout this chapter, we use a binning of

$$N_\Theta = 20 \quad \text{and} \quad N_\theta = N_\phi = N_{\overline{\theta}} = N_{\overline{\phi}} = 5 \, .$$

This choice was driven by computer CPU and memory considerations. We don't know if a a finer binning granularity can improve our results. Fig. 7 shows 95% confidence level contours for three different center-of-mass energies. The luminosities of 1.3 fb^{-1} at $\sqrt{s} = 200$ GeV, 10 fb^{-1} at $\sqrt{s} = 500$ GeV, and 44 fb^{-1} at $\sqrt{s} = 1000$ GeV were chosen to give roughly 4100 detected events at each of the three center-of-mass energies. Fig. 7(a) shows the 95% confidence level contours for $\sqrt{s} = 200$ GeV and $\sqrt{s} = 500$ GeV; the contour for $\sqrt{s} = 1000$ GeV is too small to appear on this scale. Also shown in Fig. 7(a) is an estimate[8] of the 95% confidence level contour for the SSC at 10 fb^{-1}. There is a significant improvement in the limits in going from $\sqrt{s} = 200$ GeV to $\sqrt{s} = 500$ GeV. Note that the 95% confidence limits from a $\sqrt{s} = 500$ GeV e^+e^- collider at 10 fb^{-1} are much better than the limits estimated in Ref. 8 for the SSC at 10 fb^{-1}. Fig. 7(b) shows the 95% confidence level contour for $\sqrt{s} = 500$ GeV in more detail, and it also shows the contour for $\sqrt{s} = 1000$ GeV. We can see that the individual bounds on κ_γ and λ_γ at $\sqrt{s} = 500$ GeV are $|\kappa_\gamma - 1|, \lambda_\gamma < 0.01$.

Next, we allow all four of our couplings parameters to be free as we perform a six parameter maximum log-likelihood fit of $\kappa_\gamma, \lambda_\gamma, \kappa_Z, \lambda_Z, \mathcal{L}_+, \mathcal{L}_-$. The symbols $\mathcal{L}_+$ and $\mathcal{L}_-$ refer to the luminosities of two runs at different e^- polarizations. We will first assume that $P_e = 0$ for both of these runs and that $\mathcal{L}_+ = \mathcal{L}_- = 5$ fb^{-1}. The projection onto the $(\kappa_\gamma, \lambda_\gamma)$ and (κ_Z, λ_Z) planes of the six-dimensional 95% confidence level joint probability ellipsoid is shown in Fig. 8. The two-dimensional projection of the six-dimensional ellipsoid is not a nice contour, presumably because the shape of the six-dimensional object is so complicated. Each point is generated by first choosing a direction at random in the six-dimensional space, then traveling in this direction from the standard model point to the surface of the 95% confidence level ellipsoid, and finally projecting this point onto the plane. We interpret the outer boundary of the set of points in Fig. 8(a) to be the 95% confidence level contour for κ_γ and λ_γ, independent of κ_Z and λ_Z. If we compare Figs. 7(b) and 8(a) we see that our individual bounds on κ_γ and λ_γ are degraded by a factor of about

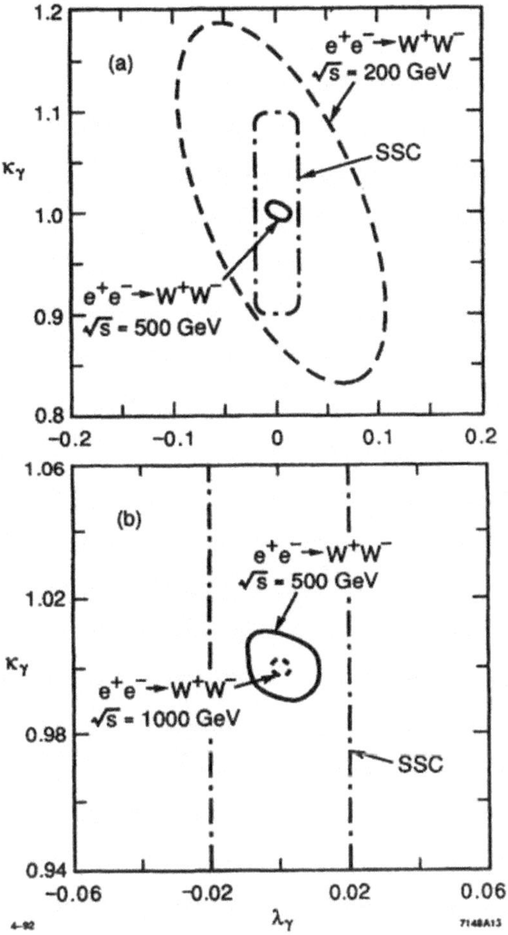

Fig. 7. Projection onto the $(\kappa_\gamma, \lambda_\gamma)$ plane of the 95% confidence level joint probability ellipsoid for a three-parameter maximum likelihood fit of $\kappa_\gamma, \lambda_\gamma, \mathcal{L}$. Standard model values are assumed for κ_Z and λ_Z. Projections for e^+e^- colliders with 1.3 fb^{-1} at $\sqrt{s}$ = 200 GeV, 10 fb^{-1} at $\sqrt{s}$ = 500 GeV, and 44 fb^{-1} at $\sqrt{s}$ = 1000 GeV are plotted, as well as an estimate of the limits from the SSC with 10 fb^{-1} (the SSC limit is independent of κ_Z and lz).

five when we allow all four of our coupling parameters to vary. The individual bounds on κ_γ and λ_γ are now $|\kappa_\gamma - 1|, \lambda_\gamma < 0.05$. Comparing Figs. 8(a) and (b) we see that the area of the excluded region in (κ_Z, lz) space is roughly the same as the excluded area in $(\kappa_\gamma, \lambda_\gamma)$ space.

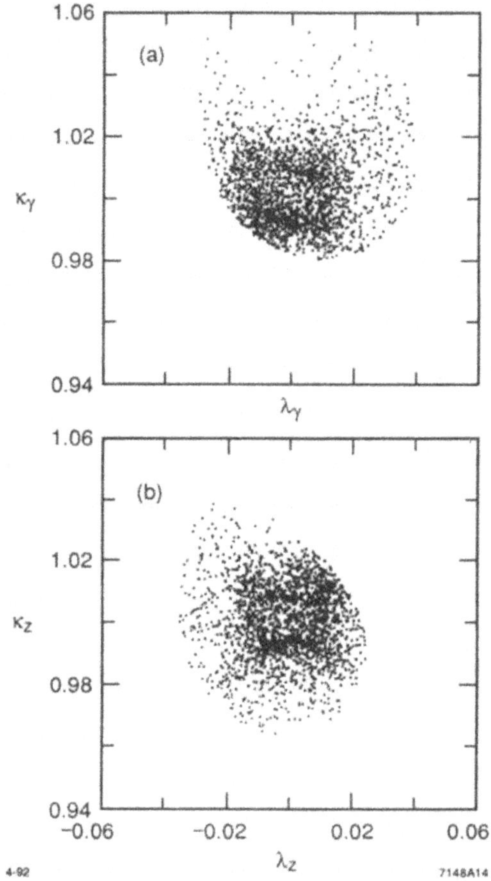

Fig. 8. Projection onto the (a) $(\kappa_\gamma, \lambda_\gamma)$ plane and (b) (κ_Z, λ_Z) plane of the 95% confidence level joint probability ellipsoid for a six parameter maximum likelihood fit of $\kappa_\gamma, \lambda_\gamma, \kappa_Z, \lambda_Z, \mathcal{L}_+, \mathcal{L}_-$. Unpolarized beams are assumed with $\mathcal{L}_+ + \mathcal{L}_- = 10\ fb^{-1}$ at $\sqrt{s} = 500$ GeV.

Given that the e^+e^-Z vertex is parity violating while the $e^+e^-\gamma$ vertex is parity conserving, it might seem plausible that polarized electron beams could help separate anomalous κ_γ and λ_γ values from anomalous κ_Z and λ_Z values. This indeed is the case. In Fig. 9 we show the projections of the six-dimensional 95% confidence level joint probability ellipsoid assuming that $\mathcal{L}_+ = 5\ fb^{-1}$ is collected with 90% right-handed electrons and $\mathcal{L}_- = 5\ fb^{-1}$ is collected with 90% left-handed electrons, all at $\sqrt{s} = 500$ GeV. $\kappa_\gamma - 1$ and λ_γ are now each bounded by ± 0.02 .

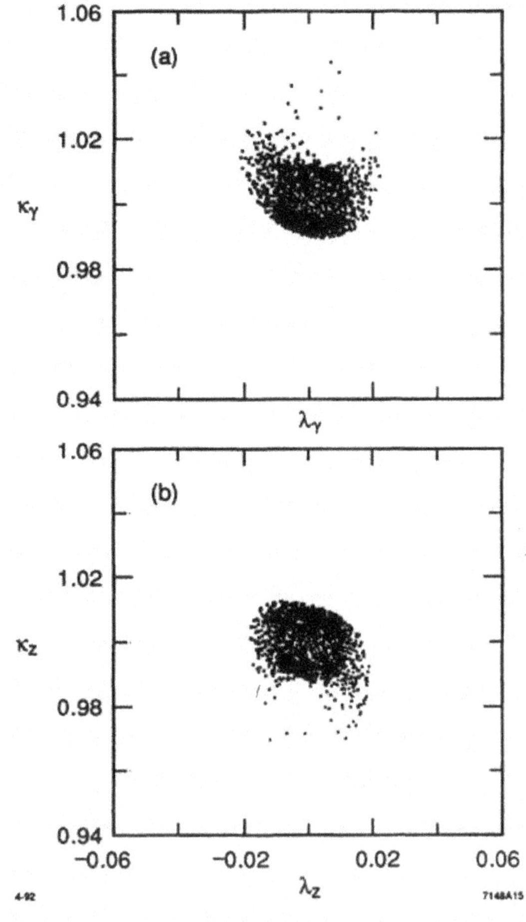

Fig. 9. Projection onto the (a) $(\kappa_\gamma, \lambda_\gamma)$ plane and (b) (κ_Z, λ_Z) plane of the 95% confidence level joint probability ellipsoid for a six-parameter maximum likelihood fit of $\kappa_\gamma, \lambda_\gamma, \kappa_Z, \lambda_Z, \mathcal{L}_+, \mathcal{L}_-$. The data set consisted of $\mathcal{L}_+ = 5\ fb^{-1}$ of 90% right-handed electrons and $\mathcal{L}_- = 5\ fb^{-1}$ of 90% left-handed electrons at $\sqrt{s} = 500$ GeV.

2.3 *A Strongly Interacting Higgs Sector*

Final state rescattering in $e^+e^- \to W^+W^-$, shown diagramatically in Fig. 10, can give us information about the $J = 1$ partial wave in the scattering process $W_L^+W_L^- \to W_L^+W_L^-$, where W_L^- denotes a longitudinally polarized W^- boson. This is discussed, for example, in Refs. 10 and 11. The effect of a 1.8 TeV techni-ρ resonance in the $W_L^+W_L^- \to W_L^+W_L^-$ channel on the $e^+e^- \to W^+W^-$ cross-section

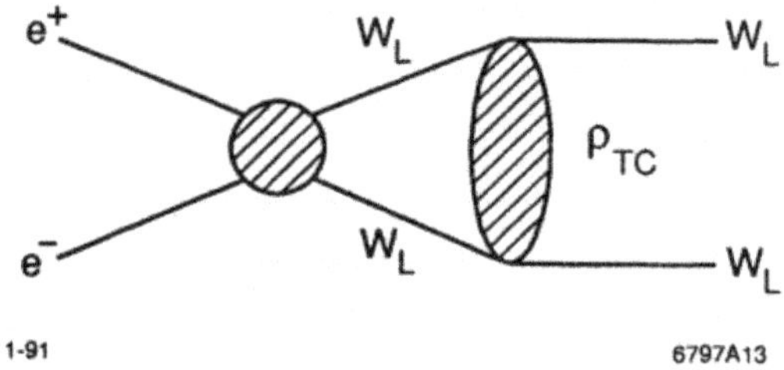

Fig. 10. Final state rescattering in $e^+e^- \to W^+W^-$

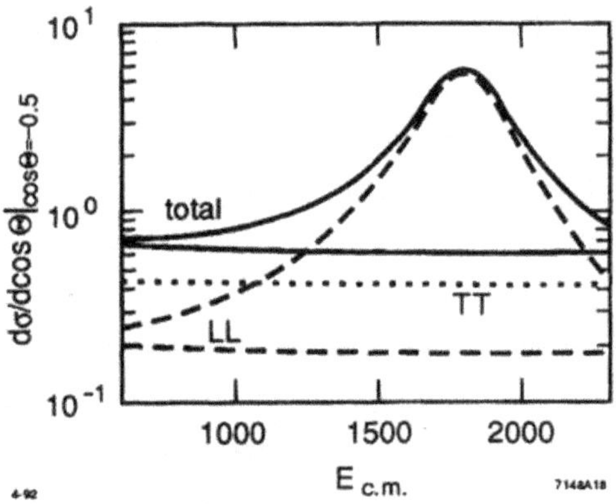

Fig. 11. The effect of a 1.8 TeV techni-ρ resonance in the $W_L^+W_L^- \to W_L^+W_L^-$ channel on the $e^+e^- \to W^+W^-$ cross-section.

is shown in Fig. 11. We incorporate this effect into our analysis by multiplying the standard model amplitude for $e^+e^- \to W_L^+W_L^-$ by the rescattering coefficient F_T defined by

$$F_T \equiv \exp\Big[\frac{1}{\pi}\int_0^\infty ds'\delta(s')\{\frac{1}{s'-s-i\epsilon}-\frac{1}{s'}\}\Big]$$

where

$$\delta(s) = \frac{1}{96\pi}\frac{s}{v^2} + \frac{3}{4}\left[\, tanh(\frac{M_\rho\Gamma_\rho}{M_\rho^2 - s}) + 1\right],$$

$v = 240$ GeV is the Higgs vacuum expectation value, M_ρ is the techni-ρ mass and Γ_ρ is the techni-ρ width. Fig. 12 shows $\mathrm{R_e(F_T)}$ and $\mathrm{I_m(F_T)}$, the real and imaginary parts of F_T for three values of the techni-ρ mass M_ρ. The shortest curve corresponds to $M_\rho = \infty$, the longest curve has $M_\rho = 2\,TeV$, and the intermediate curve has $M_\rho = 3\,TeV$. For each curve the values of $\mathrm{R_e(F_T)}$ and $\mathrm{I_m(F_T)}$ are shown

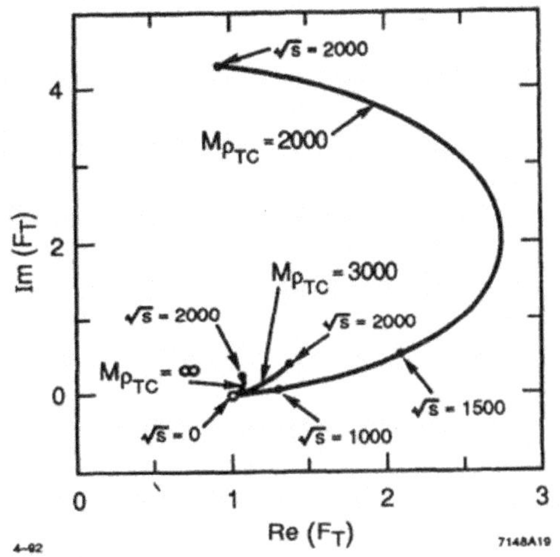

Fig. 12. $I_m(F_T)$ vs. $R_e(F_T)$ for three values of the techni-ρ mass M_ρ.

from $\sqrt{s} = 0$ to $\sqrt{s} = 2$ TeV.

To assess our sensitivity to $R_e(F_T)$ and $I_m(F_T)$ we do another maximum log likelihood analysis. This time we parameterize the log likelihood function with the variables $R_e(F_T), I_m(F_T)$, and $\mathcal{L}$, instead of the variables $\kappa_\gamma, \lambda_\gamma, \kappa_Z, \lambda_Z, \mathcal{L}$. Our log likelihood function $\ln L(R_e(F_T), I_m(F_T), \mathcal{L})$ is otherwise identical to the the one defined in Eq. (6). For the purpose of this discussion we set the coupling parameters $\kappa_\gamma, \lambda_\gamma, \kappa_Z, \lambda_Z$ equal to their standard model values. We have found, however, that this approximation will not be necessary since the variables $\kappa_\gamma, \lambda_\gamma, \kappa_Z, \lambda_Z, R_e(F_T), I_m(F_T)$ were remarkably orthogonal when simultaneous fits of all 6 variables were performed.

The 95% confidence level contours for $R_e(F_T)$ and $I_m(F_T)$ are shown in Fig. 13 for three different e^+e^- center-of-mass energies. At $\sqrt{s} = 1000$ GeV and 1500 GeV we have assumed modest luminosities of 45 fb^{-1} and of 90 fb^{-1}, respectively. We see that at these center-of-mass energies a 1.7 TeV techni-ρ resonance is easily observable. At $\sqrt{s} = 500$ GeV though, and despite the fact that we have assumed a hefty integrated luminosity of of 30 fb^{-1}, the 1.7 TeV techni-ρ is located right on the 95% confidence level contour.

Fig. 14 shows 95% confidence level contours in $R_e(F_T)$, $I_m(F_T)$ space for a larger luminosity at $\sqrt{s} = 1000$ GeV . With 200 fb^{-1} (Fig. 14) it appears possible to detect the presence of a techni-ρ resonance with a mass as large as 4 or 5 TeV.

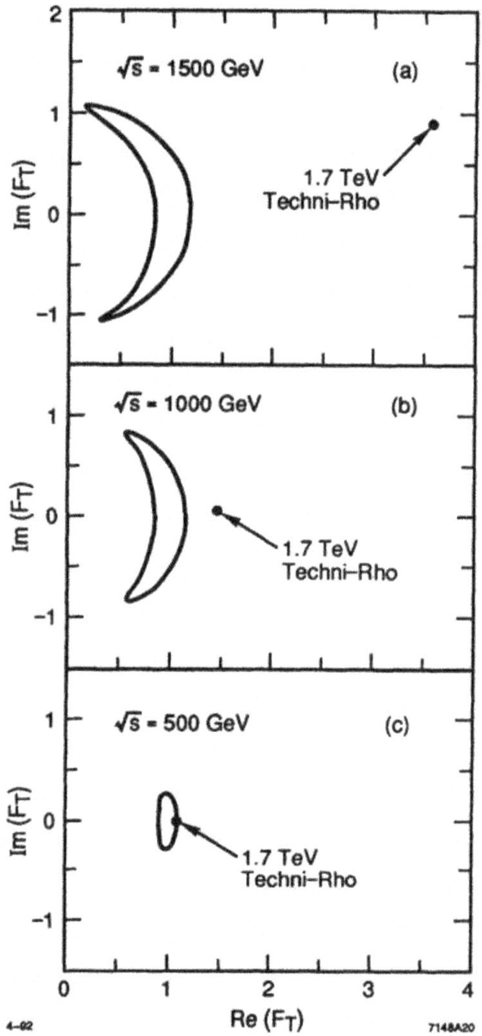

Fig. 13. The 95% confidence level contours for $R_e(F_T)$ and $I_m(F_T)$ for (a) $\sqrt{s} = 1500$, $\mathcal{L} = 90\ fb^{-1}$; (b) $\sqrt{s} = 1000$, $\mathcal{L} = 45\ fb^{-1}$; (c) $\sqrt{s} = 500$, $\mathcal{L} = 30\ fb^{-1}$

3. The Top Quark

The top quark can be studied near threshold or in the continuum. An energy scan across $t\bar{t}$ threshold probes the $t\bar{t}$ potential. The potential is determined mostly by the strong force but can also be affected by Higgs boson exchange[12]. Due to

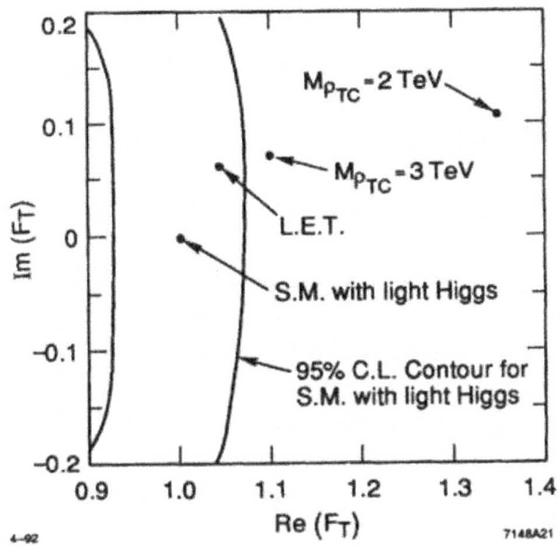

Fig. 14. The 95% confidence level contours for $R_e(F_T)$ and $I_m(F_T)$ for $\sqrt{s} = 1000$, $\mathcal{L} = 200\ fb^{-1}$

the large top mass, perturbative QCD should provide an excellent description of the $t\bar{t}$ potential. The unknown parameters which are measured with an an energy scan across the $t\bar{t}$ threshold region are the top mass M_t, the top decay width Γ_t, the strong coupling constant α_S, and, possibly, the Higgs boson mass M_{ϕ^0} and the top quark Yukawa coupling.

It has been estimated[2] that an energy scan can measure M_t, α_S, and Γ_t with errors of $\Delta M_t = 0.1$ to 0.2 GeV, $\Delta\alpha_S = 0.001$ to 0.005, $\Delta\Gamma_t/\Gamma_t < 0.2$, and that Standard Model Higgs boson exchange can be detected if the Higgs mass is less than 120 GeV or if the top Yukawa coupling is anomalously large.

Additional top physics studies can be made above the $t\bar{t}$ threshold region. An independent measurement of the top mass can be made with an accuracy of $\Delta M_t = 0.2$ GeV using beam energy-momentum constraints on the $t\bar{t}$ final state momenta. Such a measurement can be used, for example, to increase the precision of threshold measurements of α_S and Γ_t. Rare top quark decays can be searched for very effectively by selecting events in which at least one of the top quark decays via $t \to b\ l^+\ \nu$. Such a selection procedure at an e^+e^- linear collider produces a very high purity $t\bar{t}$ sample. Examples of rare t quark decays are $t \to s\ W^+$ (thought to be visible if $|V_{ts}|^2 < 0.1$), $t \to b\ H^+$ where H^+ is a charged Higgs boson, and $t \to \tilde{\chi}_1^0 \tilde{t}$ where $\tilde{\chi}_1^0$ is the lightest neutralino and $\tilde{t}$ is the stop squark.

A final state helicity analysis of the $t\bar{t}$ final state can be performed to search for structure in the $t\bar{t}\gamma$ and $t\bar{t}\ Z$ production vertices and in the $t\ b\ W^+$ decay vertex. A study of the $t\ b\ W^+$ vertex would test the expected ratio of longitudinal and transverse W gauge bosons produced in top decay and would be sensitive to a right-handed charged current. The $t\bar{t}$ helicity analysis would be very similar to the W^+W^- helicity analysis described earlier.

4. Higgs Bosons

We denote the standard model Higgs boson by ϕ^0. Large mass standard model Higgs bosons ($M_{\phi^0} > 200$ GeV) will be detected through the WW fusion process $e^+e^- \to \nu\bar{\nu}\phi^0$. The ϕ^0 decays via $\phi^0 \to W^+W^-$ or ZZ. Higgs boson with masses as as large as 500 GeV can be detected[13] with 30 fb^{-1} at a $E_{cm} = 1$ TeV.

Intermediate mass Higgs bosons will be detected using the reaction $e^+e^- \to Z\phi^0$ where the ϕ^0 decays via $\phi^0 \to b\bar{b}$ and the Z decays via $Z \to l^+l^-$, $q\bar{q}$, or $\nu\bar{\nu}$. This is how LEP searches for the standard model Higgs boson and the same methods work at NLC.

Of greater interest is how the NLC can be used to measure the branching ratios of various Higgs decay modes. For example, the branching ratios for $\phi^0 \to b\bar{b}$, $\phi^0 \to WW^*$, $\phi^0 \to \tau^+\tau^-$, and $\phi^0 \to c\bar{c} + gg$ can be be measured[14] with the errors shown in Fig. 15. The error bars are superimposed on a plot of the branching ratios of the lightest Higgs boson h^0 of the minimal supersymmetric extension to the standard model (MSSM). The branching ratios of the h^0 are plotted as a function of $\tan\beta$, the ratio of the vacuum expectation values of the two Higgs doublets of the MSSM. The branching ratios of the h^0 with small $\tan\beta$ coincide with the branching ratios of the standard model Higgs boson ϕ^0.

The $\gamma\gamma$ partial width of a Higgs boson is of special interest because it receives contributions from all electrically charged elementary particles which couple to the particular Higgs boson[15]. An elementary particle's contribution to the $\gamma\gamma$ partial width of the ϕ^0, for example, approaches a nonzero asymptotic value as the mass of the elementary particle approaches infinity. A measurement of the $\gamma\gamma$ partial width of the ϕ^0 can therefore probe physics at very large energy scales.

By colliding two beams of laser light which have been backscattered off the e^- and e^+ beams of an e^+e^- linear collider[16], the partial width $\Gamma(\phi^0 \to \gamma\gamma)$ can be measured. See Ref. 17 for a discussion of this measurement.

5. Supersymmetry

Future e^+e^- linear colliders will be in a position to study all of the supersymmetric partners of the γ, Z, and W, and of the three generations of fermions. They

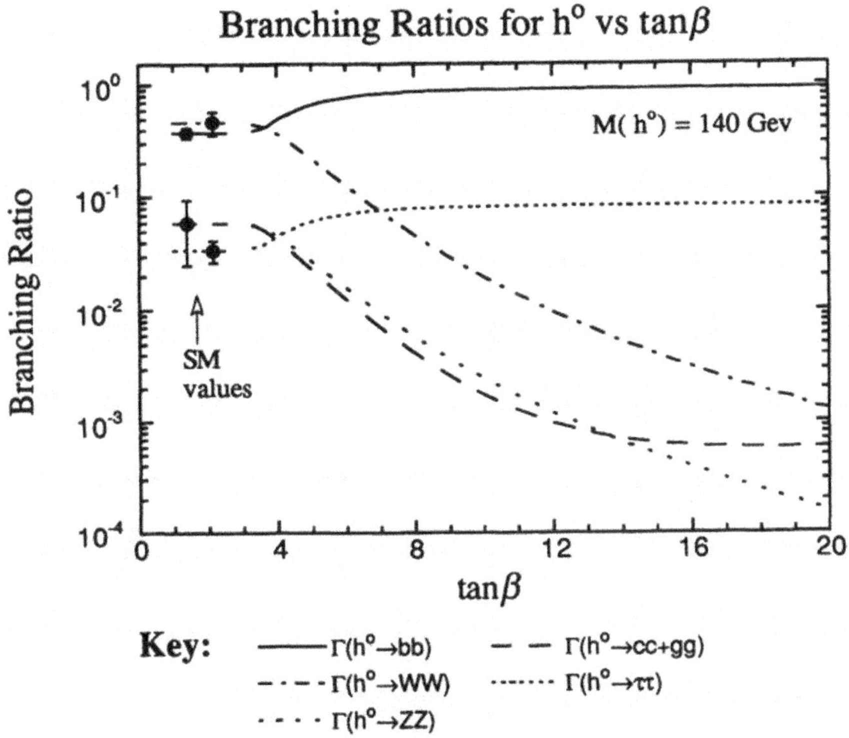

Fig. 15. Higgs branching ratio measurement errors.

will also be able to study in detail the Higgs bosons of supersymmetric theories.

We list the the additional particles required[18,19] by the MSSM:

1. Two neutral scalar Higgs particles, h^0 and H^0, a neutral pseudoscalar Higgs particle A^0, and a charged Higgs particle H^+.
2. One scalar neutrino $\tilde{\nu}_l$ for each of the three ordinary neutrinos and a left and right-handed sfermion $\tilde{f}_L$ and $\tilde{f}_R$ for each massive fermion.
3. Four neutralinos $\tilde{\chi}_1^0$, $\tilde{\chi}_2^0$, $\tilde{\chi}_3^0$, and $\tilde{\chi}_4^0$. Each neutralino is a mixture of the SUSY partners of the photon, the Z^0 and the neutral components of the Higgs doublets.
4. Two charginos $\tilde{\chi}_1^+$ and $\tilde{\chi}_2^+$. Each chargino is a mixture of the fermion partners of the $W^\pm$ and the charged components of the Higgs doublets.

5. The gluino $\tilde{g}$, which is the fermion partner of the gluon.

The Higgs mass spectrum of the MSSM is determined by two parameters which can be taken to be $\tan\beta$ and the mass M_{A^0} of the pseudoscalar Higgs. The entire parameter space of $\tan\beta$ versus M_{A^0} can be easily covered at an e^+e^- linear colliers with 10 fb^{-1} at $E_{cm} = 500$ GeV by searching for $e^+e^- \rightarrow h^0 Z$ and $e^+e^- \rightarrow H^0 Z$. In fact, less luminosity and less center–of–mass energy is required to cover the entire parameter space[20].

An e^+e^- linear collider can be used to study the spectrum of Higgs bosons in supersymmetric theories. An excellent example is given by Janot[21]. who studied the process $e^+e^- \rightarrow \gamma X Y$ where $X \rightarrow \tau^+\tau^-$, $Y \rightarrow q\bar{q}$, $X, Y = h^0, H^0, A^0, Z$, and γ is a beamstrahlung or initial state bremsstrahlung photon. Beam energy-momentum contraints are used to solve for E_γ, E_{τ^+}, E_{τ^-}, and E_Y, the energies of four of the final state particles. The momentum of the τ^+ and τ^- are determined by E_{τ^+}, E_{τ^-}, the measured directions of the $\tau^\pm$ decay products, and the τ mass. The momentum of the Y particle is determined by E_Y and the measured velocity of the system composed of the visible decay products of the Y. After imposing cuts to suppress background, the masses of the X and Y particles are plotted in the same histogram. A mass peak is seen at the Z mass from the process $e^+e^- \rightarrow ZZ$, but, in addition, mass peaks from the h^0, H^0 and A^0 appear from the processes $e^+e^- \rightarrow h^0 Z$, $e^+e^- \rightarrow H^0 Z$, $e^+e^- \rightarrow h^0 A^0$, and $e^+e^- \rightarrow H^0 A^0$.

If supersymmetric particles exist then e^+e^- linear colliders will be excellent facilities at which to study them. SUSY phenomenology will be very complicated. In general there will be $\tilde{f}_L$–$\tilde{f}_R$ mixing, intergenerational mixing, cascade decays of heavier SUSY particles such as squarks and the heavier neutralinos, and new sources of CP violation. e^+e^- linear colliders possess powerful tools with which to disentangle such a complex spectrum of states. Among the tools are initial state polarization of the e^- beam, high analyzing power of final state polarization, full reconstruction of final states with multiple numbers of leptons, quark jets, W's and Z's, and beam constrained fits of the momenta of heavy particles.

As an example we consider the process $e^+e^- \rightarrow \tilde{\chi}_1^+\tilde{\chi}_1^-$ where $\tilde{\chi}_1^+$ is the lightest chargino. There are three tree–level graphs contributing to the process $e^+e^- \rightarrow \tilde{\chi}_1^+\tilde{\chi}_1^-$: the γ s–channel annihilation, the Z s–channel annihilation, and the $\tilde{\nu}_e$ t–channel exchange graphs. The coupling of the $\tilde{\chi}_1^+$ to the photon is just given by its charge. The left and right–handed couplings of the $\tilde{\chi}_1^+$ to the Z, which we denote by O_L and O_R respectively, will be different and will depend on the relative amounts of wino and charged Higgsino in the $\tilde{\chi}_1^+$. The coupling at the e^- $\tilde{\chi}_1^+$ $\tilde{\nu}_e$ vertex will depend on the polarization of the inital state e^-, the SU(2) gauge coupling constant g, and the relative amounts of wino and charged Higgsino in the $\tilde{\chi}_1^+$. The coefficient at the e^- $\tilde{\chi}_1^+$ $\tilde{\nu}_e$ vertex which depends on the relative amounts of wino and charged Higgsino in the $\tilde{\chi}_1^+$ is denoted by V.

In the MSSM the coupling coefficients O_L, O_R, and V depend on the W^3–ino Majorana mass parameter M_2, the Higgs superpotential parameter μ and $\tan\beta$: $O_L = O_L(M_2, \mu, \tan\beta)$ $O_R = O_R(M_2, \mu, \tan\beta)$, and $V = V(M_2, \mu, \tan\beta)$. The cross–section for $e^+e^- \to \tilde{\chi}_1^+\tilde{\chi}_1^-$ will depend on the initial state e^- polarization, O_L, O_R, and $V^2/M_{\tilde{\nu}_e}^2$, where $M_{\tilde{\nu}_e}$ is the mass of the $\tilde{\nu}_e$. We assume that the $\tilde{\chi}_1^+$ decays 100% via $\tilde{\chi}_1^+ \to W^+\tilde{\chi}_1^0$ where $\tilde{\chi}_1^0$ is the lightest neutralino and the lightest supersymmetric particle. This will be the case, for example, if the only SUSY particles lighter than the $\tilde{\chi}_1^+$ are the $\tilde{\chi}_1^0$ and the the right handed selectron $\tilde{e}_R$ — a not unreasonable assumption.

We denote the masses of the $\tilde{\chi}_1^+$ and the $\tilde{\chi}_1^0$ by $M_{\tilde{\chi}_1^+}$ and $M_{\tilde{\chi}_1^0}$ respectively. Within the MSSM, we have $M_{\tilde{\chi}_1^+}=M_{\tilde{\chi}_1^+}(M_2, \mu, \tan\beta)$ and $M_{\tilde{\chi}_1}^0=M_{\tilde{\chi}_1}^0(M_1, M_2, \mu, \tan\beta)$, where M_1 is the Bino Majorana mass parameter. The energy and angle distributions of the W^+ and W^- bosons in the process $e^+e^- \to \tilde{\chi}_1^+\tilde{\chi}_2^- \to W^+W^-\tilde{\chi}_1^0\tilde{\chi}_1^0$ will depend on the inital state e^- polarization, $V^2/M_{\tilde{\nu}_e}^2$, O_L, O_R, $M_{\tilde{\chi}_1^+}$ and $M_{\tilde{\chi}_1}^0$. By collecting data half the time with right-handed longitudinally polarized electrons in the initial state, and half the time with left-handed electrons, and by binning the energy and polar angle of the $W^\pm$, the maximum likelihood method can be used to fit for $V^2/M_{\tilde{\nu}_e}^2$, O_L, O_R, $M_{\tilde{\chi}_1^+}$ and $M_{\tilde{\chi}_1}^0$.

If we assume that $\tan\beta$ has been measured using the measured Higgs boson mass M_{h^0} and the branching ratio for $h^0 \to WW^*$, then from the measured values of O_L, O_R, and $M_{\tilde{\chi}_1^+}$ we can extract M_2 and μ, and perform a MSSM consistency check since we are solving for two parameters using three measurements. Our measured value of $M_{\tilde{\chi}_1}^0$ will give us an estimate of M_1 and we can check the consistency of the Grand Unified Theory (GUT) relation between M_1 and M_2. Our measured values of M_1 and M_2 can be used to predict the mass of the gluino.

With our measured values M_2, μ and $\tan\beta$ we can solve for the parameter V. We can then use our fitted value of $V^2/M_{\tilde{\nu}_e}^2$ to estimate $M_{\tilde{\nu}_e}$ or to put a lower limit on $M_{\tilde{\nu}_e}$. For more discussion on this topic and quantitative examples see Ref. 2.

Acknowledgments

I would like to thank the organizers of this conference for an enjoyable stay at the I.C.T.P. This work was supported in part by U.S. Department of Energy contract DE-AC03-76SF00515.

REFERENCES

1. R. Ruth, in *Proc. of Workshop on Physics and Experiments with Linear Colliders*, Saariselka, Finland (1991) and SLAC-PUB-5729 (1992).
2. JLC Group, S. Matsumoto *et al.*, KEK Report 92-16 (1992)
3. K. Balewski *et al.*, preprint DESY-91-153 (1991)
4. H. Padamsee, in *Proc. of 1991 IEEE Particle Accelerator Conf.*, San Francisco, Ca. (1991) and preprint CLNS-91-1080 (1991).
5. D. Burke, in *Proc. of Workshop on Physics and Experiments with Linear Colliders*, Saariselka, Finland (1991) and SLAC-PUB-5791 (1992).
6. T. Barklow, P. Chen, and W. Kozanecki, in *Proc. of ECFA Workshop on Physics with Linar Colliders*, Hamburg, Germany (1991) and SLAC-PUB-5718 (1992).
7. K. Hagiwara, R.D. Peccei, D. Zeppenfeld, and K. Hikasa, *Nucl. Phys.* **B282,** 253 (1987).
8. G.L. Kane, J. Vidal, and C.-P. Yuan, *Phys. Rev.* **D39**, 2617 (1989).
9. A.de Rujula,M.B.Gavela,P. Hernandez, and E.Masso, preprint CERN-TH.6272/91 (1991); K.Hagiwara,S.Ishihara,R.Szalapski, D.Zeppenfeld, preprint MAD/PH/690 (1992).
10. M.E.Peskin, in *Physics in Collision 4*, A.Seiden, ed. (Editions Frontieres, Gif-sur-Yvette, 1984).
11. K.Hikasa, plenary talk at this workshop and preprint KEK-TH-319 (1992)
12. M. Strassler and M. Peskin, *Phys. Rev.* **D43**, 1500 (1991).
13. P. Burchat, D. Burke, and A. Petersen, *Phys. Rev.* **D38**, 2735 (1991).
14. M. Hildreth, SLAC-PUB-6036 (1993).
15. J. Gunion and H. Haber in *Proc. of 1990 DPF Summer Study on High Energy Physics*, Snowmass, Colorado (1990) and preprint SCIPP-90-22 (1990).
16. I.F. Ginzburg *et al.*, *Nucl. Instrum. Methods* **205**, 47 (1983); I.F. Ginzburg *et al.*, *Nucl. Instrum. Methods* **219**, 5 (1984); V.I. Telnov, *Nucl. Instrum. Methods* **A294**, 72 (1990);
17. T. Barklow in *Proc. of 1990 DPF Summer Study on High Energy Physics*, Snowmass, Colorado (1990) and SLAC-PUB-5364 (1990).
18. H. Haber and G. Kane *Phys. Rept.* **117,** 75 (1985).
19. J. Gunion and H. Haber *Nucl. Phys.* **B272,** 1 (1986).
20. P. Janot, preprint LAL 92-07 (1992).
21. P. Janot, preprint LAL 91-61 (1991).

NEW PARTICLES – POSSIBILITIES AND PROSPECTS

C. H. LLEWELLYN SMITH
Theoretical Physics,
1 Keble Road, Oxford OX1 3NP, England.

ABSTRACT

Reflections are presented on issues and questions that arose during the meetings.

1. Introduction

Rather than attempting to summarise what has been said at the meeting, which I consider unnecessary in view of the clarity of the talks, the instant availability of copies of transparencies and the ample time for questions and discussion, I shall reflect on various issues that speakers have addressed and comment on some questions that arose in the discussions. I begin with the obligatory list of "the big problems".

- **Origin of mass** ... I shall have a lot more to say about this. Here I remark only on the fact, which has not been mentioned at the meeting, that the Higgs' vacuum expectation value of the standard model generates a cosmological constant which is fifty four orders of magnitude greater than the experimental upper limit. It is unclear what we can learn from this, in the absence of a quantum theory of gravity, but it is certainly worrying.
- **Proliferation of flavours** ... As was remarked[1], the successes of two and a half millenia of science would suggest that the answer is compositeness, but it is very hard to understand how observed masses could be non-zero but much less than the scale of compositeness, and there are no plausible models. Unification of generations in grander groups is an obvious possibility, but again no good models are known. Perhaps flavour arises from the manifold structure in string theories, but such ideas are, of course, far from being established.
- **Full unification of forces** ... I shall have more to say about GUTS and here only remark that any really successful theory must incorporate a quantum theory of gravity; strings are the only candidate at present but progress in understanding string theory and linking it to observations is disappointingly slow.

- **CP violation** ... This is, perhaps, the Achilles heel of the standard model. The K decay data are consistent with the hypothesis that CP violation is entirely due to the phase in the CKM matrix but the observables are too few and the calculations so full of uncertainties that nothing more than consistency can be claimed. Furthermore the absence of strong CP violation, which is allowed by the vacuum structures of QCD, remains a mystery[2]. It can be forbidden by enlarging the symmetries of the theory but the price is the introduction of an axion. This mechanism can be adjusted to accommodate the failure to discover the axion by making it essentially invisible but this hardly enhances its plausibility. Although the arguments for an axion are good ones, and searches should not be abandoned, I think other mechanisms could account for the absence of strong CP violation and, if all else fails, it can simply be tuned to zero — a procedure regarded as unnatural although (at least in the absence of CP violating weak interactions) zero is the simplest possible value. A further twist is provided by the fact that the observed baryon number of the universe cannot be understood without non-standard model CP violation[3], except as a bizarre boundary condition in a baryon number conserving world. Further input is desperately needed from experiment — from a B factory[4] or studies of B decay at LHC or SSC[5].

The way ahead in addressing these and the other questions raised by the standard model is experiment guided by intelligent theoretical speculation. So far, of course, there is no clear evidence for any physics beyond the standard model and experimental hints are either somewhat tenuous (eg the ν_e/ν_μ ratio observed in underground experiments[6], the solar neutrino problem[7], the possible 17 KeV neutrino[8], rumours of anomalies in the CCFR neutrino experiment at Fermilab) or more convincing but less direct (eg the existence of dark matter and the baryon number of the universe). The importance of dark matter searches[9] is obvious and we can hope that the COBE data may cast light on whether dark matter is most likely to be hot or cold.

Turning to theory and phenomenology, it is obvious that although some fashionable models of non-standard physics may be correct the details should not be trusted. According to Bjorken[10] "One thing for sure is that almost all the words in the Higgs Hunter's Guide on how to find a nonstandard Higgs are wrong. There are so many mutually incompatible scenarios that generally what is on one page contradicts what is on another. Is it reasonable to average over a large number of wrong strategies in order to increase the odds of finding the right one?". Personally I think that there is no alternative and that it *is* reasonable, but the general point is well taken. Perl, quoting Francis Bacon, remarked[11] that when at sea we should not assume that there is no land because none is visible. Pursuing this metaphor, designers of experiments should certainly recall that better theory would have told Columbus that China was too far away to reach, and he would not have embarked. With this caution in mind, I shall comment on issues in and beyond the standard model and their implications for searches for new particles.

2. Precision Tests of the Standard Model and Searches for the Higgs boson

The extraordinary consistency between values of m_t that are obtained from one-loop fits to different precision measurements provides very strong support for the SU(2)×U(1) gauge theory. Using the latest LEP data, the fit of Ellis, Fogli and Lisi[12] — which is now excellent — gives m_t between 97 and 150 GeV [80 and 175] at the $1\sigma[3\sigma]$ level and M_H *less than 300 GeV at* 1σ *and 2 TeV at* 3σ.

With certain assumptions, *the limit on* M_H *can be interpreted generally as a bound on the scale of new physics* (Higgs or some alternative) *which generates masses for the* W *and* Z. Suppose we keep the standard model gauge boson-fermion couplings (which are well measured) and the triple gauge boson vertices (which is plausible[13], if not essential[14], in view of the success of LEP in fitting the precision data, and is theoretically desirable[15]), and add masses for the fermions W and Z by hand, or equivalently take the standard model Lagrangian in physical gauge and drop all terms that involve the physical Higgs field. In order to calculate one-loop radiative corrections we would have to introduce an ultraviolet cut-off Λ but the resulting formulae would be the same as in the standard model except that i) M_H would be replaced by Λ, and ii) there would be small changes in the constant (independent of m_t, M_H) terms which would have essentially no effect on the analysis. The LEP data therefore tell us that Λ is less than a few TeV. Note that Λ is a cut-off on the physics associated with boson masses and provides no information about the physics of fermion masses, which could in principle have a different origin, since the corresponding radiative corrections are small (of order $\alpha_W \frac{M_{light}}{M_W} \ln \Lambda$ where m_{light} is the mass of an external fermion, or $\alpha_W^2 \frac{m_t}{M_W} \ln \Lambda$).

I do not need to stress the importance of searching for the Higgs boson[16,17,18], or more generally investigating the mechanism that generates masses for the W and Z by studying WW scattering[17,18], to which I return in the next section. The prognosis for success with the next generation of facilities looks excellent in the light of the empirical evidence for a relatively small M_H/cut-off, which strengthens earlier theoretical arguments. These arguments follow from the fact that without the contributions from exchange of the physical Higgs boson, the standard model would predict that the Born amplitude for elastic scattering of longitudinal Ws is proportional to $\frac{g_W^2 s}{M_W^2}$ and grows one power of s faster than allowed by unitarity[19] which would fail at an energy of order 1 TeV. It follows that either perturbation theory must fail, leading to new non-perturbative phenomena, or new physics must exist which restores unitarity eg a Higgs boson, whose mass must be less than 1 TeV to ensure that perturbative unitarity holds at all energies[20], or the existence of form factors which would damp the amplitude if the W was composite. Without the

Higgs contributions, there are residual terms proportional[19] to $\frac{g^2 m_f \sqrt{s}}{M_W^2}$, where m_f is a fermion mass, but even with $m_t = 150$ GeV the unitarity limit is not reached until 20 TeV[21], so the bound on new physics associated with fermion masses is weaker.

The value of M_H can be bounded in terms of the energy domain in which it is supposed that the standard model is valid[22]. M_H is related to the quartic Higgs self interaction constant λ evaluated at a scale $\langle H\rangle$ and the Higgs field's vacuum expectation value by

$$M_H^2 = 2\lambda(\langle H\rangle)\langle H\rangle^2.$$

The renormalization group equation for λ can be solved to give, in the one loop approximation,

$$\lambda(Q) = \frac{\lambda(\langle H\rangle)}{1 - \frac{3\lambda(\langle H\rangle)}{4\pi^2}\ln\left(\frac{Q^2}{\langle H\rangle^2}\right)}.$$

Not being asymptotically free, λ increases with Q and, according to this formula, blows up at the "Landau pole" at a value of Q we will call Q_L. The one loop calculation cannot be trusted once λ become large, but lattice calculations strongly suggest[23] that indeed the theory becomes nonsense at a scale of order Q_L and that Q_L therefore gives a limit on the domain of validity of the standard model beyond which new physics must enter. The hypothesis that this limit occurs beyond the grand unification scale $M_X = 10^{15}$ GeV, gives an upper bound on $\lambda(\langle H\rangle)$ which translates into the bound[24]

$$M_H < 200 \text{ GeV}.$$

Regardless of whether the argument that the one-loop Landau pole indicates a breakdown of the theory is correct, it certainly indicates a failure of perturbation theory. It can therefore be argued that if the bound is not satisfied it would be impossible to understand the successful (?) prediction of Grand Unified theories (discussed below) that the SU(3)×SU(2)×U(1) couplings α_1, α_2, α_3 merge at M_X because this prediction depends on using perturbation theory (however this argument is not totally compelling as the influence λ on the evolution of the α_i is indirect).

As Q_L is lowered, the corresponding limit on $\lambda(\langle H\rangle)$ and hence on M_H increases until $M_H \approx Q_L \approx 650$ GeV. This indicates that, if the one-loop Landau pole indeed indicates a breakdown, the standard model makes no sense at all unless $M_H \lesssim 1$ TeV (allowing a safety factor because Q_L is calculated using perturbation theory where it is invalid; lattice calculations[24] give $M_H < 10M_W$). This is again encouraging for Higgs searchers.

3. Alternatives to Higgs?

Given that the W and Z are gauge bosons, as implied almost inescapably by the LEP data, the question arises how they acquired mass, or equivalently longitudinal degrees of freedom. These longitudinal states must either be fundamental, as in the Higgs model, or composite — presumably bound states of new fermions, as in technicolour models (for a review of recent work on technicolour and references see ref.[25]). In the latter case there must be a full spectrum of additional bound states (the techni-rho etc) lying at about 1 TeV and above, and also — in the original "extended" models introduced to attempt to explain fermion masses (unsuccessfully — see below) — relatively light pseudo Goldstone bosons, which should have been discovered at LEP.

The way to test the hypothesis that longitudinal vector bosons (V_L) are composite (apart from looking for pseudo-Goldstone bosons) is to study $V_L V_L$ scattering as discussed by Kane[17] and Pauss[18]. Even if a relatively light Higgs boson is found, it will be important to show that it really does "restore unitarity" and generate the "weak" perturbative standard model $V_L V_L$ cross-section. If no Higgs boson (or a non-standard boson) is found, it will be absolutely vital to study $V_L V_L$ scattering — perhaps it is weak after all, presumably indicating the existence of a hidden Higgs boson which escaped detection (as is possible according to supersymmetric models — see below), or perhaps it exhibits strong interactions which could reveal new underlying dynamics.

It is often suggested that in Higgs-less models there should be resonances in $W_L W_L$ scattering in the 1–2 TeV region. For example, in technicolour models the techni-rho should presumably lie in this region, and it has been argued, on the basis of the analogy between the chiral dynamics of W_L, which is a Goldstone boson, and the π in QCD, that the existence of an $I = 1$ vector resonance in this region is assured independent of the underlying dynamics. However, in an interesting recent paper Veltman and Veltman[26] have pointed out that this is wrong: it is necessary to include πN interaction terms in the chiral Lagrangian to generate the rho in the correct mass region. Indeed they show that if the effective Lagrangian that controls $W_L W_L$ scattering is obtained by taking the standard model Lagrangian and letting $M_H \to \infty$ then, if there is a rho-like resonance at all, it must lie at around 9 TeV or above. Thus the existence of a rho-like resonance in the 1–2 TeV region in Higgs-less models is by no means guaranteed, and in fact will not occur if the standard model Lagrangian correctly describes $W_L W_L$ scattering near threshold, which is therefore a very interesting quantity to study to high accuracy.

Technicolour provides an appealing way to generate masses for the W and Z. To generate fermion masses, however, it is necessary to turn to more complicated "extended technicolour models" which run into the difficulty that, unless the fermion masses are unacceptably small ($\lesssim 1$ MeV), they also produce flavour changing neutral currents (FCNCs) in contradiction to experiment. It is possible to suppress the FCNCs to an acceptable level, and also i) produce fermion masses of order a few GeV and ii) promote the pseudo Goldstone bosons out of the reach of LEP, by appealing to more complex dynamics in more complex "walking extended technicolour models". The dynamics may even make it possible to generate a mass as large as the top mass in "strong walking extended technicolour". However, this is only a hope and even if they work, the byzantine complexity of the models makes it impossible for me to take them very seriously. Furthermore, these models appear to be ruled out by the precision measurements at LEP, although proponents claim[27] that this is not so since the dynamics "is so poorly understood that nobody knows how to compute its effects on precisely-measured electroweak quantities".

Although, in contrast to technicolour models, the Higgs model can accommodate the observed fermion masses, the way it does so is not easy to take seriously.* The Higgs-fermion Yukawa couplings are simply adjusted by hand to produce the observed masses and mixings, and the values look rather unnatural eg the top coupling is more than 2×10^5 bigger than the electron coupling. There have been attempts[30] to produce empirical "Balmer formulae" for quark masses, which it is hoped might be explained later, by guessing simple forms for the mass matrix at the ground unification scale and seeing whether they can accommodate the data when extrapolated to low energy. Recent attempts have been based on minimal SUSY GUTs. One[31] utilizes seven parameters and therefore leads to six predictions for the thirteen fermion mass parameters (it is assumed that neutrino masses are zero); one is that m_t should lie in the range 174 to 183 GeV which is only just allowed by the precision LEP data. Another[32], which can accommodate m_t anywhere in the range 125 to 170 GeV, leads to seven predictions which agree with the data, including an impressively accurate prediction for the Cabibbo angle. With better data, this approach could be reversed[33] and the low energy data run up to the GUT scale to see if a simple pattern emerges.

It would seem that we may need a new idea to understand the pattern and perhaps also the origin of fermion masses. Indeed, unless or until a better understanding emerges, perhaps we should not entirely dismiss simple technicolour with one techni-quark doublet (which does not predict unwanted pseudo Goldstone bosons and, in contrast to the apparent situation in extended technicolour models, is not completely excluded by the LEP data[34]) as the source of masses for the W and Z, appealing to some new yet to be discovered idea to explain fermion masses. In any case, additional experimental information should be invaluable — more knowledge (limits or values) about neutrino masses from oscillation[35] and laboratory[8] experiments, a measurement of m_t[36], and more accurate measurements

---

* The same is true in models[28,29] in which a $t\bar{t}$ condensate plays the role of the Higgs vacuum expectation value, which are effectively equivalent to the standard model, with additional constraints, at energies well below the scale of the new unknown dynamics that must be invoked to produce the condensate. I do not dwell on these models which with a high scale for new physics i) predict $m_t \simeq 230$ GeV (this is the infrared quasi-fixed point value discussed in Section 5) which is now ruled out by the LEP data (but may[28] be avoidable), and ii) require an unnatural fine tuning to keep M_W well below the scale of the new dynamics; with a low scale, the prediction power is lost.

of the elements of the Cabibbo-Kobayashi-Maskawa matrix[11].

4. Extensions of the Standard Model

Grand Unified Theories

It is very unfortunate that, unless the proton lifetime lies in the small window between existing limits and the best limit allowed by the irreducible cosmic ray background, it would seem that GUTs can only be tested indirectly (see below). Nevertheless, they must be taken very seriously[37]. The underlying idea of unifying strong and electroweak forces, and thereby simultaneously removing any fundamental distinction between quarks and leptons, is compelling, and GUTs provide an elegant explanation for a variety of facts (eg charge quantization and the fact that quarks are heavier than the corresponding leptons).

Supersymmetry

The idea that there may be a (super) symmetry which connects fermions and bosons and the fact that local supersymmetry necessarily incorporates gravity are so appealing that the possible existence of supersymmetry must be taken extremely seriously. An additional argument for SUSY, which has the exciting by-product that it predicts that SUSY particles will be in reach of the next generation of accelerators, is that it is needed to stabilise the mass of the W in theories with elementary Higgs bosons in which the Standard Model is embedded in a more complete theory involving larger mass scales. In such theories, there are one loop contributions to M_W^2 from particles X proportional to $g_W^2 M_X^2$. This "quantum correction" is enormous if X is a heavy particle such as a GUT boson. In principle it can be cancelled against an almost equally enormous but negative zeroth order contribution, but only by fine-tuning the parameters to great accuracy in a way which is not credible. In a SUSY theory, however, the corrections are proportional to $g_W^2(M_X^2 - M_{\tilde{X}}^2)$ where $\tilde{X}$ is the superpartner of X. This constrains the mass splitting if we require the correction to be less than or of order M_W^2, so that fine tuning will be unnecessary: in particular, it requires the SUSY partners of the known particles to lie below 1 TeV.

There has been considerable discussion at this meeting[16,18] of the possible discovery of the five Higgs bosons $(h, H, A, H^{\pm})$ which occur in the minimal supersymmetric model. Optimists are pleased that one of these bosons should be found at LEP200 and/or LHC and SSC for most values of the parameters. Pessimists would respond i) that unfortunately there seems to be an inaccessible hole in parameter space (said to be small by people who seem to have privileged knowledge that the parameters are likely to be chosen with equal priori probability somewhere in the space of the two parameters that are conventionally used), and ii) that in most of the accessible region only one of the bosons will be found, which would not be enough to establish the model. However, the published analyses are mainly based on the assumption that the Higgs bosons decay into ordinary particles, whereas SUSY decay modes may be allowed in some cases which would broaden the discovery potential. This deserves further study.

The excellent discovery potential of the next generation of machines for SUSY particles has been widely discussed in the literature[18,38] and we will not dwell on it here, except to remark that most of the discussion assumes the conservation of R parity which requires that SUSY particles be pair produced and that the lightest is stable. This need not be true. Indeed the most general renormalizable Lagrangian with the minimal SUSY content contains R violating terms, so it is natural to expect them to be present[29,39,40]. They are not generally introduced because they violate lepton and baryon number, the three possible generic types of term (which enter with independent coefficients for the many different possible flavour combinations) having the structures $LL\bar{E}$, $LQ\bar{D}$ and $\bar{U}\bar{D}\bar{D}$ where L and Q represent (s)lepton and (s)quark doublets and E, D, U represent singlets. Proton stability requires that the $LQ\bar{D}$ and $\bar{U}\bar{D}\bar{D}$ terms are not present simultaneously, so presumably some symmetry must prevent this — a situation which is just as easy to image as the existence of a symmetry which banishes all three types of term. It has been argued[41] that the existence of any of these terms leads to trouble for the cosmological generation of the observed baryon number of the Universe but this is not necessarily so[42]. At present it would seem therefore that the possibility that R parity is violated should be taken very seriously. Discussion of the phenomenology is complex because there are so many options. It is different from the phenomenology of R conserving models but it is likely there will still be spectacular signatures at the next generation of colliders. R violating models may contain leptoquarks which could be found at HERA[39,43].

5. Predictions for the Masses of New Particles

It is appropriate at a meeting such as this to review predictions for undiscovered particles, beginning with the top quark and Higgs boson since they are nearly in reach according to the LEP data which already provide fairly stringent constraints. *TT* Wu[44] has presented a new approach to the calculation of m_t and M_H. It starts, following Veltman[45] but using a different calculational method, from a condition which ensures that the standard model is free of quadratic divergences. This is desirable as it stabilises M_W which, at least if the theory's ultraviolet cut-off is treated as the scale at which new physics appears, would otherwise tend to be dragged up to the cut-off. However, in contrast to SUSY which removes quadratic divergences and provides stability to *all* orders, this result is only achieved to one loop, in a seemingly *ad hoc* way (and it is not obvious that it can be consistently generalised to all orders). Nevertheless the underlying motivation is fairly strong and the resulting "Veltman relation"

$$\begin{aligned} M_H^2 &= 4m_t^2 + \text{ light fermion contributions } - 2M_W^2 - M_Z^2 \\ &= 4m_t^2 - (147 \text{ GeV})^2, \end{aligned}$$

which gives 127 GeV $< M_H <$ 261 GeV putting in the LEP 1σ limit, 97 GeV $< m_t <$ 150 GeV, is certainly interesting.

Wu and Osland[44] go further, however, requiring the absence of certain logarithmic divergences. This gives a further relationship* between the masses and fixes m_t = 120 GeV, M_H = 190 GeV. However, the calculation raises many questions, concerning the apparent need for a particular form of cut-off to get their result, whether the calculation is consistent when higher orders are taken into account, the motivation for removing logarithmic divergences in a sub-set of vertex functions — especially since others certainly cannot be removed, and the apparent inconsistency between results obtained for different leptons (Wu expresses the hope that consistency to all orders for all particles might fix further — perhaps all — parameters). Personally, I cannot take it seriously.

In my opinion, the idea of Pendleton and Ross[47,48] that some masses may be determined by infra-red fixed points for running Yukawa couplings is more promising. I shall describe it in some detail as it deserves to be better known. For those not familiar with infra-red fixed points, consider as a purely mathematical model a coupling $g(Q^2)$ which satisfies the renormalization group equation (RGE)

$$\frac{dg^2}{d\ln Q^2} = ag^2(g^2 - b),$$

where a and b are positive constants. Whatever the boundary condition $g(\bar{Q}^2) = \bar{g}$, g^2 will get driven to b as $Q^2 \to 0$ the explicit solution being

$$g^2(Q^2) = b\left[1 - \left(1 - \frac{b}{\bar{g}^2}\right)\left(\frac{Q^2}{\bar{Q}^2}\right)^{ab}\right]^{-1}.$$

Suppose $\bar{g}$ is determined by some deeper theory, in which known low energy physics is embedded, at a very large scale $\bar{Q}$ (eg the GUT scale), or rather an effective $\bar{g}$ since the low energy RGE will change as Q approaches $\bar{Q}$ and the deeper physics comes into play. Then unless $\bar{g}^2$ is very small compared to b, g^2 will be approximately

---

* Blumhofer and Stech[46] reach a different condition (which however leads to essentially the same numerical predictions for m_t and M_H) by requiring that the Higgs' vacuum expectation value is free not only of quadratic divergences (this gives the Veltman condition), but also logarithmic divergences; their approach is subject to the same criticism.

equal to b — which is determined solely by the low energy dynamics — at observable energies $\bar{Q} \ll Q$.

This phenomenon can occur in the standard model. If we keep only the two relatively large couplings g_3 (QCD) and g_t (the top Yukawa coupling) the one loop RGE equations are

$$16\pi^2 \frac{dg_3}{d\ln Q^2} = -7g_3^2$$

and

$$16\pi^2 \frac{dg_t}{d\ln Q^2} = g_t \left[\frac{9}{2}g_t^2 - 8g_3^2\right],$$

where the two terms in the second equation arise from the t loop correction to the Higgs propagator in the $Ht\bar{t}$ three point function and the gluon correction to the vertex respectively. These equations imply

$$16\pi^2 \frac{d\ln(g_t/g_3)}{d\ln Q^2} = \frac{9}{2}g_t^2 - g_3^3$$

which has an infra-red fixed point at $g_t^2 = 2g_3^3/9$. The assumption that g_t is very close to this fixed point* leads to the prediction $m_t = \sqrt{\frac{4\pi\alpha_s(m_t)}{9}}\langle H\rangle \simeq 100$ GeV which is corrected to around 135 GeV when the position of the fixed point is evaluated without neglecting g_1 and g_2. However, the structure of the complete standard model equations is such that generally g_t/g_3 is unlikely to be close to the fixed point. For $9\bar{g}_t^2 > 16\bar{g}_3^2$ the couplings become stagnant with $9g_t^2 \simeq 16g_3^2$ until Q^2 becomes very small indeed and the fixed point is finally approached[48]. The relation $9g_t^2(m_t^2) \simeq 16g_3^2(m_t^2)$ gives $m_t \simeq 230$ GeV which turns out to be an upper bound for all $\bar{g}_t/\bar{g}_3$.

The quartic Higgs self coupling satisfies the one loop RGE

$$16\pi^2 \frac{d\lambda}{dQ^2} = 6\lambda^2 + 12\lambda g_t^2 - 24g_t^4$$

where the terms are from the one loop Higgs correction to the Higgs four point function and the top loop corrections to the external propagators and to the vertex. It is easily seen that there is an infra-red fixed point at $4\lambda = \sqrt{689 - 25}\,g_t^2$ which would give $M_H \simeq 0.59m_t$. To get a realistic prediction we must again include all the couplings, and also the contributions of SUSY particles if nature is

---

* or exactly at it as supposed by Kubo Sibold and Zimmermann[49] who explored the conditions for exact scale independent relations between coupling constant to exist, which they call "Reduction of Couplings", without using the term infra-red fixed point or realising that their results had been anticipated, in a more general form, by Pendleton and Ross.

supersymmetric[50], but it turns out that in the SUSY GUT models to which we now turn this relation is approximately correct with M_H/m_t varying from 0.57 to 0.84 in the two solutions which represent the range of predictions.

6. Minimal Supersymmetric Predictions

The preceding discussion shows how the evolution of coupling constants at low (compared to the GUT scale) energy can determine the masses of particles. I shall now consider generalizations which incorporate SUSY, impose GUT scale boundary conditions suggested by SUSY GUTs — which obviously increases the predictive power, and allow for the evolution of masses as well as couplings, to produce state of the art predictions. It is by now well known that if the three gauge couplings are evolved to high energy using the standard model RGE they do not appear to converge at a grand unified scale M_X, whereas if the contributions of the minimal set of low energy SUSY particles is switched on at the expected threshold at an energy $\lesssim 1$ TeV, they do converge (see eg refs.[29,51,52]). The implications must be treated with caution because

1) Other, non-SUSY, thresholds below M_X could also produce convergence (see eg ref.[52]). Thus the most that can be concluded is that *minimal* non-SUSY GUTs are probably (see below) excluded, while minimal SUSY GUTs are allowed.
2) It is not actually true that the couplings should meet at M_X. They should only meet as $Q^2 \to \infty$ where the fact that the GUT is spontaneously broken by GUT scale masses can be neglected (this is for couplings defined by momentum subtraction[53]; with mass independent definitions, the couplings are exactly equal in the full theory but there are discontinuities[54] in matching them to the low energy couplings which do not meet exactly at M_X). This effect depends on the details of the GUT. However, it tends to be very small and is almost certainly not capable of saving minimal non-SUSY GUTs. It is ignored in the discussion of SUSY GUTs below, which is probably reasonable.
3) It is usually assumed that all particles with mass of order the GUT scale M_X have masses exactly equal to M_X although this need not be so. Differing masses near the GUT scale would put kinks in the evolution of the couplings[53], and varying these differences has much the same effect on the fit as varying the SUSY threshold[55]. Thus attempts to derive the SUSY threshold from the fit should be treated with caution. This effect is also ignored in the discussion below.
4) The assumption of a single SUSY threshold for all particles is certainly not expected to be correct. This is corrected in the discussion below.

The particular SUSY GUT predictions we shall discuss, which are due to

Roberts and Ross[56], start from a simple set of assumptions about physics at M_X and proceed by evolving couplings and masses down to low energy. The one loop top and scalar top quark contributions to the Higgs field propagator have a dominant effect because the top Yukawa coupling is large. The (fermionic) top and (bosonic) s-top loops contribute with opposite signs, but because the s-top is naturally heavier than the top in such models, the net effect is — remarkably, and rather generally — to drive the Higgs mass parameter $\mu^2(Q^2)$ negative at low Q^2 thereby automatically triggering SU(2)×U(1) breaking[57], and making the Higgs mechanism appear much less *ad hoc*.

There are eight parameters in the Lagrangian: M_X, α_X — the universal gauge coupling at M_X, $\mu_0(X)$ — the Higgs mass parameter at M_X, $m_0(X)$ — a "soft SUSY breaking" scalar mass, and $m_{1/2}(X)$ — a "soft SUSY breaking" gaugino mass, and two parameters A and B. The origin of the soft SUSY breaking terms is not specified, but the usual philosophy is to assume that primary SUSY breaking occurs among a set of "invisible" particles which communicate with the observed particles only through gravity, which transmits flavour-blind SUSY breaking. In evolving couplings and masses down from the GUT scale, Ross and Roberts allow for different SUSY thresholds and generate a self consistent set of SUSY masses. The parameters are treated as follows:

i) Values of μ_0, m_0, A and B are chosen.

ii) α_X, M_X and $m_{1/2}$ are adjusted to produce the observed low energy values of the three gauge couplings α_i.

iii) g_t is adjusted to produce the observed value of M_W (which is produced by the breakdown of SU(2)×U(1) that is driven by the top and s-top quarks thanks to the relatively large value of g_t, as discussed above).

All other masses (m_t, M_H, the different SUSY masses) are then predicted. It turns out that there are consistent solutions for SUSY scales up to 90 TeV. However, when the SUSY scale is much larger than M_W, g_t has to be fine tuned to produce the observed value of M_W. This is of course precisely the fine tuning problem discussed earlier, which generally suggests that SUSY masses cannot plausibly be much greater than M_W/g_W. Restricting $\mu_0(X)$ and $m_0(X)$ so that the "no fine tuning" condition $\delta M_W^2/M_W^2 < 10\delta g_t^2/g_t$ is satisfied, it turns out that the supersymmetric particles are indeed relatively light. Roberts and Ross present two solutions, representing the range of predictions for which the no fine tuning condition is satisfied, with the following predictions for masses (in GeV):

m_t	160	100
$\tilde{\gamma}$	57	83
$\tilde{Z}$; $\tilde{W}$	99; 99	120; 112
$\tilde{g}$	354	559
$\tilde{\ell}_L$	220	206
$\tilde{\ell}_R$	195	146
$\tilde{u}_L, \tilde{c}_L; \tilde{d}_L, \tilde{s}_L$	365; 373	511; 517
$\tilde{u}^c_L, \tilde{c}^c_R$	359	495
$\tilde{d}^c_R, \tilde{s}^c_R, \tilde{b}_R$	358	491
$\tilde{t}_L$; $\tilde{b}_L$	325; 335	491; 497
$\tilde{t}^c_R$	273	452
H^0	91, 264	84, 221
$H^\pm$; A^0	276; 264	232; 218
$\tilde{H}^0$	205, 225	139, 226
$\tilde{H}^\pm$	229	227

It is unclear how seriously the details of such predictions should be taken as they not only depend on very strong assumptions about the underlying physics, but also involve technical assumptions eg that particles with GUT scale masses are degenerate (point 3) above). Nevertheless this is the best that can be done at present and, at the least, this exercise produces a plausible suggestion for the ordering and order of magnitude of the masses of SUSY particles. However, one experimental number would be worth more than any theoretical speculation!

7. Concluding Remarks

As we have heard in so many excellent talks at this conference, existing data — Janus-like — give excitingly accurate support for the basic gauge structure of the standard model, while frustratingly refusing to give hints of the unknown physics which must exist beyond the standard model. However, we may not have to wait long for a new discovery as existing experiments are capable of finding a large range of sought for new particles and phenomena eg

- The top quark — seems likely to be found at FNAL[36], and could still be in reach of LEP200 if it has non-standard decays.
- Supersymmetry — could be discovered at FNAL, or at LEP200[38] through the production of a pair of neutralinos.
- A Leptoquark — could be discovered at HERA[43].
- A heavy Majorana neutrino — could be discovered at HERA[43].
- Higgs — H^0 and $H^\pm$ could be discovered at LEP200[16], while $H^\pm$ could be discovered at FNAL in the decay $t \to H^+ b$.

- A non-zero neutrino mass — could be inferred from the study of solar[7] or laboratory neutrinos[35], or measured in the 17 KeV experiments[8].

In any case it is clear that the energies of LHC[18] and SSC[58] provide a sufficient safety margin for exploring the new physics which theory predicts must exist below 1 TeV that we can be confident that they will provide a real breakthrough in understanding the origin of mass. Of course the Higgs boson may be found first at LEP but it would then be odds on that nature is supersymmetric, in my opinion; LHC and SSC will be needed to explore the SUSY spectrum, even if the lightest SUSY states are found earlier, and to look for the other Higgs bosons required by SUSY, in which task a linear e^+e^- collider could also play an important role[59]. In conclusion, the prospects for finding new particles and deepening our understanding of nature in the coming years are excellent, and we can hope that this excellent meeting will be the first of a long series in which, however, the phrase "The Search for" can be dropped from the title.

8. References

1. M. Bardadin, these proceedings.
2. For a review see R.D. Peccei, p.503 in "CP Violation" ed. C. Jarlskog, World Scientific 1989.
3. For a review see P. Langacker, p.552 in "CP Violation" ed. C. Jarlskog, World Scientific 1989.
4. For a list of proposed B factors see P. Kim, these proceedings.
5. See presentations by P. Schlein, G. Carboni and T. Nakada in "Towards the LHC Experimental Programme" (Evian Meeting) CERN, 1992 and talks in "Large Hadron Collider Workshop" CERN90-10/ECFA 90-133.
6. The ν/e ratio in underground experiments is found to be about 1.2 while the expected value is about 2. This has been interpreted in terms of a dearth of ν_μ and $\bar{\nu}_\mu$ due to $\nu_\mu \to \nu_\tau$ oscillations, but uncertainties in the flux calculations allow the possibility that it is due to an excess of positrons, which might result from the decay $p \to \nu\bar{\nu}e^+$. For a discussion of this possibility and references to the experiments see W.A. Mann, T. Kafka and W. Leeson *Phys. Lett. B* (1992) in press.
7. D. Sciama,, these proceedings.
8. J. Jaros, these proceedings.
9. S. Cooper, these proceedings.
10. J.D. Bjorken, SLAC-PUB-5673 (1991).
11. M. Perl, these proceedings.
12. G.L. Fogli, these proceedings.
13. A. de Rujula, M.B. Gavela, P. Hernandez and E. Masso, CERN-TH 6972/91.

14. K. Hagiwara, S. Ishihara, R. Szalapski and D. Zeppenfeld, *Phys. Lett.* **B283** (1992) 353.
15. C. Latourre and G. Moultaka, Montpellier preprint PM/92-15.
16. M. Felcini, these proceedings.
17. G. Kane, these proceedings.
18. F. Pauss, these proceedings.
19. C.H. Llewellyn Smith, *Phys. Lett.* **46B** (1973) 233. J. Cornwall, D. Levin and G. Tiktopoulos, *Phys. Rev. Lett.* **30** (1973) 1268 (E:31, 572).
20. D.A. Dicus and V.S. Mathur, *Phys. Rev.* **D7** (1973) 3111. B.W. Lee, C. Quigg and H.B. Thacker, *Phys. Rev. Lett.* **38** (1977) 883 and *Phys. Rev.* **D16** (1977) 1519. For improvements on the original bounds see: M. Luscher and P. Weisz, *Phys. Lett.* **212B** (1988) 472, and L. Durand, J.M. Johnson and J. Lopez, *Phys. Rev. Lett.* **64** (1990) 1215.
21. T. Appelquist and M. Chanowitz, *Phys. Rev. Lett.* **59** (1987) 2405.
22. R. Dashen and H. Neuberger, *Phys. Rev. Lett.* **50** (1983) 1897.
23. A. Hazenfratz and T. Neuhaus, *Nucl. Phys.* **B297** (1988) 205. J. Kuti, L. Lin and Y. Shen, *Phys. Rev. Lett.* **61** (1988) 678. M. Luscher and P. Weisz, *Phys. Lett.* **B212** (1989) 472. A. Hasenfratz *et al*, *Nucl. Phys.* **B317** (1981) 81.
24. L. Maiani, G. Parisi and R. Petronzio, *Nucl. Phys.* **B136** (1978) 115. N. Cabibbo, L. Maiani, G. Parisi and R. Petronzio, *Nucl. Phys.* **B158** (1979) 295.
25. K. Lane, Proc. Joint International Lepton-Photon Symposium and European Conference on High Energy Physics, ed. S. Hegarthy *et al*, World Scientific 1992, Vol.I p675.
26. H. Veltman and M. Veltman, DESY 91-050.
27. K. Lane, Proc. Joint International Lepton-Photon Symposium and European Conference on High Energy Physics, ed. S. Hegarthy *et al*, World Scientific 1992, Vol.2 p210.
28. M. Lindner, Proc. Joint International Lepton-Photon Symposium and European Conference on High Energy Physics, ed. S. Hegarthy *et al*, World Scientific 1992, Vol.I p679.
29. G.G. Ross, Proc. Joint International Lepton-Photon Symposium and European Conference on High Energy Physics, ed. S. Hegarthy *et al*, World Scientific 1992, Vol.II p181.
30. H. Fritzsch, *Phys. Lett.* **70B** (1977) 436 and **73B** (1978) 317. F.J. Gilman and Y. Nir, *Ann. Rev. Nucl. Part. Sci.* **40** (1990) 213.
31. S. Dimopoulous, L.J. Hall and S. Raby, *Phys. Rev. Lett* **68** (1992) 1984, and *Phys. Rev.* **D45** (1992) 4192. V. Barger, M.S. Berger, T. Han and M. Zralek, *Phys. Rev. Lett.* **68** (1992) 3394.
32. G.F. Giudice, University of Texas preprint UTTG-5-92.
33. G.G. Ross, private communication.

34. R.N. Cahn and M. Suzuki, *Phys. Rev.* **D44** (1991) 3641.
35. F. Vannucci, these proceedings.
36. F. Snider, these proceedings.
37. Not only the canonical, "SU(5) type", GUTs, but also the GUTs of the type originally proposed by Pati and Salam — see R. Pati, Maryland preprint UMDHEP92-210, for a recent review.
38. A. Lopez-Fernandes, these proceedings.
39. For the phenomenology of models in which R parity is explicitly broken. See H. Dreiner and G.G. Ross, OUTP-91-15P, J. Butterworth and H. Dreiner, OUTP-92-01, to be published in Proc. 2nd HERA Workshop on Physics. H. Dreiner and R.J.N. Phillips, *Nucl. Phys.* **B367** (1991) 591. H. Dreiner and S. Lola, OUTP-92-02, to be published in Proc. EE500 Workshop, Hamburg 1991, and references therein.
40. R parity may also be broken spontaneously — see M.C. Gonzalez Gania, J.C. Romao and J.W.F. Valle, FTUV/91-42, IFIC/91-41, MAD/PH/678 and references therein.
41. B.A. Campbell *et al*, CERN TH 6208/91.
42. H. Dreiner, Proc. Joint International Lepton-Photon Symposium and European Conference on High Energy Physics, ed. S. Hegarthy *et al*, World Scientific 1992, Vol.I p702, and H. Dreiner and G.G. Ross, OUTP-92-08P.
43. C. Berger, these proceedings.
44. T.T. Wu, these proceedings. P. Osland and T.T. Wu, CERN-TH 6385/92, 6386/92, 6387/92.
45. M. Veltman, *Acta Physica Polonica* **B12** (1981) 437.
46. A. Blumhofer and B. Stech, Heidelberg preprint HD-THEP-92-24.
47. B. Pendleton and G.G. Ross, *Phys. Lett.* **98B** (1981) 291.
48. C.T. Hill, *Phys. Rev.* **D24** (1982) 691.
49. J. Kubo, K. Sibold and W. Zimmermann, *Nucl. Phys.* **B259** (1985) 331 and *Phys. Lett.* **B220** (1989) 191. K. Sibold and W. Zimmermann, *Phys. Lett.* **B191** (1987) 427. J. Kubo, *Phys. Lett.* **B262** (1991) 472.
50. C.T. Hill, C.N. Leung and S. Rao, *Nucl. Phys.* **B262** (1985) 517.
51. J. Ellis, Proc. Joint International Lepton-Photon Symposium and European Conference on High Energy Physics, ed. S. Hegarthy *et al*, World Scientific 1992, Vol.II p27.
52. S.L. Glashow, Proc. Joint International Lepton-Photon Symposium and European Conference on High Energy Physics, ed. S. Hegarthy *et al*, World Scientific 1992, Vol.II p467.
53. C.H. Llewellyn Smith, G.G. Ross and J.F. Wheater, *Nucl. Phys.* **B177** (1981) 263.
54. S. Weinberg, *Phys. Lett.* **B91** (1980) 51.
55. R. Barbieri and L.J. Hall, *Phys. Rev. Lett.* **68** (1992) 752.
56. G.G. Ross and R.G. Roberts, RAL-92-005, *Nucl. Phys. B* (1992) in press.

57. L.E. Ibanez and G.G. Ross, *Phys. Lett.* **B110** (1982) 215. K. Inone *et al*, *Prog. Theor. Phys.* **68** (1982) 927. L. Alvariz-Gaumé, J. Polchinsky and M. Wise, *Nucl. Phys.* **B221** (1983) 495. For a recent review of supersymmetric model building see L.E. Ibanez and G.G. Ross, CERN-TH 6412/92, to appear in "Perspectives in Higgs Physics", ed. G. Kane.
58. B. Grinstein, these proceedings.
59. T. Barklow, these proceedings.

www.ingramcontent.com/pod-product-compliance
Lightning Source LLC
Chambersburg PA
CBHW041304210926
51851CB00002B/2
9789810212360